Embedded Devices and Internet of Things

The text comprehensively discusses machine-to-machine communication in real-time, low-power system design and estimation using field programmable gate arrays, PID, hardware, accelerators, and software integration for service applications. It further covers the recent advances in embedded computing and IoT for healthcare systems. The text explains the use of low-power devices such as microcontrollers in executing deep neural networks, and other machine learning techniques.

This book:

- Discusses the embedded system software and hardware methodologies for system-on-chip and FPGA.
- Illustrates low-power embedded applications, AI-based system design, PID control design, and CNN hardware design.
- Highlights the integration of advanced 5G communication technologies with embedded systems.
- Explains weather prediction modeling, embedded machine learning, and RTOS.
- Highlights the significance of machine-learning techniques on the Internet of Things (IoT), real-time embedded system design, communication, and healthcare applications, and provides insights on IoT applications in education, fault attacks, security concerns, AI integration, banking, blockchain, intelligent tutoring systems, and smart technologies.

It is primarily written for senior undergraduates, graduate students, and academic researchers in the fields of electrical engineering, electronics and communications engineering, and computer engineering.

Embedded Devices and Internet of Things

Technologies and Applications

Edited by
Adesh Kumar, Surajit Mondal,
Gaurav Verma and Prashant Mani

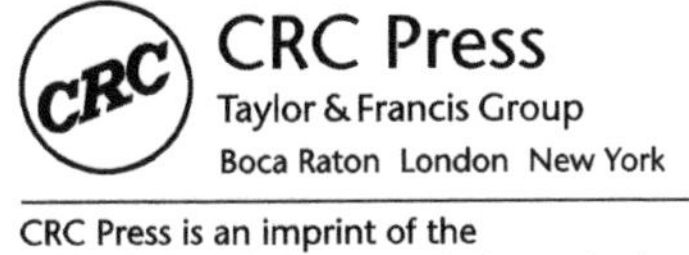

CRC Press
Taylor & Francis Group
Boca Raton London New York

CRC Press is an imprint of the
Taylor & Francis Group, an **informa** business

Designed cover image: fixed image template

MATLAB® and Simulink® are trademarks of The MathWorks, Inc. and are used with permission. The MathWorks does not warrant the accuracy of the text or exercises in this book. This book's use or discussion of MATLAB® or Simulink® software or related products does not constitute endorsement or sponsorship by The MathWorks of a particular pedagogical approach or particular use of the MATLAB® and Simulink® software.

First edition published 2025
by CRC Press
2385 NW Executive Center Drive, Suite 320, Boca Raton FL 33431

and by CRC Press

4 Park Square, Milton Park, Abingdon, Oxon, OX14 4RN
CRC Press is an imprint of Taylor & Francis Group, LLC

ISBN: 9781032606002 (hbk)
ISBN: 9781032836461 (pbk)
ISBN: 9781003510420 (ebk)

DOI: 10.1201/9781003510420

Typeset in Sabon
by Deanta Global Publishing Services, Chennai, India

Contents

14 The efficiency of blending AI technology to enhance behavior intention and critical thinking in higher education

242

AHMAD AL YAKIN, MUTHMAINNAH, AHMED J. OBAID, EKA APRIANI, SOUVIC GANGULI, AND ABDUL LATIEF

15 Embedded machine learning

267

SHWETA GUPTA AND ABHIGNA RAGALA

About the Editors

Adesh Kumar is a Professor with the Department of Electrical & Electronics, Engineering, School of Advanced Engineering, UPES, Dehradun, India. He earned a BTech in electronics and communication engineering from Uttar Pradesh Technical University Lucknow, India, in 2006; an MTech (Hons) in embedded systems technology, from SRM University, Chennai, in 2008; and a PhD in electronics engineering from the University of Petroleum and Energy Studies (UPES), Dehradun, India, in 2014. He has also worked as a senior engineer in TATA ELXSI Limited Bangalore and a faculty member at ICFAI University, Dehradun, India. His areas of interest are VLSI design, embedded systems design, signal processing, and digital image analysis. He has published more than 100 research papers in international peer-reviewed journals (SCI/Scopus) and conferences (h-index-20, i-10 index-40, citations > 2000). He has contributed as conference secretary to ICICCD-2016 and ICICCD-2017, and editor in ICICCD 2018, 2022, and 2024. He has supervised ten PhD scholars, and five candidates are doing PhD research under his guidance.

Surajit Mondal is an Assistant Professor– Selection Grade with the Department of Electrical & Electronics, Engineering, University of Petroleum and Energy Studies, Dehradun, India. He did his BTech in electrical engineering from Asansol Engineering College, West Bengal, India; MTech in energy systems from UPES, Dehradun, India; and PhD in electrical engineering from the University of Petroleum and Energy Studies (UPES), Dehradun, India, in 2020. His areas of interest are wireless power transformation, renewable energy, waste to energy, alternate energy, ESG, and carbon footprint. He has published more than 35 research papers in international peer-reviewed journals (SCI/Scopus) and conferences, published more than 85 patents (25 granted), and published three books with eminent publishers. He has also handled many government-funded projects as a lead or principal investigator. He received the prestigious Young Scientist award from the DST, Government of India, New Delhi. Three students are pursuing their PhD research under his guidance.

Gaurav Verma earned BTech, MTech, and PhD degrees in electronics and communication engineering in 2005, 2012, and 2018, respectively. He earned his BTech from COER (UP Technical University, Lucknow, India), MTech from IIT Kharagpur, West Bengal, India, and PhD from Jaypee University, Noida, India. In addition, he earned an MBA degree in marketing management from IMT, Ghaziabad, India, in 2008. Since 2013, Verma has been an Associate Professor (senior grade) and head of research group (IoT, embedded, and AI) in the Department of Electronics & Communication Engineering, Jaypee Institute of Information Technology, Sector-62 campus, Noida (UP), India. He started his career as a trainer in VLSI, embedded domain at CIPL, Noida, in 2005, and was promoted to the position of assistant manager. From 2007 to 2013, he served at different engineering colleges in different positions such as DBIT, Dehradun (2007–2008), DIT Dehradun (2008–2010), and VIT, Greater Noida (2012–2013). He has about 15 years of experience in training, teaching, research, and administration. He has multidisciplinary abilities in teaching and training in the latest technologies. His current research focuses on the development of IoT and embedded-based solutions for the benefit of society and power estimation of FPGA implementations using machine learning. He is an active researcher with about 25 publications in reputed SCI and SCOPUS journals and about 40 publications in conferences of IEEE and Springer repute. He has also authored a book on embedded system design for students and published more than five book chapters for Springer. He is a senior member of IEEE and served as a TPC member, session chair, and keynote speaker at reputed conferences in India and abroad. He also serves as a member of the Website and Portal Management Sub-Committee, IEEE UP Section (R-10), India; and as an editor, advisor, and reviewer of several well-known international journals published by IEEE, Taylor & Francis, Springer, and Elsevier. He was also awarded the Best Doctoral Thesis award in RACE-2019 at Kasetsart University, Bangkok, Thailand.

Prashant Mani is an Associate Professor and Principal at the Government Engineering College Aurangabad Bihar, India. He has worked for 14 years in the Department of Electronics & Communication Engineering, SRM University, NCR campus. He holds a PhD in VLSI from the Faculty of Engineering and Technology, SRM University, Chennai, India. The focus of his research is related to nanoscale SOI devices. He has multi-discipline/cross-disciplined ability in teaching, and he has been teaching both core nanoelectronics engineering subjects, such as, nanoscale devices, microelectronics, VLSI devices and design, VLSI testing, nano-technology, digital image processing, and real-time and embedded systems. He is the deputy head of Research & Publication, SRM University Campus. He is also the author of several research articles indexed in SCOPUS/SCI journals.

Contributors

Prateek Agarwal
Gurukula Kangri (Deemed to be
  University), Haridwar, India

M. Anand
Dr. M.G.R Educational
  and Research Institute,
  Maduravoyal, Chennai, India

T. Anirudh
Vignan Institute of Technology and
  Science, Hyderabad, India

Eka Apriani
Institut Agama Islam Negeri
  (IAIN), Curup, Indonesia

Piyush Bagla
Graphic Era Hill University,
  Dehradun, India

Pritish Bhanja
GITA Autonomous College,
  Madanpur, Bhubaneswar,
  Odisha, India

K. Thomas Alwa Edison
SRM Institute of Science and
  Technology, Kattankulathur,
  Chennai, Tamilnadu, India

Souvic Ganguli
Thapar Institute of Engineering
  and Technology, Patiala, Punjab,
  India

Tanuj Kumar Garg
Gurukula Kangri (Deemed to be
  University), Haridwar, India

Umang Garg
Krishana Institute of Engineering
  and Technology Ghaziabad,
  India

R. Gowri
Graphic Era Hill University,
  Dehradun, India

Monika Gupta
Veer Madho Singh Bhandari
  Uttarakhand Technical
  University, Dehradun India

Shweta Gupta
Woxsen University, Hyderabad,
  India

Raj Gusain
Graphic Era (Deemed to be
  University), Dehradun, India

Naga Durga Saile. K
VNR Vignana Jyothi Institute of
  Engineering and
  Technology, Hyderabad,
  Telangana, India

Priyanka Kaushik
CHRIST (Deemed to be
  University), Bangalore, India

Adesh Kumar
UPES, Dehradun, India

Akula Praveen Kumar
Woxsen University, Hyderabad,
  India

Amit Kumar
Graphic Era University Dehradun,
  Uttarakhand

B. Vinay Kumar
Vignan Institute of Technology and
  Science, Hyderabad, India

M. Lenin Kumar
Vignan Institute of
  Technology and Science,
  Hyderabad, India

Mukul Kumar
Veer Madho Singh Bhandari
  Uttarakhand Technical
  University, Dehradun, India

N. Dinesh Kumar
Vignan Institute of Technology and
  Science, Hyderabad, India

Abdul Latief
Universitas Al Asyariah Mandar,
  Sulawesi, Barat, Indonesia

V. Lokesh
Vignan Institute of Technology and
  Science Hyderabad, India

Alagesan M.
SRM Institute of Science and
  Technology, Kattankulathur,
  Chennai, Tamilnadu, India

Bishnu Paramguru Mahapatra
SRM Institute of Science and
  Technology, Kattankulathur,
  Chennai, Tamilnadu, India

V. Malathy
SR University, Warangal, India

Amit Kumar Mishra
Graphic Era Hill University,
  Dehradun, Uttarakhand, India

Rasabihari Mishra
GITA Autonomous College,
  Madanpur, Bhubaneswar,
  Odisha, India

Muthmainnah
Universitas Al Asyariah Mandar
  Sulawesi Barat, Indonesia

L. Kavitha Nair
SRM Institute of Science and
  Technology, Kattankulathur,
  Chennai, Tamilnadu, India

Ahmed J. Obaid
University of Kufa, Iraq

Prasanth P.S.
SRM Institute of Science and
  Technology, Kattankulathur,
  Chennai, Tamilnadu, India

Rupendra Kumar Pachauri
UPES, Dehradun, India

Neeraj Kumar Pandey
Graphic Era (Deemed to be
  University), Dehradun, India

Pankaj Singh Panwar
Veer Madho Singh Bhandari
    Uttarakhand Technical
    University, Dehradun, India
Shivalik College of Engineering,
    Dehradun, Uttarakhand, India

Rishi Prakash
Graphic Era (Deemed to be
    University), Dehradun, India

Abhigna Ragala
Woxsen University, Hyderabad,
    India

M. Rohith Reddy
Vignan Institute of Technology and
    Science, Hyderabad, India

M. Sujeeth Reddy
Vignan Institute of Technology and
    Science, Hyderabad, India

Bukka Sanjay
Woxsen University, Hyderabad,
    India

Priyanka Sharma
Dev Bhoomi Uttarakhand
    University, Dehradun, India

Shashikant
Babu Banarasi Das University,
    Lucknow, India

Sushant Shekhar
Graphic Era (Deemed to be
    University), Dehradun, India

F. B. Shiddanagouda
Vignan Institute of Technology and
    Science, Hyderabad, India

Devesh Pratap Singh
Graphic Era (Deemed to be
    University), Dehradun, India

Jai Govind Singh
Asian Institute of Technology,
    Pathumthani, Thailand

Ninni Singh
CMR Institute of Technology,
    Hyderabad, Telangana

Rohit Tanwar
UPES, Dehradun, India

Neha Tripathi
Graphic Era (Deemed to be
    University), Dehradun, India

Deval Verma
Bennett University, Greater Noida,
    India

Anurag Vidyarthi
Graphic Era (Deemed to be
    University), Dehradun, India

Vijai
Vel Tech Rangarajan Dr.
    Sagunthala R&D Institute
    of Science and Technology,
    Chennai, India

Worakamol Wisetsri
King Mongkut's University of
    North Bangkok (KMUTNB),
    Thailand

Ahmad Al Yakin
Universitas Al Asyariah Mandar
    Sulawesi Barat, Indonesia

Chapter 1

Low-power embedded system design applications using FPGAs

Prateek Agarwal, Tanuj Kumar Garg, and Adesh Kumar

1.1 INTRODUCTION

Field programmable gate arrays (FPGAs) are programmable devices that are utilized by end users to create digital circuits. Configurable logic blocks (CLBs), also known as FPGA building blocks, are arranged in a two-dimensional array and linked to one another via a routing network (Verma et al. 2017). The communication-oriented structure supporting multiple resources and their networks on a single chip is known as NoC communication architecture (Jain et al. 2022). The NoC offers a solution that enables the resources to function independently as building blocks and as components of a specific network. Reconfigurable computing is an architecture that arranges highly flexible computing hardware and software components on processing platforms like FPGAs (Babu and Parthasarathy 2021). After production, they are reprogrammed to a particular application based on their functional requirements. This distinguishing feature of FPGAs sets them apart from ASICs, which are created specifically for a given purpose or application. The capacity of FPGAs to upgrade or reuse hardware designs after implementation. FPGAs provide several benefits over ASICs or full-custom designs, including lower silicon chip costs, improved performance, and low power consumption. On the other side, the manufacturers of these FPGAs, such as Xilinx, Altera, and Atmel, are examined and contrasted using different metrics. FPGA reconfigurable logic devices are used in digital design to increase design flexibility and speed up the design-to-implementation process (Rajaei and Gholipour 2018). Their flexibility and efficiency in configuration due to their high performance and versatility, SRAM-based FPGAs find widespread usage in digital design. However, SRAM-based FPGAs lose their settings whenever the power is switched off since SRAM is volatile. Non-volatile and often slow configuration memory is necessary for SRAM-based FPGAs to store their configuration pattern, which will be delivered via serial connection into the FPGA during power-up. Despite the existence of two additional non-volatile FPGA types – flash and anti-fuse-based – many high-performance applications cannot be supported by their performance. Due to their economic power usage and

DOI: 10.1201/9781003510420-1

quick execution times, FPGAs are widely employed for applications that require a lot of computing (Skhiri et al. 2019). For instance, the sliding-window application in signal processing demonstrates that the FPGA suggests a higher energy economy and can increase the rate by up to 57× and 11× in comparison to multicores and GPUs. Additionally, FPGA may be a preferable option for specific applications of image processing, such as stereo vision. The usual design approach for an FPGA architecture, however, is time-consuming and necessitates knowledge of hardware tools and languages like HDL. High-level synthesis (HLS) is a process used to create designs for FPGAs (Reyes et al. 2020). HLS converts an algorithmic explanation into a synthesized RTL netlist, which is also referred to as behavioral synthesis or architectural synthesis. By defining the hardware description in high-level languages like C/C++, HLS enables engineers to operate at an advanced degree of abstraction. An acyclic-directed graph is typically used to internally express this description (as an intermediary structure). It identifies the data dependencies noted in the design's input/output relationships and data flow. Different RTL implementations may exist for the same behavioral description. In general, hardware and software implementations of block ciphers are equivalent (Mohd et al. 2016). Hardware realizations are quicker and use less power than software systems. One method for putting hardware concepts into action is by using FPGAs. A flexible and inexpensive development time is two of the many advantages of an FPGA design. Additionally, the FPGA's reconfigurability feature makes it a desirable choice for implementing cryptographic algorithms. Algorithm agility (changing ciphers while in use), upload (upgrading to a new cipher), and modification are made easier by reconfigurability. It has been hypothesized that the hardware's reconfigurability characteristic will develop alongside algorithms and so become more resistant to fresh attacks. FPGAs are reprogrammable, adaptable, and low-cost circuits with a simple and rapid design process (Alaei et al. 2021). Physical layout, clock distribution, manufacturing constraints, and testing are often controlled via mapping tools and synthesis for FPGA engineers. Designers focus on improving the functionality of the plans and the algorithms they use. For special-purpose and embedded systems, like NoC-oriented SoCs, FPGAs provide rapid, simple, and easily accessible devices, they are constrained in their design area and resource availability and suffer from excessive power consumption. FPGAs are very flexible and reconfigurable, but they also have low operating frequencies and high power dissipation. When synthesizing and executing processing elements (PEs) and routers for FPGA-oriented NoCs, several FPGA constraints and characteristics must be taken into consideration. The advancement of FPGA technology is often claimed (Rodríguez et al. 2015). The ability to modify an arrangement while the remaining circuit is running is known as run-time reconfiguration (RTR) (Ram et al. 2020). It is typically used in FPGAs; configurable circuits are available thanks to Xilinx Altera.

For more than a decade, FPGAs have offered partial reconfiguration, which modifies only a portion of the hardware at run-time. Current SRAM-based FPGAs generate a significant amount of dual port (DP). The main benefits of hardware reconfiguration are increased hardware flexibility and better use of hardware space, which result in lower power consumption and lower production costs.

1.2 LITERATURE REVIEW

In-depth testing is done on the performance of two different dynamic threshold MOS (DTMOS) arrangements used to build 4-input multiplexer switches (Kumar et al. 2010). The switch design's DTMOS transistors were all scaled accurately and efficiently, which improved latency and power-delay product (PDP) performance. Improvements in the delay of 12.32% and 11.19%, respectively, as 8.29% and 8.26% in the optimal PDP for Virtex-4 cost-effective 90 nm FPGA are seen when secured voltage transistor DTMOS-oriented switch layouts and transistor DTMOS-based switch designs with transistors are upgraded. As the design of an FPGA contains hundreds of 4-input multiplexers, a reduction in latency and PDP will have a substantial impact on the improvement of high-speed, low-power FPGAs. The calculation only uses a 4-input multiplexer switch, however, the results also apply to larger multiplexers with 5 and 6 inputs. Diminish the assessment period of an FPGA multiplexer-oriented connection by using an efficient built-in self-test (BIST) technique that makes use of the design of partly self-configurable structures and the cost-effective test point insertion (Jianfeng et al. 2011). It is shown that the minimum number of TCs needed for connectivity in the Virtex-2/Spartan-3 FPGA may be lowered from 32 to 8 by using the provided strategy and technique. Consequently, the test duration is cut in half while the overhead area only grows by around 1.2%. A substantial impact on evaluating Flash-based FPGAs in addition to adapting to SRAM-based FPGAs. Matrix multiplication serves as the kernel function in most procedures used in high-performance super-computing, digital signal processing (DSP), image and video processing, computer graphics, and vision applications (Qasim et al. 2010). Two alternative matrix multiplier architecture examples are used where speed is the primary restriction. The performance of the first design, which computes dense matrix–vector multiplication, is assessed by calculating the execution time of the design on the Xilinx Virtex-4 FPGA. The throughput of 16970 frames per second, which is suitable for many image and video processing applications, can be shown from hardware simulation outcomes. The second design uses systolic arrays and is realized on the Spartan-3 and Virtex-2 Pro platforms, respectively, for the multiplication (Kumar, Joshi, Patkar, & Narayanan (2010) of three matrices. Findings of the simulation show that FPGAs are appropriate for use in these kinds of applications. The memory-based structure is extremely effective for multiplying big matrices, and systolic array approaches are quite effective for multiplying small and medium-sized matrices, as shown by the simulation outcomes.

For DSP applications, creating a high-speed, low-power arithmetic circuit is a constant issue (Bhattacharjee et al. 2011). The FPGA has been used to compare the fixed-point multiplier and adder circuits. In peer-reviewed articles, the power consumption of arithmetic circuits was not considered; only the delay and area were compared. The FPGA has been used for the power and delay utilization of several types of arithmetic circuits. It offers specific characteristics to enable arithmetic tasks, a direction that appears to be prevalent now for FPGA devices. The smaller delays in the circuits also result in lower power consumption, in addition to designs with fewer circuits, or slices. The importance of learning how to effectively implement floating point units on FPGAs has increased as the need for such features has grown (Hemmert and Underwood 2010). FPGA technology is used to implement two different floating-point algorithms. In comparison to earlier options, the outcomes are designs that are substantially smaller, quicker, and have shorter pipelines. An adder structure may run at 271 MHz utilizing 507 slices on a Virtex4 with fewer than nine cycles of delay and full IEEE, double precision authorization. This reduces latency by five cycles, space by 30%, and clock rate by almost nothing compared to the previous best plan. The multiplier may operate at 274 MHz in just 737 slices using a 14-stage pipeline. It can get amazing area and latency reductions with just very few changes: the adder and multiplier both experience a two-cycle decrease in latency, while the adder and multiplier both experience an increase of around 20% and 30%, respectively. The hardware execution of multipliers on FPGA devices using Verilog HDL is described. The Spartan 3E (xc3s500e4ft256), Virtex 4 (xc4vlx15-10-sf363), Virtex 5 (xc5vlx30-1-ff324), and Virtex 6 (xc5vlx30-1-ff324) were used to execute the design (Anitha and Bagyaveereswaran 2011). When compared to all existing multipliers, the suggested multiplier exhibits lower consumption. The Virtex-6 low-power FPGA chip has decreased average pin latency and combinational path time. Hence, in comparison to other FPGA devices, the Virtex-6 low power received the best outcome.

The physical creation and execution of a parallel FPGA design for conventional and truncated multipliers using VHDL are provided (Rais 2010). A comparison of the conventional and truncated multipliers will be shown. In comparison to the regular multiplier, the truncated multiplier exhibits a significantly greater decrease in device consumption. The IEEE 754 quadruple precision floating-point multiplication on FPGAs is implemented using an effective design (Jaiswal and Cheung 2012). It has demonstrated the effective use of larger multipliers, which improves the use of the quadruple precision (QP) multiplier. As compared to the array multiplier, booth multiplier, and traditional Vedic multiplier execution on FPGA, the architecture's attempt to minimize propagation delay resulted in a 45% improvement in the minimization of delay (Kumar et al. 2013). Future work on this specific project may include designing the arithmetic logic units (ALUs) for reduced instruction set computer (RISC) processors. Designing a low-power, high-speed finite impulse response (FIR) filter is a persistent task for applications of DSP (Bhattacharjee et al. 2013). It compares the various FIR filter architectures used in Virtex-6 FPGA implementation. In peer-reviewed research, the latency, the area, and the power for pipelined FIR filters have been compared, but the power dissipation of non-pipelined FIR filters has not been considered. But in cases when power consumption and resource use are more important considerations than DSP application speed, we can utilize a non-pipeline FIR filter or a simple DF FIR filter.

A novel robust physical unclonable function (PUF) architecture based on the FPGA is suggested (Hatti and Paramasivam 2022). It is a collection of two comparable sets of feed-forward MUX chains. The response bits are produced based on the race situation between two similar latency pathways. The Virtex-6 FPGA is used to build the three alternative arbiter PUFs (APUFs), which are explored for the double APUF, feed-forward APUF, and suggested double feed-forward XOR APUF (DFFX APUF). These three APUFs have distinct values of 43%, 43%, and 48%, respectively. On the Virtex-6 FPGA, three PUF topologies were evaluated for dependability at various temperatures. The results were 80%, 82%, and 97%, respectively. It may be inferred from the results above that the suggested DFFX APUF reports average distinctiveness and dependability that are 5% and 8% greater than those of the other two designs, respectively. In comparison with FF-MPUF, the suggested PUF design takes up a lot more space. The higher privacy level and resistance to machine learning threats make up for it. The suggested DFFX APUF may be utilized for safety purposes since it is more distinctive and dependable than previous delay-based PUF systems. Since the DFFX APUF arrangement uses more FPGA resources than the other two PUF systems, it must be revised to consume less space and be implemented on higher-end FPGA boards (7 series). Further, it will examine various modeling assaults and protection methods to determine the suggested structure's prediction rate. The 3D multilayer mesh NoC architecture developed in Xilinx 14.2 using VHDL is effectively emulated by ModelSim 10.0 software with 8, 16, 32, 64, and 128-bit data transmission (Kumar et al. 2019). Internode communication is made possible by the scalable architecture, which is beneficial for larger networks like wireless sensor networks. Future studies might focus on hierarchical NoC architecture with a multilayer system environment and network security integration. This work introduces pipelined and non-pipelined strategies for generating high-performance JH hash function FPGA implementations (Athanasiou et al. 2013). The suggested designs greatly outperform the existing ones in terms of frequency and throughput/area. The upcoming task will involve optimizing the architecture to provide low-power designs.

A high-speed R2DIF SDF pipelined FFT processor was implemented for 1024-point FFT calculations on a Xilinx Virtex-7 FPGA (Nguyen et al. 2017). The effectiveness of the suggested proposal is assessed in terms of hardware problem, speed, and accuracy. The difficult shift-add operations-based twiddle factor multiplication is optimized in the proposed design to take full advantage of the pipelined R2DIF SDF architecture while also boosting performance and resource utilization. In comparison to conventional designs, it provides superior performance and accuracy while utilizing fewer resources, which lowers cost and power. The IEEE 754 double precision floating-point multiplication is implemented on FPGAs utilizing efficient topologies (Jaiswal and Cheung 2013). Comparing the suggested three architectures to other earlier ideas, they perform better. In contrast, 3-PKM (Design-2) has the same advantages for the number of multipliers and suffers no accuracy loss. By combining the two proposed systems, we get large resource savings, accuracy assurance, and performance improvement. Furthermore, the third design boasts promising space and performance characteristics, without any pipeline stalls, and reconfigurable dual capability, including both dual single and double precision as well as their extended precision multiplication. The building blocks for applications that use floating-point multipliers are in an effective and condensed manner. Future work will involve measuring the benefits of our method and incorporating it into specific applications such as FFT and matrix multiplication. Moreover, work can be done on more optimization to further understand the benefits of DSP, and the design-3 ASIC-optimized design will also be investigated.

Particularly for portable battery-operated gadgets like mobile phones, power consumption is a significant concern (Verma et al. 2017). Therefore, pipelining and parallelism, two low-power approaches, have been suggested and tested on three alternative realizations of an ALU intended for an FPGA in this study. The findings of the pipelined and parallel ALU realization using the X-power analyzer tool are considered important. These realizations can be used to create CPUs that are communication-centric for wireless applications. Employing the Han-Carlson, Weinberger, and Ling adder with a binary to excess-1 converter (BEC) circuit, this study suggests and constructs a hybrid adder-based multiplier (CSELA) (Thamizharasan and Kasthuri 2021). Verilog HDL is used in Xilinx ISE 12.1, where the simulation is run. It may also be expanded to include the development of various input-size (input bit) multipliers. It offers a mechanism for the per-device setup of PUFs as well as a novel PUF on Xilinx FPGAs based on the Anderson PUF architecture (Usmani et al. 2018). Instead of the traditional design, which employs the more uncommon SLICEMs, our new PUF uses the more common Slices. The PUF architecture and technique may be utilized to generate trustworthy lookup table (LUT)-based PUF keys on FPGAs and have a wide range of applications in future reconfigurable systems.

In a hardware description language environment, a three-stage multistage telecommunication network is implemented on a hardware device (Kumar, Kuchhal, and Singhal 2015). The Virtex-5 made with a Digilent FPGA kit is used to validate the results. A cluster arrangement (64 × 64) is used for the chip's inlets and outputs. To enhance speed, decrease traffic congestion, and minimize delays, the switching system's FPGA design aims to make the entire switching system programmable. It represents a sizable contribution to the complete digitization and programmability of switching systems and will undoubtedly be beneficial for the VLSI design industry. It may be conducted on the incorporation of security measures for data transit between inlets and outlets using encryption and decryption. Future chip implementations may support up to five switching phases, four stages, or even more stages with a configurable number of inlets and outlets. The DSP48E1 Slices included in current Virtex-6 FPGA processors seem to be the finest resource for building correlation-based frame synchronizers (Pham et al. 2012). For synchronization in IEEE 802.16 OFDM systems, simpler multiplier-less architectures provide comparable synchronization performance. Higher clock rates can be achieved with the DSP48E1-based correlators, but only with a carefully thought-out pipelined architecture. Additionally, their resource and power utilization are significantly higher. Another benefit is that multiplier correlation can be used in any FPGA design. The efficiency of synchronization is affected by very low quantization precision. The 128-bit compact AES encryption method is made and used to make safety better (Arul Murugan et al. 2020). In the suggested design, CFA makes use of the S-box operation. In the mix-columns transformation, the Vedic multiplier is used to balance the speed and further minimize the area. The Virtex 4 FPGA chip is used to implement the complete design. In comparison to other current conventional procedures, the area is decreased by 67% with a 69% increase in latency, according to the data. The trade-off between memory and bandwidth is graphically shown in two matrix multiplication designs (Kumar et al. 2010). The designs; simplicity and usage of Xilinx Coregen's off-the-shelf floating-point units make it simple to reproduce them, portable across FPGA generations, and maintainable. They also provide greater IEEE compliance and features like configurable precision. The designs scale effectively over many FPGAs with only a 1% performance loss thanks to design II. An effective design for boosting the AES algorithm's throughput has been demonstrated (Rahimunnisa et al. 2013). The proposed PSP architecture has been compared against the loop unrolled, pipelined, parallel, and parallel pipelined architectures using a prototype built on an FPGA Virtex-6 XC6VLX75T device. It has also been prototyped using 0.13 and 0.18 micron ASIC technology. The parallel sub-pipelining design outperforms all other AES algorithm structural designs in terms of throughput. The space and power dissipation of the suggested parallel sub-pipelining architecture can be decreased by using optimization techniques. The Zynq 7000 All Programmable SOC uses the greatest power, whereas the Artix-7, Kintex-7, and Ultrascale FPGAs dissipate the least amount of power (Pandey et al. 2017). Ultrascale FPGA is also suitable for packet processing in 100G networking and heterogeneous wireless infrastructure. The hyper-scale FPGA is the optimum architecture for implementing any communication design on FPGA in an energy-efficient manner. From a seven-series architecture based on a 28 nm process technology to an ultrascale architecture based on a 20 nm process technology, there is a 47.74% reduction in latency. Any communication architecture, including this FIR filter and many others, will function more effectively with less latency.

An innovative Xilinx System Generator (XSG) model-based low-power and high-performance FPGA execution for the 4D memristor chaotic system with cubic nonlinearity was proposed by Hagras and Saber (2020). In addition, the pseudo-random sequence of the 4D memristor chaotic arrangement is shown using the 15 randomizations evaluated in the SP 800-22 standard. Additionally, the cubic nonlinear 4D memristor chaotic system-based gray image encryption approach has been proposed. The mechanisms of confusion and dispersion are utilized to increase the cryptosystem's strength. The strength of the suggested gray image encryption system is evaluated using statistical testing. Reconfigurable infinite impulse response (IIR) filters are designed and implemented (Datta and Dutta 2021). To assess performance, the creation of several IIR filter types has been effectively implemented in FPGA. The reaction time findings from the IIR filter tests are quite good, highlighting the processing levels of reconfigurable architectures, and pipeline and parallel technologies yield greater overall performance. The suggested FIR-based IIR filter has reduced power consumption while simultaneously accelerating sampling. According to comparative statistics, the recommended FIR-based IIR filter has the best power and area with a maximum operating frequency of 285.105 MHz. The recommended solutions can be produced via real-time signal processing systems. It is proposed to use a low-power, scalable CBMW multiplier (Solanki et al. 2021). It enables the Wallace tree multiplier to be easily and effectively implemented on both the FPGA and ASIC platforms. The recommended multiplier uses a more efficient 7:3 counter than other 7:3 counter designs already in use, which are composed of multiplexers and ex-or-gate-based adders. One 7:3 counter is utilized at each stage, which reduces the amount of hardware needed and reduces power consumption significantly owing to the enhanced area. The performance of the MAC unit created with the suggested CBMW is assessed. The suggested CBMW may be employed in low-power applications since it has improved PDP performance.

The distorted power computation is provided using an FPGA-based optimized technique that assumes a distinct power triangle for each frequency

(Jarrah et al. 2022). A real-time power analysis system on the FPGA platform is proposed, simulated, and experimentally evaluated using the Vivado HLS tool. Multiple optimization approaches are used to function hardware circuits that are synthesized from a variety of power components using the Vivado HLS. The suggested architecture significantly cuts down on computing time. Additionally, the improved design reduces power line losses and addresses voltage distortion-related problems. The entire process is carried out and an effective design is produced using the Vivado HLS tool. The suggested method can accelerate computations by a factor of up to 15 and is capable of accurately estimating the total apparent power for all harmonics. For greater utilization of all power components, more money should be invested in finding solutions to the proposed coupled problems. The FDFM technique was used to create a three-layer perceptron in an FPGA of the Xilinx Virtex-6 series (Ago et al. 2013). Our solution can operate at incredibly high speeds and throughput, according to experimental results. The goal of this effort is to create and construct a portable, power-efficient DPSK modem (Bag et al. 2014). The CORDIC method has been utilized to replace the expensive multipliers, which also includes additional hardware, increased power consumption, and noise effects. The noise issue is also resolved via carrier generation on a semiconductor. Because the CORDIC algorithm's operations are straightforward and well-suited for VLSI implementation, it has been used. The shortest processor delay is 0.669 ns, and 0.029 mW is the least amount of dynamic power dissipation. How well a modem functions is determined by the software, not by how complicated the hardware is. For a given realistic performance level, the DPSK modem needs fewer calculations per baud. It is better suited for an FPGA implementation. The standard IIR filter, fast IIR filter using levels 1 and 2, typical moving average (MA) FIR filter, fast MA FIR filter using levels 1 and 2, as well as FPGA implementation, have all been thoroughly described (Seshadri and Ramakrishnan 2021). The results of the simulation demonstrate that the fast IIR and fast MA FIR filters perform better than the conventional filters. According to the simulation findings, the first-order IIR conventional filter performs at 153.4 MHz, 189.79 MHz with look-ahead level 1, and 199.96 MHz with level 2 for the fast filter. A comparable filter using a level 1 look-ahead operates at 167.22 MHz, and a similar filter using a level 2 look-ahead operates at 168.56 MHz. When putting cryptographic algorithms on a hardware platform, Montgomery modular multiplication is crucial (Khan et al. 2018). The implementation of an effective full-word Montgomery modular multiplication may be used to speed up the operation of the RSA and ECC cryptographic algorithms. Comparisons and implementation outcomes demonstrate that the suggested system is superior to alternative schemes. The suggested Montgomery multiplier is also quite adaptable and may be used well in cryptographic processors for pairing-based and elliptic curve cryptography. Interleaved modular multipliers for any prime number p, both serially and simultaneously, were developed in the article by Javeed et al. (2015). The radix-4 interleaved multiplication technique serves as the foundation for both modular multipliers. Utilizing the Montgomery power laddering technique, a parallelism has been added to the parallel interleaved multiplier. In comparison to other existing designs, each of these modular multipliers performs multiplication in 50% fewer clock cycles. The parallel interleaved modular multiplier speeds up the existing platform-independent procedures by 62% and 49% because of the suggested parallelism technique.

A more secure and dependable basic true random number generator (TRNG) is created by using all the entropy seeds provided by ADPLL, Ring Oscillator, flip-flop, and other primitives (Bharat Meitei and Kumar 2022). ADPLL-based TRNG has a promising future as a security solution, making it a reliable and more secure choice. The utilization of WSNs in several industries and sectors has continuously increased, especially in the monitoring of machinery, sensors, and equipment (Mishra and Kumar 2022). In comparison to industrial monitoring and control systems based on microcontrollers and microprocessors, the main benefit of adopting FPGA is that it offers quick switching. Using specialized ZigBee networks with mesh chip designs can remotely monitor and manage any plant.

1.3 DISCUSSION

Using Vivado Design Suite 2018.3, the recommended cubic design is successfully synthesized, simulated, and realized on the Kintex-7 FPGA board (Reddy et al. 2022). The suggested work is programmed into an FPGA and executed on an ASIC. The data confirmed the efficacy of the recommended cubic design by outperforming predictions for performance improvement and area reduction. Because of this, the recommended cubing architecture performs well for high-performance processors. A 28 nm Artix-7 FPGA for green communication is built (Haripriya et al. 2022). The Xilinx design suite is applied to execute the analysis and simulation, and the X-power analyzer is used to calculate power. The aim is to reduce UART's power consumption in organizations that use an FPGA device. It has been noted that while many various strategies have been used by researchers to minimize power usage, it may still be decreased by up to 0.080 W. The power usage in our suggested design is found to be lowered by up to 58.75% when our findings are compared. In the future, sophisticated ultrascale and ultrascale+ FPGAs will be able to implement UART. These ideas may later be transformed into ASIC designs, which are more practical and portable than FPGAs. Table 1.1 lists the low-power design applications with tool descriptions and FPGA hardware.

Table 1.1 Low-power design applications with tool description and FPGA hardware

Reference	Platform/hardware	Simulation environment	Application
Kumar et al. (2010)	FPGA	Virtex 4	Low-power and high-speed FPGA design
Jianfeng et al. (2011)	FPGA	Virtex-II and Spartan-3	A cost-efficient self-configurable BIST technique
Qasim et al. (2010)	FPGA(Virtex4)	Spartan-3 and Virtex-II	Matrix multiplier architectures for image and signal processing
Bhattacharjee et al. (2011)	FPGA	Xilinx XC5VLX30 (Virtex-5)	Evaluation of power-efficient adder and multiplier circuits
Hemmert and Underwood (2010)	FPGA	Xilinx Virtex 4	Fast, efficient floating-point adders and multipliers
Anitha and Bagyaveereswaran (2011)	FPGA	Spartan-3E, Virtex-4, Virtex-5 and Virtex-6	Braun's multiplier implementation with bypassing techniques
Rais (2010)	FPGA	Spartan-3A, Virtex-4 and Virtex-5	Hardware implementation of truncated multipliers
Jaiswal and Cheung (2012)	FPGA	Virtex 4	Quadruple precision floating point multiplier
Kumar et al. (2013)	FPGA	Xilinx Spartan-6 family xc6s1x75T-3-fgg676	High-speed 8-bit Vedic multiplier using barrel shifter
Bhattacharjee et al. (2013)	FPGA	Virtex-6	Power efficient FIR filter evaluation for DSP
Hatti et al. (2022)	FPGA	Xilinx Virtex-6	Design and implementation of enhanced PUF architecture
Kumar et al. (2019)	FPGA	Virtex 5	3D multilayer mesh NoC communication and FPGA synthesis
Athanasiou et al. (2013)	FPGA	Xilinx Virtex-4, Virtex-5 and Virtex-6, ALTERA Stratix-III and Stratix-IV	Cryptographic hash function and security
Nguyen et al. (2017)	FPGA	Xilinx Virtex-7	Pipelined FFT processor for high-speed signal processing applications

(Continued)

Table 1.1 (Continued) Low-power design applications with tool description and FPGA hardware

Reference	Platform/hardware	Simulation environment	Application
Jaiswal and Cheung (2013)	FPGA	Virtex-4 and Virtex-5	Area-efficient topologies for double-precision multipliers with dual single-precision support that may be changed at run time
Verma et al. (2017)	Xilinx ISE	Spartan 3E XC3S250E FPGA	Analysis of low power consumption techniques for wireless devices
Thamizharasan and Kasthuri (2021)	Xilinx ISE 12.1	Virtex-6 FPGA devices (Device. XC6VLX75T, Package FF484, Speed -3)	Design of proficient two operands adder using hybrid carry select adder
Usmani et al. (2018)	Zynq ZedBoard	Virtex-7 family	Efficient PUF-based key generation using per-device configuration
Kumar, Kucchal, and Singhal (2015)	Xilinx ISE 14.2	Virtex- 5	Three-stage telecommunication switching design and synthesis in an HDL Environment
Pham et al. (2012)	FPGA	Xilinx Virtex-6 and Spartan-6	Low-power correlation for OFDM synchronization in IEEE 802.16
Arul et al. (2020)	FPGA	Xilinx Spartan 3, Virtex-4 and Virtex-5	Hardware architecture with AES encryptor using sub-pipelined S-box techniques
Kumar et al. (2010)	FPGA	Virtex-5 SX240T	High-performance double-precision matrix multiplication
Rahimunnisa et al. (2013)	FPGA	Virtex XC6VLX75T	Parallel sub-pipelined architecture for high throughput AES
Pandey et al. (2017)	FPGA (Ultrascale FPGA, Artix-7, Kintex-7, and Zynq)	Vivado HLS 2016.2	Performance evaluation of FIR filter and its utilization in communication and network
Hagras and Saber (2020)	FPGA	Xilinx Spartan-6 X6SLX45	4D memristor chaotic system for image encryption
Datta and Dutta (2021)	FPGA	Xilinx Virtex-5	High performance IIR filter

(Continued)

Table 1.1 (Continued) Low-power design applications with tool description and FPGA hardware

Reference	Platform/hardware	Simulation environment	Application
Solanki et al. (2021)	Xilinx ISE design suite 14.7	Spartan 3E FPGA	Wallace tree multiplier architecture with low power and modular design
Jarrah et al. (2022)	FPGA	Xilinx Vivado	High-performance implementation of power components
Ago et al. (2013)	FPGA	Xilinx Virtex-6	Neural networks with the OFDM processor core approach
Bag et al. (2014)	Kintex-7 FPGA	Xilinx block and Spartan-6	Design of a DPSK modem using the CORDIC algorithm
Seshadri and Ramakrishnan (2021)	AlteraEP4CE115F29C7 FPGA	Quartus II 13.1	Fast digital FIR and IIR filters
Khan et al. (2018)	Xilinx ISE 14.1 Design Suite	Virtex-5, Virtex-6 and Virtex-7	Full-word Montgomery multiplier implementation on a fast FPGA for ECC
Javeed et al. (2015)	Xilinx ISE 14.2 Design suite.	Virtex-6	Serial and parallel interleaved modular multipliers
Bharat Meitei and Kumar (2022)	Artrix-7 (XC7A35T-CPG236-1)	Vivado v.2015.2	True random number generator architecture using all digital phase-locked loop
Mishra and Kumar (2022)	Xilinx ISE 14.7	Virtex-5	Performance of IEEE 802.15.4 ZigBee wireless sensor nodes for monitoring and automating industrial plants
Reddy et al. (2022)	Vivado Design Suite 2018.3	Kintex-7 FPGA	Implementation of cubing architecture
Haripriya et al. (2022)	Artix-7 FPGA	XILINX design suite	Dynamic voltage scaling for green communication in the industrial sector with energy-efficient UART design
Gupta et al. (2021)	Virtex-5	Xilinx ISE 14.7 ModelSim 10.0	DSDV and OLSR routing protocols implementation for wireless sensor networks
Kumar, Rastogi & Srivastava (2015)	Virtex-5	Xilinx ISE 14.2 ModelSim 10.0	Discrete wavelets transform (DWT)-based hardware chip design and image processing applications
Ompal et al. (2021)	Virtex-5	Xilinx ISE 14.7	Zigbee-based WSN application using star, mesh, and cluster tree topological communication

1.4 CONSTRAINTS IN EMBEDDED SYSTEM DESIGN

The design of low-power embedded systems that make use of FPGA involves several challenges and limits that engineers are required to handle. The following is a list of important constraints that must be considered while designing low-power embedded systems with FPGAs. Power consumption constraints are related to dynamic power and static power. FPGAs exhibit power consumption in both static and dynamic states. Minimizing power consumption is a significant priority concerning dynamic power, and designers must optimize the switching activity and clock frequency to achieve this. Static power pertains to the amount of power that is consumed by the FPGA when it is not undergoing any changes or activities. To reduce static power consumption, it is necessary to employ power gating techniques and ensure that any unused components are off. Timekeeping and frequency limitations are applicable to decrease power consumption, and one can reduce the clock frequency or implement clock gating techniques. Nevertheless, it is crucial to maintain a balance between these considerations and the performance demands of the embedded system. The embedded systems have limitations on the size or extent of an area. FPGAs possess constrained resources, encompassing LUTs, flip-flops, and block RAMs (Kumar, Joshi, Patkar & Narayanan 2010). Optimal exploitation of these resources is essential to decrease both the physical space required and the amount of power consumed.

Power source and activation levels and decreasing the supply voltage have the potential to decrease power usage, but they also have an impact on the general functionality of the FPGA. Designers must carefully weigh the balance between power conservation and ensuring consistent performance when operating at reduced voltages. The system design also has input/output constraints. Power minimization requires careful consideration of input and output constraints. Optimizing power consumption can be achieved by configuring I/O pins, employing strategies such as controlling the slew rate, and activating tri-state buffers when they are not in use. The deviation in manufacturing processes in the semiconductor industry can result in disparities in power consumption. Designers must consider these variances and employ resilient strategies to guarantee continuous low power consumption across many manufacturing batches. Embedded systems frequently depend on constrained energy sources, such as batteries or energy harvesting. Efficient energy management, which encompasses the use of power-aware algorithms and duty cycling, is essential for prolonging the operating lifespan of these systems. As the functionality of embedded systems grows, the task of managing power becomes increasingly difficult. Coordinating power management across numerous components and subsystems necessitates the utilization of sophisticated techniques to overcome cost constraints.

Utilizing low-power components and design techniques may result in increased expenses. The issue lies in finding the optimal balance between

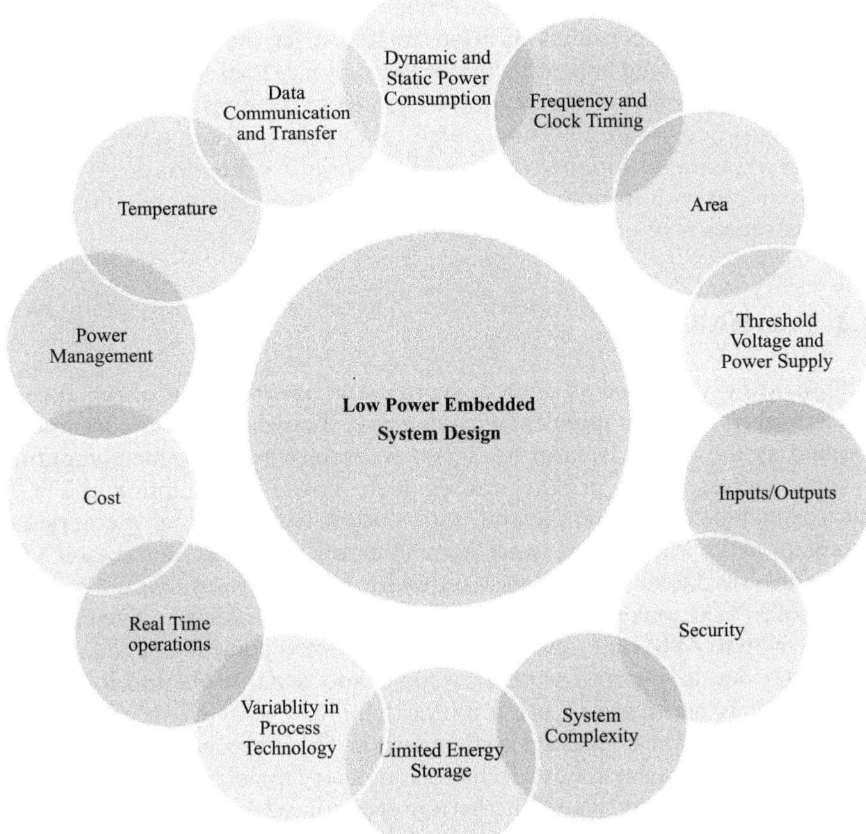

Figure 1.1 Constraints in low-power embedded system design.

power efficiency and system cost, particularly in situations where cost is a significant factor. Data transfer and communication is one of the challenges in embedded systems. Reducing the transfer of data between components can conserve energy. To minimize power consumption during data transfer, one can utilize methods such as data compression, effective data encoding, and optimizing communication protocols. FPGA functions within predetermined temperature thresholds. Excessive heat might result in higher energy usage and decreased dependability. Hence, it is imperative to take thermal limitations into account during the entire design process. It is crucial to use efficient power management strategies, such as dynamic voltage and frequency scaling (DVFS) and power gating. These strategies enable the system to adjust its power usage according to the workload. Adhering to time limitations is crucial for the optimal operation of the embedded system. Designers must strategically optimize the design to fulfill timing requirements while considering the influence on power usage. Some embedded systems have security limitations that restrict the implementation of specific power-saving approaches. Power analysis attacks can limit the utilization of dynamic voltage scaling in some applications. Certain embedded systems necessitate real-time operation, hence restricting the ability to use power-saving techniques that could potentially cause delays. Achieving a harmonious equilibrium between energy conservation and immediate responsiveness is of utmost importance.

1.5 SUMMARY

FPGAs are widely used as semiconductors and are the result of rising consumer electronics and mobile device demand. This market will continue to expand as the rising demand for smart wearables and scalable computing systems increases. The primary causes of the power consumption of FPGA devices include both dynamic and static power, which is notably composed of components like the hardware logic components, clock signal, and I/O. The ultra-low latency and large bandwidth for data transmission requirements of FPGAs make them suitable for 5G communication applications. In comparison to ASICs, FPGAs provide dynamic reconfigurability to facilitate the better development of wireless systems and scalable 5G and IoT architectures. It is necessary to use a multidisciplinary strategy to address the issues in low-power embedded system design. This approach should involve collaboration between system architects, industry experts, and designers of both hardware and software. Furthermore, continual improvements in semiconductor technologies and design approaches contribute to the process of solving these obstacles in the field of low-power embedded system design.

REFERENCES

Alaei, M., & Yazdanpanah, F. (2021). A high-performance FPGA-based multicrossbar prioritized network-on-chip. *Concurrency and Computation: Practice and Experience*, 33(6), e6055. https://doi.org/10.1002/cpe.6055.

Athanasiou, G. S., Michail, H. E., Theodoridis, G., & Goutis, C. E. (2013). High-performance FPGA implementations of the cryptographic hash function JH. *IET Computers and Digital Techniques*, 7(1), 29–40. https://doi.org/10.1049/iet-cdt.2012.0070.

Anitha, R., & Bagyaveereswaran, V. (2011). Braun's multiplier implementation using FPGA with bypassing techniques. *International Journal of VLSI Design & Communication Systems*, 2(3), 201.

Arul Murugan, C., Karthigaikumar, P., & Sathya Priya, S. (2020). FPGA implementation of hardware architecture with AES encryptor using sub-pipelined S-box techniques for compact applications. *Automatika*, 61(4), 682–693. https://doi.org/10.1080/00051144.2020.1816388.

Ago, Y., Ito, Y., & Nakano, K. (2013). An FPGA implementation for neural networks with the FDFM processor core approach. *International Journal of Parallel, Emergent and Distributed Systems*, 28(4), 308–320. https://doi.org /10.1080/17445760.2012.684686.

Bag, J., Roy, S., Dutta, P. K., & Sarkar, S. K. (2014). Design of a DPSK modem using CORDIC algorithm and its FPGA implementation. *IETE Journal of Research*, 60(5), 355–363,doi: https://doi.org/10.1080/03772063.2014 .961979.

Babu, P., & Parthasarathy, E. (2021). Reconfigurable FPGA architectures: A survey and applications. *Journal of the Institution of Engineers (India): Series B*, 102(1), 143–156. https://doi.org/10.1007/s40031-020-00508-y.

Bharat Meitei, H., & Kumar, M. (2022). FPGA implementation of true random number generator architecture using all digital phase-locked loop. *IETE Journal of Research*, 68(3), 1561–1570. https://doi.org/10.1080/03772063 .2021.1963333.

Bhattacharjee, S., Sil, S., Basak, B., & Chakrabarti, A. (2011, December). Evaluation of power efficient adder and multiplier circuits for FPGA based DSP applications. In *International Conference on Communication and Industrial Application* (pp. 1–5). IEEE. https://doi.org/10.1109/ICCIndA .2011.6146691.

Bhattacharjee, S., Sil, S., & Chakrabarti, A. (2013). Evaluation of power efficient FIR filter for FPGA based DSP applications. *Procedia Technology*, 10, 856– 865. https://doi.org/10.1016/j.protcy.2013.12.431.

Datta, D., & Dutta, H. S. (2021). High performance IIR filter implementation on FPGA. *Journal of Electrical Systems and Information Technology*, 8(1), 1–9. https://doi.org/10.1186/s43067-020-00025-4.

Gupta, N., Jain, A., Vaisla, K. S., Kumar, A., & Kumar, R. (2021). Performance analysis of DSDV and OLSR wireless sensor network routing protocols using FPGA hardware and machine learning. *Multimedia Tools and Applications*, 80(14), 22301–22319. https://doi.org/10.1007/s11042-021-10820-4.

Hagras, E. A., & Saber, M. (2020). Low power and high-speed FPGA implementation for 4D memristor chaotic system for image encryption. *Multimedia Tools and Applications*, 79(31–32), 23203–23222,doi: https://doi.org/10 .1007/s11042-019-08517-w.

Haripriya, D., Kumar, K., Shrivastava, A., Al-Khafaji, H. M. R., Moyal, V., & Singh, S. K. (2022). Energy-efficient UART design on FPGA using dynamic voltage scaling for green communication in industrial sector. *Wireless Communications and Mobile Computing*, 2022. https://doi.org/10.1155 /2022/4336647.

Hatti, K., & Paramasivam, C. (2022). Design and implementation of enhanced PUF architecture on FPGA. *International Journal of Electronics Letters*, 10(1), 57–70. https://doi.org/10.1080/21681724.2020.1859141.

Hemmert, K. S., & Underwood, K. D. (2010). Fast, efficient floating-point adders and multipliers for FPGAs. *ACM Transactions on Reconfigurable Technology and Systems (TRETS)*, 3(3), 1–30. https://doi.org/10.1145/1839480.1839481.

Jain, A., Dwivedi, R. K., Alshazly, H., Kumar, A., Bourouis, S., & Kaur, M. (2022). Design and simulation of ring network-on-chip for different configured nodes. *Computers, Materials and Continua*, 71(2), 4085–4100. https:// doi.org/10.32604/cmc.2022.023017.

Jaiswal, M. K., & Cheung, R. C. (2012, May). Area-efficient FPGA implementation of quadruple precision floating point multiplier. In *26th International Parallel and Distributed Processing Symposium Workshops & PhD Forum* (pp. 376–382). IEEE. 10.1109/IPDPSW.2012.46.

Jaiswal, M. K., & Cheung, R. C. (2013). Area-efficient architectures for double precision multiplier on FPGA, with run-time-reconfigurable dual single precision support. *Microelectronics Journal, 44*(5), 421–430. https://doi.org/10.1016/j.mejo.2013.02.021.

Jarrah, A., Haymoor, Z. S., Al-Masri, H. M., & Almomany, A. (2022). High-performance implementation of power components on FPGA platform. *Journal of Electrical Engineering and Technology, 17*(3), 1555–1571, doi: https://doi.org/10.1007/s42835-022-01005-6.

Javeed, K., Wang, X., & Scott, M. (2015, September). Serial and parallel interleaved modular multipliers on FPGA platform. In *25th International Conference on Field Programmable Logic and Applications (FPL)* (pp. 1–4). IEEE. 10.1109/FPL.2015.7293986.

Jianfeng, Z., Hu, H., Dong, W., & Liyang, P. (2011). A cost-efficient self-configurable BIST technique for testing multiplexer-based FPGA interconnect. *Journal of Electronic Testing, 27*(5), 647–655. https://doi.org/10.1007/s10836-011-5238-3.

Khan, S., Javeed, K., & Shah, Y. A. (2018). High-speed FPGA implementation of full-word Montgomery multiplier for ECC applications. *Microprocessors and Microsystems, 62*, 91–101. https://doi.org/10.1016/j.micpro.2018.07.005.

Kumar, A., Kuchhal, P., & Singhal, S. (2015). Design and FPGA synthesis of three stage telecommunication switching in HDL environment. *Procedia Computer Science, 48*, 454–460. https://doi.org/10.1016/j.procs.2015.04.119.

Kumar, A., Rastogi, P., & Srivastava, P. (2015). Design and FPGA implementation of DWT, image text extraction technique. *Procedia Computer Science, 57*, 1015–1025. https://doi.org/10.1016/j.procs.2015.07.512.

Kumar, A., Verma, G., Gupta, M. K., Salauddin, M., Rehman, B. K., & Kumar, D. (2019). 3D multilayer mesh NoC communication and FPGA synthesis. *Wireless Personal Communications, 106*(4), 1855–1873. https://doi.org/10.1007/s11277-018-5724-3.

Kumar, D., Kumar, P., & Pattanaik, M. (2010). Performance analysis of dynamic threshold MOS (DTMOS) based 4-input multiplexer switch for low power and high-speed FPGA design. In *Proceedings of the 23rd Symposium on Integrated Circuits and System Design* (pp. 2–7). https://doi.org/10.1145/1854153.1854156.

Kumar, U. P., Goud, A. S., & Radhika, A. (2013, April). FPGA Implementation of high speed 8-bit Vedic multiplier using barrel shifter. In *International Conference on Energy Efficient Technologies for Sustainability* (pp. 14–17). IEEE. 10.1109/ICEETS.2013.6533349.

Kumar, V. B., Joshi, S., Patkar, S. B., & Narayanan, H. (2010). FPGA based high performance double-precision matrix multiplication. *International Journal of Parallel Programming, 38*(3–4), 322–338. https://doi.org/10.1007/s10766-010-0131-8.

Mishra, V. M., & Kumar, A. (2022). FPGA integrated IEEE 802.15.4 ZigBee wireless sensor nodes performance for industrial plant monitoring and automation. *Nuclear Engineering and Technology, 54*(7), 2444–2452. https://doi.org/10.1016/j.net.2022.01.011.

Mohd, B. J., Hayajneh, T., Khalaf, Z. A., & Ahmad Yousef, K. M. (2016). Modeling and optimization of the lightweight HIGHT block cipher design with FPGA implementation. *Security and Communication Networks, 9*(13), 2200–2216. https://doi.org/10.1002/sec.1479.

Nguyen, N. H., Khan, S. A., Kim, C. H., & Kim, J. M. (2017, April 3–7). An FPGA-based implementation of a pipelined FFT processor for high-speed signal processing applications. In *Applied Reconfigurable Computing Proceedings of the 13: 13th International Symposium* (pp. 81–89). ARC, Springer International Publishing. https://doi.org/10.1007/978-3-319-56258-2_8.

Ompal, Mishra, V. M., & Kumar, A. (2021). Zigbee internode communication and FPGA synthesis using mesh, star and cluster tree topological chip. *Wireless Personal Communications, 119*(2), 1321–1339. https://doi.org/10.1007/s11277-021-08282-w.

Pandey, B., Das, B., Kaur, A., Kumar, T., Khan, A. M., Akbar Hussain, D. M., & Tomar, G. S. (2017). Performance evaluation of FIR filter after implementation on different FPGA and SOC and its utilization in communication and network. *Wireless Personal Communications, 95*(2), 375–389. https://doi.org/10.1007/s11277-016-3898-0.

Pham, T. H., Fahmy, S. A., & McLoughlin, I. V. (2012). Low-power correlation for IEEE 802.16 OFDM synchronization on FPGA. *IEEE Transactions on Very Large Scale Integration (VLSI) Systems, 21*(8), 1549–1553. https://doi.org/10.1109/TVLSI.2012.2210917.

Qasim, S. M., Telba, A. A., & AlMazroo, A. Y. (2010). FPGA design and implementation of matrix multiplier architectures for image and signal processing applications. *International Journal of Computer Science and Network Security, 10*(2), 168–176.

Ram, R. S., Prabhaker, M. L. C., Suresh, K., Subramaniam, K., & Venkatesan, M. (2020). Dynamic partial reconfiguration enhanced with security system for reduced area and low power consumption. *Microprocessors and Microsystems, 76*, 103088. https://doi.org/10.1016/j.micpro.2020.103088.

Rajaei, R., & Gholipour, A. (2018). Low power, reliable, and nonvolatile MSRAM cell for facilitating power gating and nonvolatile dynamically reconfiguration. *IEEE Transactions on Nanotechnology, 17*(2), 261–267. https://doi.org/10.1109/TNANO.2018.2792782.

Rahimunnisa, K., Karthigaikumar, P., Christy, N. A., Kumar, S. S., & Jayakumar, J. (2013). PSP: Parallel sub-pipelined architecture for high throughput AES on FPGA and ASIC. *Central European Journal of Computer Science, 3*(4), 173–186. https://doi.org/10.2478/s13537-013-0112-2.

Rais, M. H. (2010). Hardware implementation of truncated multipliers using spartan-3an, virtex-4 and virtex-5 FPGA devices. *American Journal of Engineering and Applied Sciences, 3*(1), 201–206.

Reyes Fernandez de Bulnes, D., Maldonado, Y., & Trujillo, L. (2020). Development of multiobjective high-level synthesis for FPGAs. *Scientific Programming, 2020*. https://doi.org/10.1155/2020/7095048.

Reddy, B. N. K., Seetharamulu, B., Krishna, G. S., & Vani, B. V. (2022). An FPGA and ASIC implementation of cubing architecture. *Wireless Personal Communications, 125*(4), 3379–3391. https://doi.org/10.1007/s11277-022-09715-w.

Rodríguez-Andina, J. J., Valdes-Pena, M. D., & Moure, M. J. (2015). Advanced features and industrial applications of FPGAs-A review. *IEEE Transactions on Industrial Informatics*, 11(4), 853–864. https://doi.org/10.1109/TII.2015.2431223.

Seshadri, R., & Ramakrishnan, S. (2021). FPGA implementation of fast digital FIR and IIR filters. *Concurrency and Computation: Practice and Experience*, 33(3), e5246. https://doi.org/10.1002/cpe.5246.

Skhiri, R., Fresse, V., Jamont, J. P., Suffran, B., & Malek, J. (2019). From FPGA to support cloud to cloud of FPGA: State of the art. *International Journal of Reconfigurable Computing*, 2019, 1–17. https://doi.org/10.1155/2019/8085461.

Solanki, V., Darji, A. D., & Singapuri, H. (2021). Design of low-power wallace tree multiplier architecture using modular approach. *Circuits, Systems, and Signal Processing*, 40(9), 4407–4427. https://doi.org/10.1007/s00034-021-01671-3.

Thamizharasan, V., & Kasthuri, N. (2021). High-speed hybrid multiplier design using a hybrid adder with FPGA implementation. *IETE Journal of Research*, 1–9. https://doi.org/10.1080/03772063.2021.1912655.

Usmani, M. A., Keshavarz, S., Matthews, E., Shannon, L., Tessier, R., & Holcomb, D. E. (2018). Efficient PUF-based key generation in FPGAs using per-device configuration. *IEEE Transactions on Very Large Scale Integration (VLSI) Systems*, 27(2), 364–375. https://doi.org/10.1109/TVLSI.2018.2877438.

Verma, G., Kumar, M., Khare, V., & Pandey, B. (2017). Analysis of low power consumption techniques on FPGA for wireless devices. *Wireless Personal Communications*, 95(2), 353–364. https://doi.org/10.1007/s11277-016-3896-2.

Chapter 2

CNN-embedded hardware chip design and simulation

Mukul Kumar, Monika Gupta, and Adesh Kumar

2.1 INTRODUCTION

Convolutional neural networks (CNNs) are a standard multilayer neural network. Deep learning technology is constantly evolving, and it is currently being used extensively in voice analysis and machine vision. Convolutional neural networks use central processing units (CPUs) to carry out their computations. It is challenging to achieve real-time computation requirements with such calculations since they are slow and inefficient. Thus, graphics processing unit (GPU)-based CNNs are popular. The literature on CNN open-source GPU projects revealed GPU concerns including excessive cost and power consumption. A popular method of designing digital circuits is field programmable gate arrays (FPGAs), which are hardware circuit structures that may be customized for programming. The computational features of CNNs are compatible with the parallel computing mode offered by FPGA. In addition, the reprogrammable properties of FPGAs are appropriate for the flexible network architecture of neural networks. As a result, the FPGA-based design of CNNs has drawn a lot of attention (Kumar et al. 2018). An FPGA-based deep CNN accelerator is suggested. To achieve a high-efficiency CNN, a deep pipeline FPGA cluster is created.

CNNs have produced incredibly precise results in a variety of application areas, as was discussed in the previous section. This kind of neural network advancement is happening quickly. State-of-the-art CNNs are continually being pushed by more potent hardware, bigger datasets, bigger models, new algorithms, and superior network topologies. The major goal of this thesis is to have a deeper understanding of the field of efficient deep neural network architecture design and evaluation. This research suggests an FPGA-based CNN accelerator intending to lower the computational cost in terms of time performance and power consumption. More specifically, the work discussed in this study is concentrated on the creation of a CNN architecture for high-performance computing on an FPGA. CNNs are the most common type of multilayer neural network. They are commonly engaged in the areas of voice analysis and machine vision due to the ongoing development of deep learning technology. Calculations using CPUs are carried

DOI: 10.1201/9781003510420-2

out by conventional convolutional neural networks. Real-time computation demands are challenging to meet with such calculations because they are slow and inefficient. So convolutional neural networks based on graphics processing units are frequently used. The literature analyzed CNN open-source techniques that rely on GPUs and identified several drawbacks associated with GPUs, such as substantial power consumption and exorbitant costs. The FPGA is a widely used method for building digital circuits due to its ability to enable customized programming of hardware circuitry. The FPGA's parallel processing mode is compatible with the computational properties of convolutional neural networks. The FPGA's reconfigurable characteristics are well-suited for accommodating the dynamic network architectures of neural networks simultaneously. The CNN architecture implemented on an FPGA has generated significant interest. This chapter introduces an FPGA-based deep CNN accelerator. A deep pipeline FPGA cluster is developed for the development of a very efficient CNN. Many studies have already been published since deep learning was first mentioned in the literature. Thousands of these publications covered numerous frameworks, network architecture, databases, training, and validation software implementations for different applications. Of these hundreds of studies, only a small number discuss hardware implementation and acceleration. CNNs have been used widely for the assessment of brain activity (Goel et al. 2023, Kumar 2023).

2.2 LITERATURE REVIEW

Huang et al. (2021) presented a high-throughput, highly resource-efficient CNN hardware accelerator built on FPGA. Due to the ineffective computational resource mapping technique, the data supply issue, and other factors, the accelerator's real throughput is increasing, but it is still significantly lower than the theoretical throughput. This chapter suggests a revolutionary composite hardware CNN accelerator architecture to address these issues. A brand-new multi-central-point enhanced (CE) architecture built on a row-level pipelined streaming approach is suggested to effectively execute the convolution layer (CL). A continuous data supply and efficient data system are created to prevent the idle state of the CE, and an optimum mapping mechanism is proposed for each CE to increase its computing resource usage ratio. A weighted data allocation approach is also recommended to ease the strain on off-chip bandwidth. Kyriakos et al. (2019) introduced a high-performance accelerator for CNN applications. Accelerators for AI that are implemented in hardware offer the required performance gain for applications in particular fields, such as robotics, autonomous systems, and IoT. An FPGA-based CNN accelerator is presented in the current study. The MNIST dataset is used to train the CNN model, and the VHDL design only uses on-chip memory while aiming for high throughput and low power. The Xilinx Virtex VC707 architectural implementation validates the outcomes. Wu et al. (2019) investigated a CNN processor for MobileNets based on FPGA. With only a slight loss in accuracy, modern CNN MobileNet has significantly decreased operations and parameters by converting from normal convolution to depth-wise separable convolution. This research proposes a high-performance CNN processor based on FPGA.

Liu et al. (2021) introduced an edge computing system using FPGA-based CNN accelerators to enhance energy efficiency. Conventional CNNs are generally unsuitable for large-scale, time-sensitive failure tolerance (FT) systems due to their substantial energy consumption and demanding processing needs. To achieve rapid and energy-efficient failure tolerance, create an edge-cloud collaborative computing (ECCC) system using AI and the IoT (AIoT). This system is built on energy-efficient FPGAs that employ CNN accelerators. The AIoT-enabled ECCC system is structured with four main components: an application subsystem, an edge-cloud collaborative subsystem, an IoT subsystem, and an intelligent computing subsystem. After that, a hardware accelerator is built that is based on FPGA technology and is specifically designed for the tiny MobileNet CNN. This was accomplished by applying hardware design approaches such as systolic array, matrix tiling, and parallelism. The scheduling technique was implemented for this diverse system that is both conscious of delays and economical in its use of energy. The energy consumption and processing time of CNNs can be significantly decreased using the previously outlined hardware and software codesign approach. Lu et al. (2021) demonstrated an FPGA-based sparse CNN hardware accelerator. As CNN models become more complex, compressing CNN to sparse by removing redundant network connections is a popular way to save computing and memory. Recent research has revealed that FPGAs can accelerate CNN inference. Dense CNN models dominate modern FPGA architectures. Modern sparse FPGA accelerators just focus on fully connected layers. Next, develop an FPGA architecture that manages input-weight and weight-output links. We develop a tile look-up table for the input-weight connection to eliminate compressed weight runtime indexing matches. We also design a weight arrangement for high on-chip memory access. Shamma et al. (2020) boosted CNN performance based on an FPGA accelerator. CNNs are commonly used for image recognition due to their accuracy. Simulating human visual nerves achieves this accuracy. The rapid development of deep learning applications has enhanced research and development. An FPGA-based architecture for deep CNN accelerators is planned due to its fast development cycles, reconfigurability, and high performance. The FPGA is faster than the CPU due to its parallel mechanism and low energy consumption. Mittal (2020) described FPGA-based CNN accelerators using deep CNNs, which have shown great accuracy in various thought tasks. CNN performance depends on specialized hardware accelerators because they process a lot. FPGA's energy efficiency, computing capability, and reconfigurability make it a good platform for CNN hardware acceleration. The study should benefit hardware architecture, system design, and AI researchers. See the review on FPGA-based computing architecture for CNN inference by Xiyuan et al. (2021). CNN models are computationally complex and have many parameters, making them unsuitable for common processors. An FPGA-based computing architecture may boost CNN inference performance. Software-hardware codesign improves inference performance, processing overhead, and accuracy. The research covers CNN structure design methods and offers an example to demonstrate how they can improve CNN inference performance. Discussion of difficulties and prospective research directions encourages this area of research. Hu and Chen et al (2019) presented an FPGA CNN accelerator. CNN accuracy is record-setting across all picture interpretation metrics, but computational complexity is considerable. Today's high-end FPGA generations have several hardened functional units for rapid and efficient implementation of common functions. Additionally, these generations feature hundreds of thousands of programmable logic blocks. Many researchers have proposed FPGA-based CNN accelerators. However, contemporary approaches have not adequately addressed data reliance. Data dependency greatly affects accelerator performance. Current techniques address data dependency by adding hardware modules to FPGAs. Unfortunately, contemporary approaches have not overcome the data dependence problem. Data dependence severely affects accelerator performance. Adding hardware modules to FPGAs solves data dependency. This method is useless and complicates hardware. The data dependency of CNNs could be addressed by rearranging the data. The reorganized data is saved in a hardware-compatible format. The accelerator will use pipeline technology better than current designs. Wen et al. (2020) proposed an efficient FPGA accelerator optimized for CNN inference. By compressing CNN models using pruning techniques, the enormous quantity of labor required by large-scale CNNs can be reduced. This is accomplished by setting the weights for unimportant components to zero. However, because pruned weights are distributed at random, it is extremely difficult for hardware architectures to load and process nonzero data effectively and with high parallelism. It is suggested to use a sparsity-aware CNN accelerator to process the irregularly trimmed CNN models to overcome this issue. A candidate pool architecture is used to select only the nonzero-weighted, randomly required activations, and the proposed sparsity-aware CNN accelerator is shown to improve throughput by up to 89.7% over the baseline design for several widely used CNN models. Hamdan and Rover (2017) presented a VHDL generator for a high-performance CNN FPGA-based accelerator. Modern CNNs are computationally demanding, but because of their parallelism and modularity, systems like FPGAs are well-suited for the accelerating process. This study introduced a tool that empowers developers to generate VHDL code for a specific CNN model in a robotic fashion using a configurable user interface. The resulting code or architecture is modular, fully pipelined, scalable, reconfigurable, and adaptable to a wide variety of CNN models. By developing LeNet, a small-scale CNN model, and AlexNet, a large-scale CNN model, we demonstrate the versatility of the autonomous VHDL generator.

Saglam et al. (2019) presented an FPGA implementation of a CNN algorithm to recognize blood cells infected with malaria. It involves the completion of CNN implementation on FPGA for categorizing cells infected with malaria. The hardware is developed and implemented using Xilinx Zynq-7000 FPGA and VHDL. Image processing employs the CNN classification technique to facilitate expert analysis and commentary on diseased cells. The classification procedure allows us to deliver a more straightforward interpretation by categorizing intricate pictures This breakthrough makes it easier to use image processing for early diagnosis in the medical field, which immediately lowers death and treatment expenses. The experimental results show that for 200 8 × 8 binary images, the accuracy rate for locating a malaria-infected cell using the CNN approach is 94.76%. Shahshahani et al. (2018) introduced a memory optimization technique for FPGA-based CNN implementations. Speech recognition, picture classification, and natural language processing have all benefited from deep learning. These learning algorithms are typically executed on groups of CPUs and GPUs. However, the models built on CPUs and GPUs are not scalable as data sizes grow. FPGA is useful in this situation. The development of computer-aided design (CAD) tools for FPGAs has eliminated the requirement for designers to develop network topologies at the RTL level using HDLs like Verilog and VHDL. When creating the models with software like Xilinx Vivado HLS, they can use high-level languages like C or C++. Additionally, the power usage of FPGA-based. The storage of weights and pictures requires a significant amount of memory for the CNN architectures. Bouaafia et al. (2022) proposed an FPGA-based hardware acceleration of CNN. CNNs have proven beneficial in image processing and computer vision applications. However, they do necessitate intensive CPU operations and memory bandwidth, which hinder general-purpose CPUs from achieving the desired levels of speed. Hardware accelerators like GPUs, FPGAs, and ASICs have been used to increase CNN throughput. Due to their capacity to maximize parallelism and power economy, FPGAs have recently been utilized to speed up the creation of deep learning networks. Many deep learning applications have made extensive use of CNNs. The large number of convolutional operations in convolutional neural networks makes real-time performance extremely difficult. Due to their great performance and energy efficiency, FPGA implementations of convolutional accelerators have attracted a lot of attention nowadays. In this study, using a systolic array architecture and a Xilinx ZedBoard device, we construct an accelerator for convolutional processes. The results of the experiments demonstrate that the accelerators we created can achieve performance densities of up to 0.032 Gop/s/DSP.

Linares et al. (2021) suggested the use of FPGAs as accelerators for CNNs in the context of visual categorization. CNN exhibits high accuracy in classification tasks. In recent years, there has been a proliferation of hardware accelerators aimed at enhancing CPU- or GPU-based systems. Before contemplating the production of an ASIC for large-scale manufacture, it is common practice to prototype and evaluate this technology using FPGAs. The performance of these systems in high-speed applications is limited using commercially available cameras. By utilizing a 30 frames per second (fps) dynamic vision sensor (DVS), the computation of a frame can be achieved at the maximum rate supported by a CNN accelerator. This study presents a VHDL/HLS representation of an FPGA pipelined system designed to collect events from an address-event-representation (AER) DVS retina. The system aims to provide a normalized histogram that can be utilized by a specialized CNN accelerator called Null Hop. The VHDL language is used to describe the circuit responsible for normalizing a frame in CNN, while the computation blocks are specified using high-level synthesis (HLS). The results exhibit superior performance in terms of latency and power consumption compared to previous endeavors including frame gathering and normalization using ARM processors running at 800 MHz on a Zynq 7100. Alaeddine et al. (2021) presented a highly effective network structure on an FPGA accelerator. Utilizing FPGA-based acceleration appears to be the optimal choice for improving the energy efficiency and performance of computing CNN. A hardware and software accelerator has been developed to effectively speed up the performance of FPGA-based accelerators. This accelerator utilizes parallelism and weight reorganization to maximize the computing capability of the FPGA. In addition, the accelerator allows for the calculation of MLPs and 33 convolutional layers without relying on the CPU. The TensorFlow deep learning framework incorporates this accelerator to provide software developers with a straightforward interface via which they may define a network and utilize an FPGA engine. Utilizing the PYNQ-Z1 platform and a Xilinx Zynq SoC, this system achieves a frame rate of 5.91 frames per second using 16-bit fixed point arithmetic. Additionally, it exhibits an energy efficiency that is 186.25 times greater than a CNN processed by an Intel Xeon on FPGAs. The input data is initially partitioned using tiling methodologies. Subsequently, the suggested accelerator is integrated into the TensorFlow deep learning framework. The VHDL autogeneration tool is employed to enhance the hardware acceleration of CNN on FPGA. The CNN, a popular machine learning technique, has been used in handwritten digit identification, visual perception, and image categorization. Empirical data has proven that this method is highly accurate and effective. Modern CNNs require significant processing resources. Nevertheless, their intrinsic parallelism and adaptability render them highly suitable for integration with platforms like FPGAs to achieve optimal acceleration. The developed architecture or code is highly optimized, exhibiting modularity, high parallelism, reconfigurability, scalability, pipelining, and adaptability to CNN models. The adaptability of the automatic VHDL creation tool is demonstrated by implementing two CNN models, LeNet-5 and AlexNet, which differ in size. Unlike the conventional approach used for large-scale models, the code for the small-scale model does not incorporate any external memory management for the CNN parameters. Korol and Moraes (2019) presented an FPGA-based multilayer architecture for CNNs. Convolutional neural networks are used to address issues across a variety of disciplines, including natural language processing. To accelerate the performance of CNNs, specialized architectures have been proposed in response to advancements in the learning and inferencing algorithms for CNNs. The CNNs' demands for bandwidth and processing power present a challenge to engineers, who must develop architectures appropriate for ASICs and FPGAs. CNNs could be useful in embedded applications like IoT, medical equipment, smartphones, and other battery-powered devices. That requires a distinct approach for the CNN design, where the cost function is a compact area footprint and lower power usage. This work takes a step in that direction by suggesting an architecture for the central components of contemporary CNNs. The Alexnet CNN is used as a case study in the proposal, which is directed toward Xilinx FPGA devices. Results show a reduction in the number of required DSP modules of up to nine times when compared to the literature (Lu et al. 2019).

2.3 CNN DESIGN

In recent years, different researchers have done sustainable work in the direction of hardware chip design and accelerators for specific applications. The fundamental layers of the hierarchical CNN model are input layers, convolution layers, pooling layers, fully connected layers, and output layers (Dhyani et al. 2023). The primary characteristic is that the pooling and convolution layers alternate, with the local feature of the earlier layer acquiring the properties of the subsequent convolution layer via the convolution's weight. Images increasingly transition from surface to deep characteristics through the connections of numerous convolutional layers with pooling layers (Kumar 2023). The features are then categorized by the fully connected layer and the output layer, and the images are finally split into classes. Therefore, the CNN may be separated into two components based on the purpose of each layer: the feature extractor is made up of the input layer, the convolution layer, and the pooling layer, while the classifier is made up of the fully connected layer and the output layer. Figure 2.1 depicts the fundamental structure of the CNN proposed for the chip

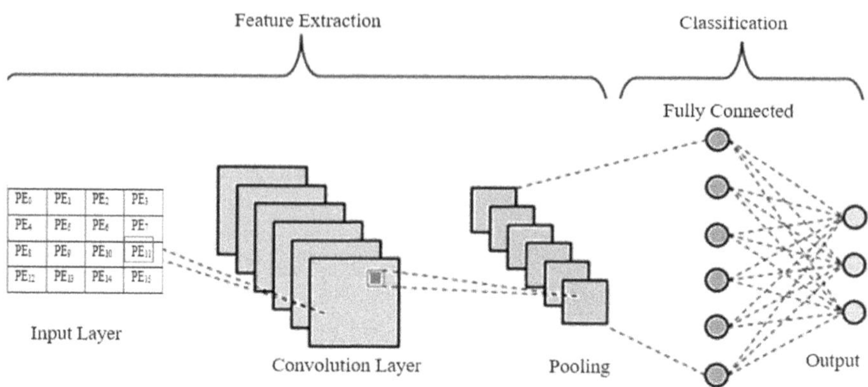

Figure 2.1 **The basic structure of the convolution neural network.**

design, in which the input data can be processed in 16 processing units in the input layer.

A CNN is comprised of three distinct layers: convolutional, pooling, and fully connected layers. These layers are all interconnected with one another. The formation of a CNN architecture occurs when these layers consist of weights. In addition to these three layers, the activation function and the dropout layer are two more significant aspects that are taken into consideration (Rawat et al. 2018).

Convolutional layer: This is the first layer that was utilized in the process of extracting the numerous attributes from the images that were sent in. At this level, a mathematical operation known as convolution is carried out between the picture that is being input and a filter that has the sizes (M × M). The calculation of the spot product between the filter and the areas of the picture regarding the filter size (M × M) is accomplished by moving the filter across the input image (Wu et al. 2023).

Pooling layer: A pooling layer is often placed after a convolutional layer in the neural network. One of the primary goals of this layer is to reduce the size of the convolved feature map to reduce the number of computational resources that are required. By removing the linkages between layers and doing this action independently on each feature map, this is accomplished. There are many kinds of pooling procedures, and each one is determined by the mechanism that is used. In its most basic form, it is an introduction to the characteristics that are generated by a convolution layer (Gupta & Kumar 2023).

Fully connected layer: Connecting neurons that are sandwiched between two layers, the fully connected (FC) layer also includes biases and weights in addition to the neurons themselves. Typically, the output layer comes before the last few layers of CNN architecture.

This step feeds the FC layer a compressed version of the input picture from the preceding layers. Mathematical function operations often take place in successive FC levels, which are applied to the compressed vector. The process of categorization begins now.

Dropout: Typically, overfitting in the training dataset might occur if all characteristics related to the FC layer were used (Dhyani et al. 2023). The term "overfitting" describes what happens when a model does exceptionally well on training data but poorly on new data. To tackle this challenge, we employ a dropout layer during training, which reduces the size of the model by removing some neurons from the neural network. After the dropout threshold is crossed, 35% of the nodes in the neural network are removed without purpose.

2.4 RESULTS AND DISCUSSION

The suggested design has been implemented using ISE Design Suite 14.7. Xilinx has created ISE Design Suite 14.7, a software that enables the synthesis and analysis of HDL designs, supporting both VHDL and Verilog. VHDL programming is used for the modeling of the chip. The model was subjected to timing and behavior simulations using ISE Design Suite 14.7 software and ModelSim software. The RTL and internal schematics architecture of a simple CNN accelerator are depicted in Figure 2.2 and Figure 2.3, respectively. The CNN accelerator chip consists of seven pins, with six

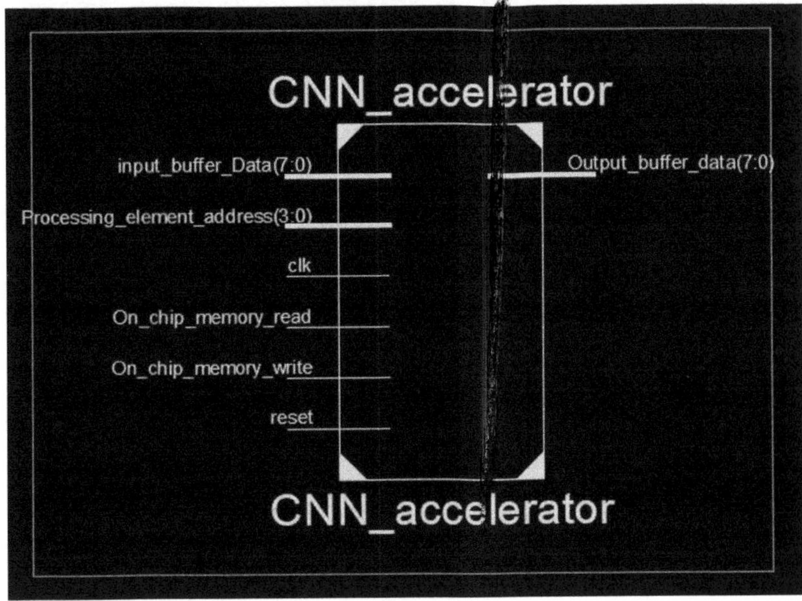

Figure 2.2 RTL of simple CNN accelerator.

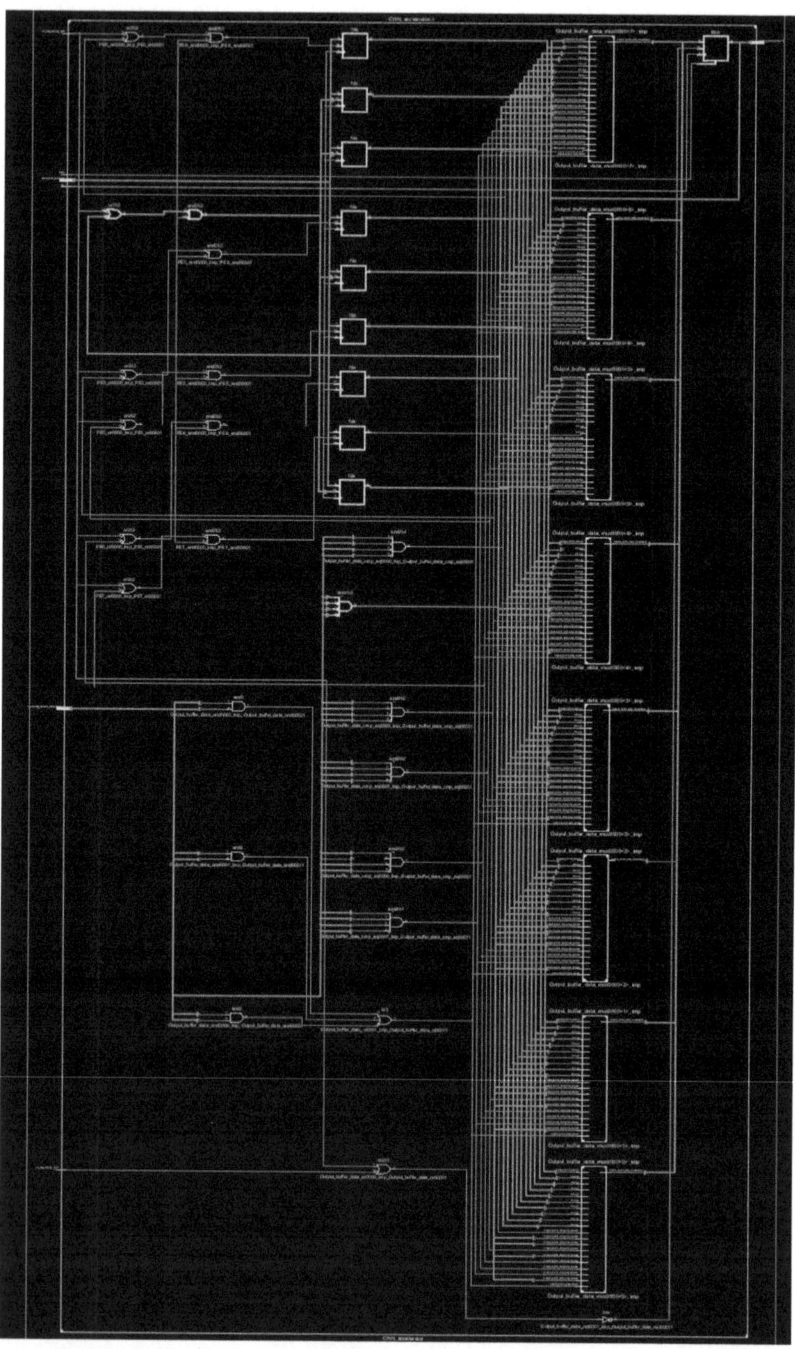

Figure 2.3 Internal schematic view of chip.

CNN_accelerator Project Status (06/10/2023 - 14:53:28)			
Project File:	CNN_accelerator.xise	Parser Errors:	No Errors
Module Name:	CNN_accelerator	Implementation State:	Synthesized
Target Device:	xc5vlx20t-2ff323	• Errors:	No Errors
Product Version:	ISE 14.7	• Warnings:	2 Warnings (0 new)
Design Goal:	Balanced	• Routing Results:	
Design Strategy:	Xilinx Default (unlocked)	• Timing Constraints:	
Environment:	System Settings	• Final Timing Score:	

Device Utilization Summary (estimated values)				[-]
Logic Utilization	Used	Available	Utilization	
Number of Slice Registers	80	12480		0%
Number of Slice LUTs	42	12480		0%
Number of fully used LUT-FF pairs	13	109		11%
Number of bonded IOBs	24	172		13%
Number of BUFG/BUFGCTRLs	1	32		3%

Figure 2.4 Estimated hardware parameters for CNN accelerator.

pins serving as input pins and one pin serving as an output pin to receive and save data in the input buffer.

The details of the RTL pins are given to understand the functionality of the design. Input_buffer_data (7:0) presents the 8-bit data considered for the processing elements and the corresponding output is taken based on output_buffer_data (7:0). The identification of the processing_element_address (3:0) is considered to address 16 processing element inputs of the CNN model. The accelerator has the storage and processing capability, and the data is written and read using on_chip_memory_write and on_chip_memory_read control pins. The clock and reset serve as the default inputs due to the sequential structure of the architecture (Ompal et al. 2021). The FPGA hardware synthesis report is generated by Xilinx ISE 14.7 according to our designed CNN_ACCELERATOR and shows a device utilization summary of the proposed CNN accelerator. Figure 2.4 presents the estimated hardware parameters for the CNN accelerator. The processing elements are placed in the mesh network design (4 × 4) (Jain et al. 2022) and then processed using different layers of the CNN based on their address. The address in the mesh-based design and processing elements are recognized based on row and column address.

The simulation of the CNN accelerator is shown in Figure 2.5 and Figure 2.6 in which the data is written in all the processing elements based on their address. The following steps are used to sum the output.

2.4.1 Test bench inputs and outputs for test 1

Step 1: Assign reset = 1 , then Run the operation , output_buffer_data will get the value = 00000000.

Step 2: Assign reset = 0, and Clock will be assigned any positive edge of clock, then on_chip_memory_write = 1 and on_chip_memory_read = 0, After that assign the value of input output_buffer_data = 10101010, Then

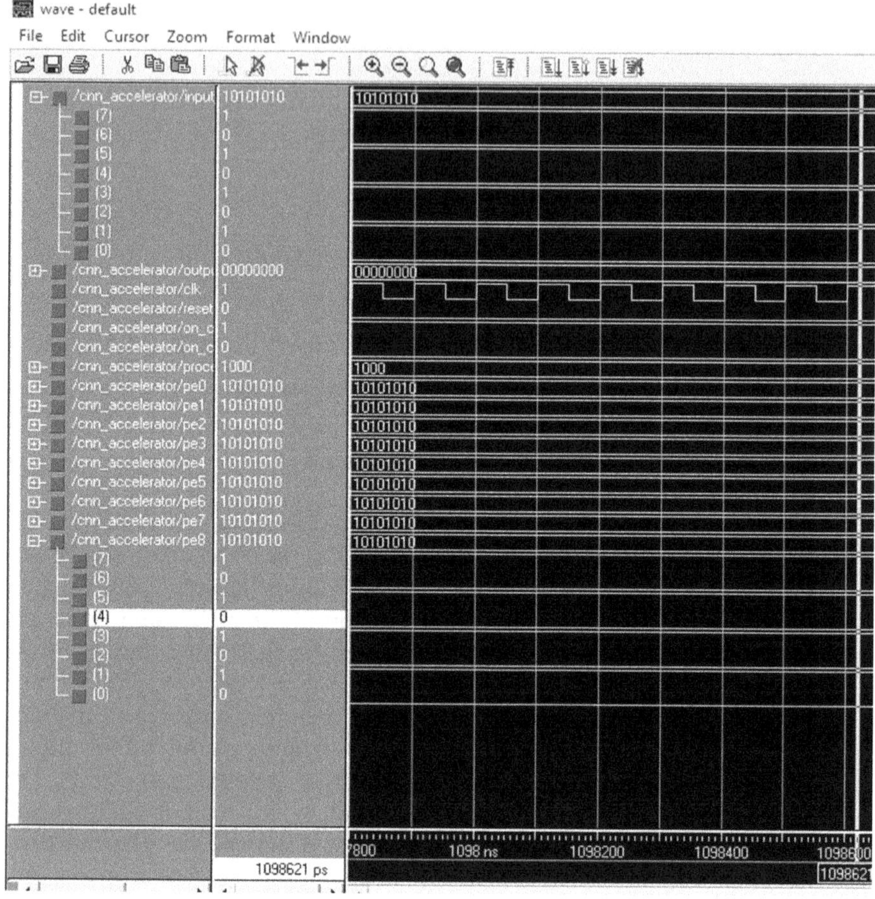

Figure 2.5 ModelSim simulation waveform output for CNN accelerator.

processing_element_address = 0000 the PE_0 data = 10101010, processing_element_address = 0001 the PE_1 data = 10101010, processing_element_address = 0010 the PE_2 data = 10101010, processing_element_address = 0011 the PE_3 data = 10101010, processing_element_address = 0100 the PE_4 data = 10101010, processing_element_address = 0101 the PE_5 data = 10101010, processing_element_address = 0110 the PE_6 data = 10101010, processing_element_address = 0111 the PE_7 data = 10101010, processing_element_address = 1000 the PE_8 data = 10101010, processing_element_address = 1001 the PE_9 data = 10101010,, processing_element_address = 1010 the PE_{10} data = 10101010, processing_element_address = 1011 the PE_{11} data = 10101010, processing_element_address = 1100 the PE_{12} data = 10101010, processing_element_address = 1101 the PE_{13} data = 10101010,

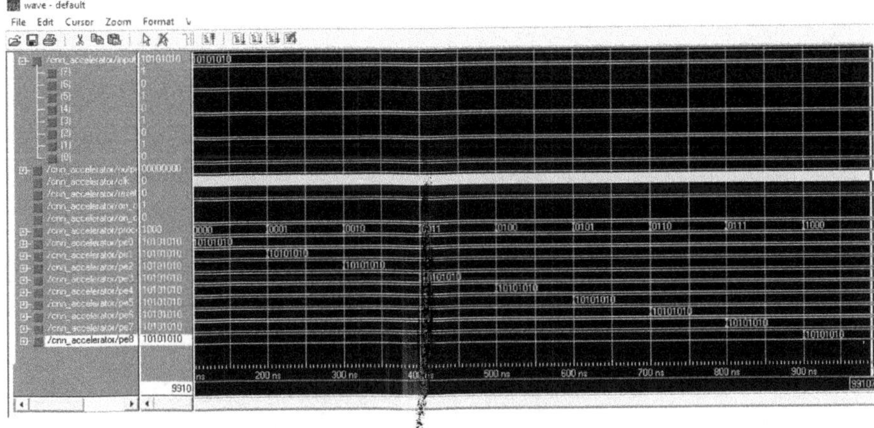

Figure 2.6 CNN accelerator output waveform of write data input in ModelSim.

processing_element_address = 1110 the PE$_{14}$ data = 10101010, and processing_element_address = 1111 the PE$_{15}$ data = 10101010.

Step 3: Assign reset = 0, and the Clock will be any positive edge of the clock, then on-chip memory write = 0 and on-chip memory read = 1. After the assign the value of with input buffer data = 10101010. Then processing_element_address = 0000 the output_buffer_data = 10101010 for PE$_0$, processing_element_address = 0001 the output_buffer_data = 10101010 for PE$_1$, processing_element_address = 0010 the output_buffer_data = 10101010 for PE$_2$, processing_element_address = 0011 the output_buffer_data = 10101010 for PE$_3$, processing_element_address = 0100 the output_buffer_data = 10101010 for PE$_4$, processing_element_address = 0101 the output_buffer_data = 10101010 for PE$_5$, processing_element_address = 0100 the output_buffer_data = 10101010 for PE$_6$, processing_element_address = 0111 the output_buffer_data = 10101010 for PE$_7$, processing_element_address = 1000 the output_buffer_data = 10101010 for PE$_8$, processing_element_address = 1001 the output_buffer_data = 10101010 for PE$_9$, processing_element_address = 1010 the output_buffer_data = 10101010 for PE$_{10}$, processing_element_address = 1011 the output_buffer_data = 10101010 for PE$_{11}$, processing_element_address = 1100 the output_buffer_data = 10101010 for PE$_{12}$, processing_element_address = 1101 the output_buffer_data = 10101010 for PE$_{13}$, processing_element_address = 1110 the output_buffer_data = 10101010 for PE$_{14}$, and processing_element_address = 1111 the output_buffer_data = 10101010 for PE$_{15}$.

The simulation of the CNN accelerator is verified for another test data as shown in Figure 2.7 in which the data is written in all the processing elements based on their address. The following steps are used to sum the output. Figure 2.8 depicts the CNN accelerator out waveform of read data output for another test data.

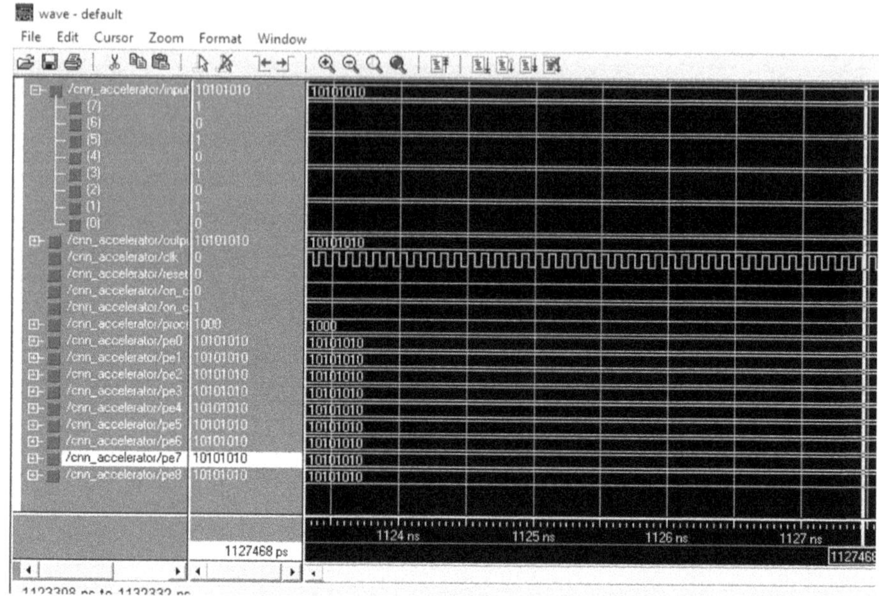

Figure 2.7 CNN accelerator output waveform of write data input in ModelSim.

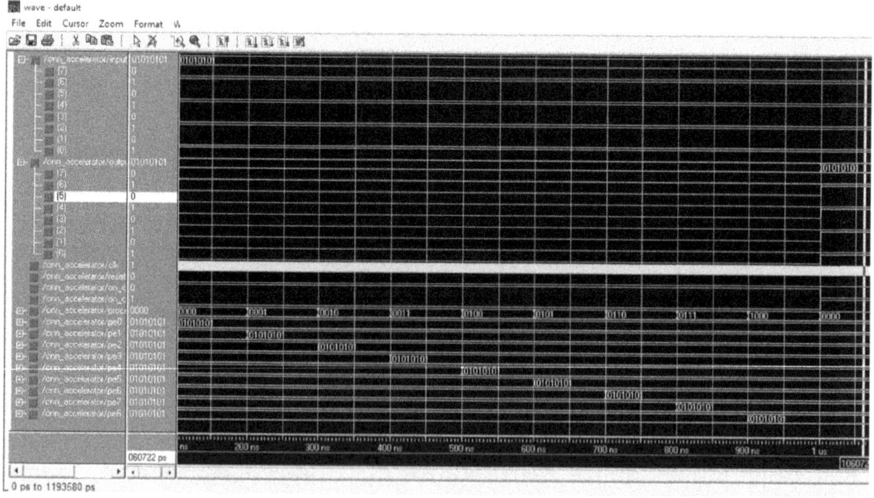

Figure 2.8 CNN accelerator out waveform of read data output for another test data.

2.4.2 Test bench inputs and outputs for test 2

Step 1: Assign Reset value = 1, then Run the operation, output_buffer_data will get the value = 00000000.

Step 2: Assign reset = 0, and the clock will be assigned any positive edge of the clock, then on_chip_memory_write = 1 and on_chip_memory_read = 0. After that assign the value of input output_buffer_data = 01010101. Then processing_element_address = 0000 the PE_0 data = 01010101, processing_element_address = 0001 the PE_1 data = 01010101, processing_element_address = 0010 the PE_2 data = 01010101, processing_element_address = 0011 the PE3 data = 01010101, processing_element_address = 0100 the PE_4 data = 01010101, processing_element_address = 0101 the PE_5 data = 01010101, processing_element_address = 0110 the PE_6 data = 01010101, processing_element_address = 0111 the PE_7 data = 01010101, processing_element_address = 1000 the PE_8 data = 01010101, processing_element_address = 1001 the PE_9 data = 01010101, processing_element_address = 1010 the PE_{10} data = 01010101, processing_element_address = 1011 the PE_{11} data = 01010101, processing_element_address = 1100 the PE_{12} data = 01010101, processing_element_address = 1101 the PE_{13} data = 01010101, processing_element_address = 1110 the PE_{14} data = 01010101, and processing_element_address = 1111 the PE_{15} data = 01010101.

Step 3: Assign reset = 0, and Clock will be any positive edge of the clock, then on-chip memory write = 0 and on-chip memory read = 1. After the assign the value of with input buffer data = 01010101,Then processing_element_address = 0000 the output_buffer_data = 0101010, processing_element_address = 0001 the output_buffer_data = 01010101, processing_element_address = 0010 the output_buffer_data = 01010101, processing_element_address = 0011 the output_buffer_data = 01010101, processing_element_address = 0100 the output_buffer_data = 01010101, processing_element_address = 0101 the output_buffer_data = 01010101, processing_element_address = 0100 the output_buffer_data = 01010101, processing_element_address = 0111 the output_buffer_data = 01010101.

The performance of the designed chip is accessed based on the comparison of the existing work of Lu et al. (2021) and Jameil and Al-Raweshidy (2022) with different parameters such as memory, DSP elements, delay, LUTs, and frequency. Table 2.1 presents the comparison with existing work on CNN chips.

The number of GOP/S in our intended design is 24, which is lower than other designs, and our KLUT is 0.575, which is also lower than other designs. Therefore, as compared to other existing CNN designs, ours has less device consumption and is more optimized. So, we're on the right track. The frequency of our design is 582.683 MHz, which is higher than the existing design. So, our design is fast and has high switching capability.

Table 2.1 Comparison with existing work on CNN chips

Parameter	Lu et al. (2021)	Jameil and Al-Raweshidy (2022)	Purposed work
Method	I_D CNN	I_D CNN	I_D CNN
FD	Zyxc7z 045	Zyxc7z 045	Virtex-5
Power (W)	—	0.176	0.168
Frequency (MHz)	200	442.94	582.683
GOP/s	25.6	30	24
Data size (bit)	Fixed 16	Fixed 16	Fixed 8 and 16
Parameter	11065	3878178	118048
KLUT	1.538	1.067	0.575
DSP	80	9	9
GOP/S/KLUT	16.71	28	18

2.5 SUMMARY

This chapter presents a concept for chip design that can be synthesized on a universal FPGA and can accommodate 16 processing elements in a mesh arrangement. A 1-D convolutional neural network technique may achieve an average accuracy of about 99% when tested against the projected model, far higher than the other models. Furthermore, the model that was recommended as having the best average accuracy of a 1-D CNN accelerator is put into action. A resource-efficient accelerator is produced by the recommended 1-D CNN accelerator, which has an energy efficiency of 24 GOP/s/W and 18 GOP/s/KLUT, which is an improvement over the previous model. In contrast to other existing designs, the KLUT of our proposed design is 0.575, which is lower. Moreover, the CNN offers a lower and more appropriate device usage as compared to earlier CNN chip designs. The frequency of the design is 582.683 MHz, which is greater than the frequency of the current design. As a consequence of this, our design is not only speedy but also a highly alterable, platform-based one-dimensional CNN with an accelerator based on Xilinx Virtex-5 FPGA. Subsequent investigations will prioritize cutting-edge technologies that enhance the accuracy, velocity, and effectiveness of categorization while minimizing the utilization of hardware resources. There are plans to develop an accelerator that minimizes power consumption while enabling the integration of additional applications, evaluating speed, and facilitating execution at any given moment. The CNN accelerator will be further utilized for many biological applications through the implementation of diverse signaling and processing approaches.

REFERENCES

Al-Shamma, O., Fadhel, M. A., Hameed, R. A., Alzubaidi, L., & Zhang, J. (2020). Boosting convolutional neural networks performance based on FPGA accelerator. Intelligent Systems Design and Applications: 18th International Conference on Intelligent Systems Design and Applications (ISDA 2018) Vellore, India, December 6–8, 2018, Volume 1, pp. 509–517. Springer International Publishing, doi: 10.1007/978-3-030-16657-1_47.

Alaeddine, H., Jihene, M., & Khemaja, M. (2021, September). An efficient deep network in network architecture for image classification on FPGA accelerator. 2021 International Conference on Cyberworlds (CW), pp. 72–77. IEEE, doi: 10.1109/CW52790.2021.00018.

Bouaafia, S., Messaoud, S., Khemiri, R., & Sayadi, F. E. (2022, May). An FPGA-SoC-based hardware acceleration of convolutional neural networks. 2022 IEEE 9th International Conference on Sciences of Electronics, Technologies of Information and Telecommunications (SETIT), pp. 537–542. IEEE, doi: 10.1109/SETIT54465.2022.9875430.

Dhyani, S., Kumar, A., & Choudhury, S. (2023a, November). Review of analysis of ECG-based arrhythmia detection system using machine learning. AIP Conference Proceedings, Vol. 2930, No. 1. AIP Publishing, doi: 10.1063/5.0178062.

Dhyani, S., Kumar, A., & Choudhury, S. (2023b). Arrhythmia disease classification utilizing ResRNN. *Biomedical Signal Processing and Control*, 79, 104160, doi: 10.1016/j.bspc.2022.104160.

Goel, A., Goel, A. K., & Kumar, A. (2023a). The role of artificial neural network and machine learning in utilizing spatial information. *Spatial Information Research*, 31(3), 275–285, doi: 10.1007/s41324-022-00494-x.

Goel, A., Goel, A. K., & Kumar, A. (2023b). Performance analysis of multiple input single layer neural network hardware chip. *Multimedia Tools and Applications*, 1–22, doi: 10.1007/s11042-023-14627-3.

Gupta, N., & Kumar, A. (2023). Study on the wireless sensor networks routing for Low-Power FPGA hardware in field applications. *Computers and Electronics in Agriculture*, 212, 108145, doi: 10.1016/j.compag.2023.108145.

Hamdan, M. K., & Rover, D. T. (2017, December). VHDL generator for a high-performance convolutional neural network FPGA-based accelerator. 2017 International Conference on ReConFigurable Computing and FPGAs (ReConFig), pp. 1–6. IEEE, doi: 10.1109/RECONFIG.2017.8279827.

Hu, W., Chen, S., Li, Z., Liu, T., & Li, Y. (2019, December). Data optimization CNN accelerator design on FPGA. IEEE International Conference on Parallel & Distributed Processing with Applications, Big Data & Cloud Computing, Sustainable Computing & Communications, Social Computing & Networking (ISPA/BDCloud/SocialCom/SustainCom), pp. 294–299. IEEE, doi: 10.1109/ISPA-BDCloud-SustainCom-SocialCom48970.2019.00051.

Huang, W., Wu, H., Chen, Q., Luo, C., Zeng, S., Li, T., & Huang, Y. (2021). FPGA-based high-throughput CNN hardware accelerator with high computing resource utilization ratio. *IEEE Transactions on Neural Networks and Learning Systems*, 33(8), 4069–4083, doi: 10.1109/TNNLS.2021.3055814.

Jameil, A. K., & Al-Raweshidy, H. (2022). Efficient CNN architecture on FPGA using high-level module for healthcare devices. *IEEE Access*, 10, 60486–60495, doi: 10.1109/ACCESS.2022.3180829.

Jain, A., Dwivedi, R. K., Alshazly, H., Kumar, A., Bourouis, S., & Kaur, M. (2022). Design and simulation of ring network-on-chip for different configured nodes. *Computers, Materials and Continua*, 71(2), 4085–4100, doi: 10.32604/cmc.2022.023017.

Korol, G., & Moraes, F. G. (2019, August). A FPGA parameterizable multi-layer architecture for CNNs. Proceedings of the 32nd Symposium on Integrated Circuits and Systems Design, pp. 1–6, doi: 10.1145/3338852.3339840.

Kyriakos, A., Kitsakis, V., Louropoulos, A., Papatheofanous, E. A., Patronas, I., & Reisis, D. (2019, July). High-performance accelerator for CNN applications. 29th International Symposium on Power and Timing Modeling, Optimization and Simulation (PATMOS), pp. 135–140. IEEE, doi: 10.1109/PATMOS.2019.8862166.

Kumar, A. (2023). Study and analysis of different segmentation methods for brain tumor MRI application. *Multimedia Tools and Applications*, 82(5), 7117–7139, doi: 10.1007/s11042-022-13636-y.

Kumar, A., Sharma, P., Gupta, M. K., & Kumar, R. (2018). Machine learning-based resource utilization and pre-estimation for network on chip (NoC) communication. *Wireless Personal Communications*, 102(3), 2211–2231. https://doi.org/doi:10.1007/s11277-018-5376-3.

Linares-Barranco, A., Rios-Navarro, A., Canas-Moreno, S., Piñero-Fuentes, E., Tapiador-Morales, R., & Delbruck, T. (2021, July). Dynamic vision sensor integration on FPGA-based CNN accelerators for high-speed visual classification. International Conference on Neuromorphic Systems, pp. 1–7, doi: 10.1145/3477145.3477167.

Liu, X., Yang, J., Zou, C., Chen, Q., Yan, X., Chen, Y., & Cai, C. (2021). Collaborative edge computing with FPGA-based CNN accelerators for energy-efficient and time-aware face tracking system. *IEEE Transactions on Computational Social Systems*, 9(1), 252–266, doi: 10.1109/TCSS.2021.3059318.

Lu, J., Liu, D., Liu, Z., Cheng, X., Wei, L., Zhang, C., ... Liu, B. (2021). Efficient hardware architecture of convolutional neural network for ECG classification in wearable healthcare device. *IEEE Transactions on Circuits and Systems. Part I: Regular Papers*, 68(7), 2976–2985.

Lu, L., Xie, J., Huang, R., Zhang, J., Lin, W., & Liang, Y. (2019, April). An efficient hardware accelerator for sparse convolutional neural networks on FPGAs. 27th Annual International Symposium on Field-Programmable Custom Computing Machines (FCCM), pp. 17–25. IEEE, doi: 10.1109/FCCM.2019.00013.

Mittal, S. (2020). A survey of FPGA-based accelerators for convolutional neural networks. *Neural Computing and Applications*, 32(4), 1109–1139, doi: 10.1007/s00521-018-3761-1.

Ompal, Mishra, V. M., & Kumar, A. (2021). Zigbee internode communication and FPGA synthesis using mesh, star, and cluster tree topological chip. *Wireless Personal Communications*, 119(2), 1321–1339, doi: 10.1007/s11277-021-08282-w.

Rawat, A. S., Rana, A., Kumar, A., & Bagwari, A. (2018). Application of multi-layer artificial neural network in the diagnosis system: A systematic review. *IAES International Journal of Artificial Intelligence*, 7(3), 138, doi: 10.11591/ijai.v7.i3.pp138-142.

Saglam, S., Tat, F., & Bayar, S. (2019, November). FPGA implementation of CNN algorithm for detecting malaria diseased blood cells. International Symposium on Advanced Electrical and Communication Technologies (ISAECT), pp. 1–5. IEEE, doi: 10.1109/ISAECT47714.2019.9069724.

Shahshahani, M., Goswami, P., & Bhatia, D. (2018, November). Memory optimization techniques for FPGA based cnn implementations. 13th Dallas Circuits and Systems Conference (DCAS), pp. 1–6. IEEE, doi: 10.1109/ISAECT47714.2019.9069724.

Wen, J., Ma, Y., & Wang, Z. (2020, December). An efficient FPGA accelerator optimized for high throughput sparse CNN inference. Asia and the Pacific Conference on Circuits and Systems (APCCAS), pp. 165–168. IEEE, doi: 10.1109/APCCAS50809.2020.9301696.

Wu, D., Zhang, Y., Jia, X., Tian, L., Li, T., Sui, L., Xie, D., & Shan, Y. (2019). A high-performance CNN processor based on FPGA for MobileNets. 29th International Conference on Field Programmable Logic and Applications (FPL), pp. 136–143. IEEE, doi: 10.1109/FPL.2019.00030.

Wu, R., Liu, B., Fu, P., & Chen, H. (2023). An efficient lightweight CNN acceleration architecture for edge computing based on FPGA. *Applied Intelligence*, 53(11), 13867–13881, doi: 10.1007/s10489-022-04251-3.

Xiyuan, P., Jinxiang, Y., Bowen, Y., Liansheng, L., & Yu, P. (2021). A review of FPGA-based Custom computing architecture for convolutional neural network inference. *Chinese Journal of Electronics*, 30(1), 1–17, doi: 10.1049/cje.2020.11.002.

Chapter 3

Imperative role of embedded system in solar photovoltaic-based stand-alone applications

Rupendra Kumar Pachauri, Jai Govind Singh, Priyanka Sharma, and Shashikant

3.1 INTRODUCTION

Energy conservation has taken a very vital space in the current generation where everyone is working to generate energy from natural sources rather than using the conventional methods of burning fossil fuels. Renewable sources of energy have become important enough to be discussed on a global level. The older ways of generating electricity have some major drawbacks that are very harmful to biodiversity (Koutroulis et al., 2001). They not only increase pollution levels but also have many adverse effects on the farming sector. Among all the available sources of energy, the solar photovoltaic (PV) system is being used by many people around the world. Energy availed from sun rays is cost-friendly and the system is easily accessible for commercial and industrial purposes. Solar-powered appliances like geysers, heaters, streetlights, and generators are popular in rural areas or places with poor electrical facilities. Similarly, urban areas are also focusing on shifting from conventional systems to roof-top solar panels.

Thus, shifting attention toward renewable sources of energy will ultimately help in living life. Although solar energy has many benefits, the system is also equipped with many drawbacks. To start, the installation cost is very high as the setup requires expensive high-grade tools. Further, the semiconductor material required to convert sunlight to electricity is placed inside the solar panel which increases its price many folds (Brunelli et al., 2009). Not only is the setup cost high, but the solar array system requires proper cleaning and hygienic environmental conditions, which further leads to many expenses. On the other hand, solar radiation is very unpredictable and in many parts of the world, the sun doesn't shine continuously throughout the year. These are some of the major drawbacks that lead to unpredictable outputs and further leads to many energy losses.

There are several ways to monitor the performance of the system electronically. There are several sensors, microcontrollers, transmitters-receivers, and other embedded circuits that can be installed along with the solar panels and their performance can be monitored from any place in the world. Keeping an eye on the measured data helps the operator to further make

DOI: 10.1201/9781003510420-3

some required changes to increase the efficiency of the system (Al-Qubaisi et al., 2009).

3.2 LITERATURE REVIEW

3.2.1 Performance improvements of PV system

Due to several instances, researchers have designed novel embedded-based systems for performance improvement of solar PV systems. Comprehensive studies are available in the literature regarding the development of novel approaches to harvest maximum power or performance improvement of solar PV systems.

Koutroulis et al. (2001) developed a novel maximum power point tracking (MPPT) to achieve the global power of a PV system by adjusting the performance voltage and current parameters regardless of the ambient temperature and sun irradiance. The schematic diagram of the MPPT-assisted PV module is shown in Figure 3.1.

Brunelli et al. (2009) developed a cost-effective PV harvester for low-power and environmental embedded systems. An experimental setup with an MPPT-assisted system was developed for maximum power harvesting from a solar PV system. The developed system is shown in Figure 3.2.

Al-Qubaisi et al. (2009) integrated analog sensors and designed a microcontroller-controlled actuator system for dust cleaning of the solar PV module. The developed model could predict the dust based on power generated by the PV module. The microcontroller decides whether to activate the motor-assisted wiper to remove the dust particles from the PV surface. Chow and Abiera (2013) developed a prototype experimental setup of a

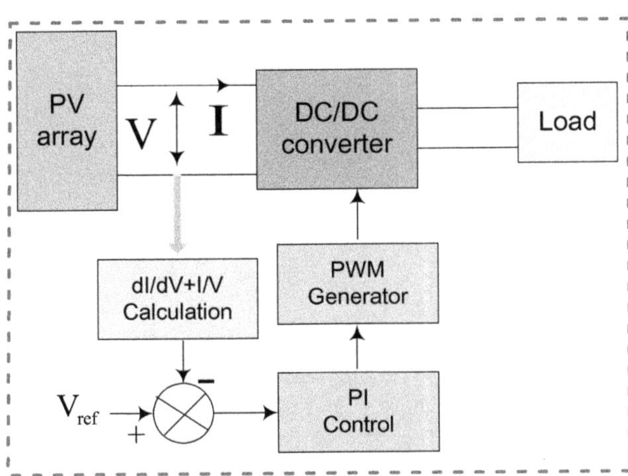

Figure 3.1 Schematic diagram of MPPT-assisted PV module (Koutroulis et al., 2001).

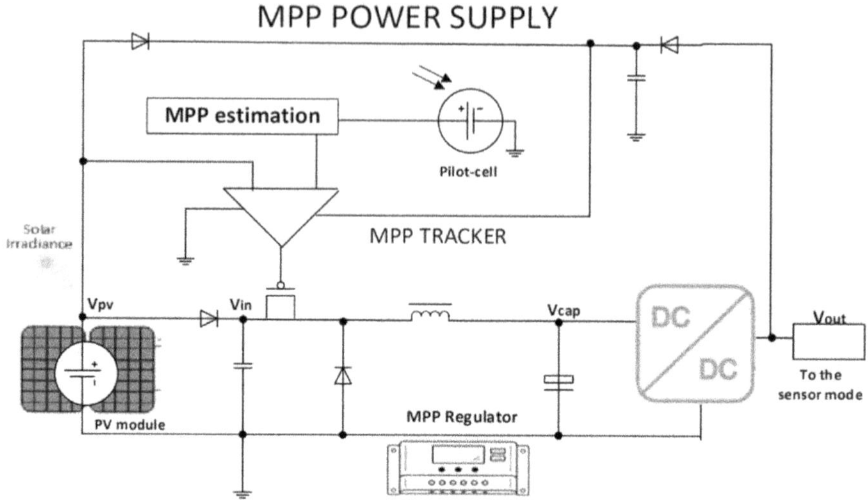

Figure 3.2 Circuit diagram: embedded platform powered by the solar harvester (Brunelli et al., 2009).

sun-tracking system to enhance the PV system performance. The developed system is comprised of an Arduino Uno ATmega-328 microcontroller, light-dependent resistor (LDR), and servomotor to drive the gear assembly for harvesting maximum solar irradiation from the sun. Moreover, a MATLAB-based graphical user interface (GUI) system was developed to log the solar irradiation from horizontally and vertically installed sensors. Azimuth and tilt angles are continuously observed for performance monitoring and the developed model is shown in Figure 3.3.

Chang et al. (2014) have presented a ZigBee wireless sensor node with four key modules such as signal processing, energy harvesting, and storage

Figure 3.3 Prototype of the optimized solar tracker system (Chow and Abiera, 2013).

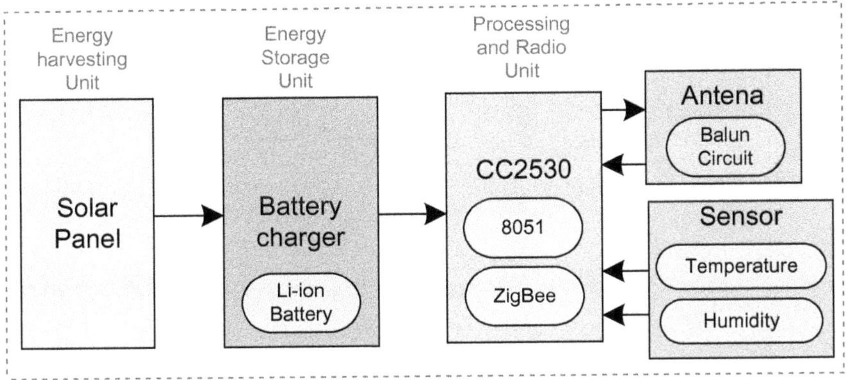

Figure 3.4 Schematic diagram of ZigBee sensor-assisted energy storage system (Chang et al., 2014).

system. A small amount of solar power is supplied for the functioning of the energy harvesting unit. Moreover, the energy storage unit has a 1200 mAh Li-ion battery system with a charge controlling unit. At the sensor unit, temperature and humidity sensors are attached to the system for continuous monitoring. The developed system is shown in Figure 3.4.

Evans and Garcia (2014) have DC motor systems powered by three 12V solar PV systems with the on/off functioning of a solid-state relay (SSR) system. A microprocessor using an optical sensor (TEMT6000) regulates the SSR system. A data storage card with a real-time clock module is used for data storage and developed system. Munoz-Garc et al. (2014) have a goal to create a cheap sensor that can detect irradiance deep inside a forest canopy. For the same purpose, two data acquisition systems were also deployed and tested. The obtained data from the low-cost designed system, which comprised an irradiation sensor and controller, is compared for two consequent years from September 2011 to September 2012. The developed system is shown in Figure 3.5.

Barsoum et al. (2016) and Zakariah et al. (2015) designed a solar tracking system based on LDR sensors. Four sensors are used for the tracking functioning of the developed system. Two motors are used to move the solar module in vertical and horizontal directions for sun tracking and the rotation of the motor is based on the LDR sensors. The circuit diagram of the overall system is shown in Figure 3.6.

For efficiency improvement of solar PV modules, a robotic arm-based wiper system was developed and called in a solar panel cleaning robotic arm (SPCRA) in (Mondal, et al. and Bansal, 2015). The designed system is used to clean the dust from the solar PV surface and obstructions of the sun irradiation are removed to enhance the system's performance. A weekly comparative study is carried out in terms of output current generated from

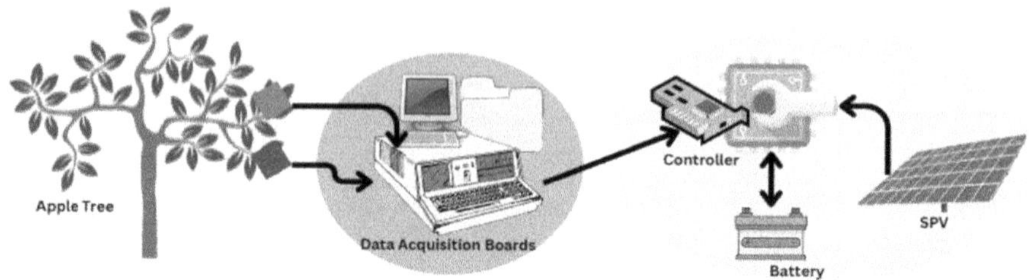

Figure 3.5 Schematic setup of the PC-based data acquisition system (Munoz-Garc et al., 2014).

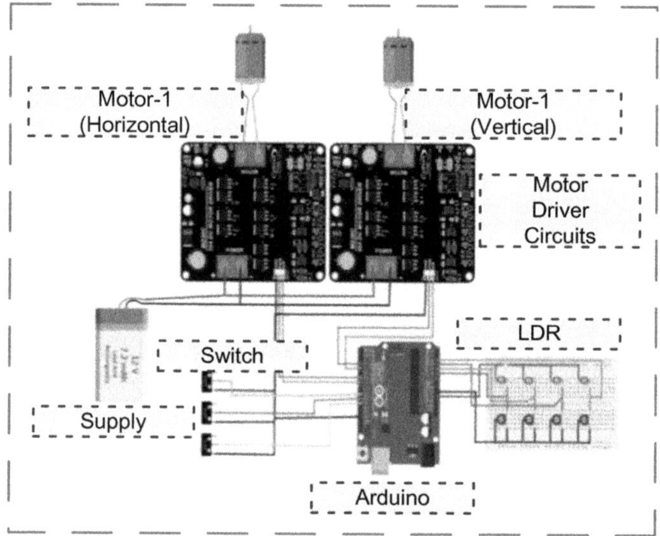

Figure 3.6 Circuit design of overall system (Barsoum et al., 2016; Zakariah et al., 2015).

the dusty and clean solar PV system. In Visconti et al. (2015), a graphical user interface (GUI) PC program is designed and tested to handle the electrical system that controls and drives the bi-axial solar trackers of a PV plant. Elevation and azimuth angles of solar PV modules are controlled by a microcontroller-assisted motor. For the linear Fresnel reflector solar concentrator (LFRSC), Babu et al. (2016) presented an embedded system-based solution for the one-axis sun-tracking mechanism. The location of the sun is monitored by sensors and a geared stepper motor controlled by an Arduino microcontroller. The Arduino controller coordinates the tracking of the sun using the LFRSC method. Two photoresistors are used to fine-tune the solar tracker's location. Fathabadi (2016) developed a solar tracker system with two stepper motors for controlling altitude and azimuth motion. The

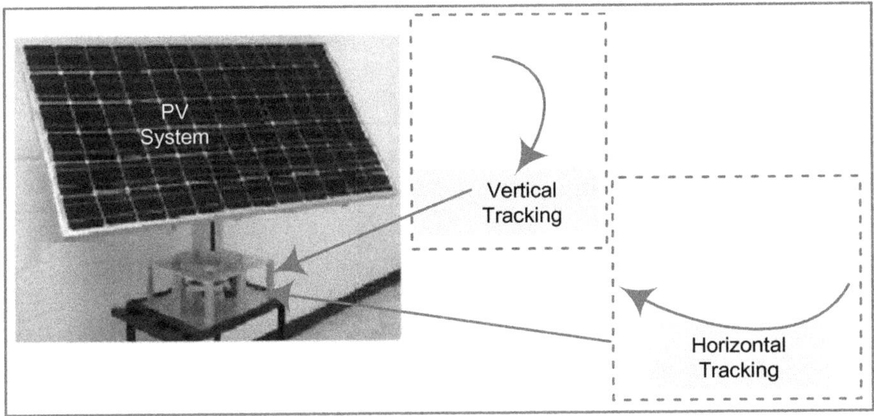

Figure 3.7 Solar tracking system (Fathabadi, 2016).

entire system is developed to enhance the system's performance. A prototype model of the system is shown in Figure 3.7.

In Kaur et al. (2016), a dual-axis solar tracker is developed to harvest maximum sun irradiation from the sun. For performance investigation of the developed model, a Lab VIEW-based GUI system is developed to analyze the system behavior. Mojallizadeh et al. (2016) developed a duty cycle controlling for the Boost converter using the ATmega 2560 (Arduino-Mega) microcontroller to achieve the maximum power point of a solar PV system. A variable programmable resistive load from 50 ohm to 100 ohm is considered for performance investigation under variable solar irradiation. In Poulek et al. (2016), solar cells with two faces were used to create a self-sustaining solar tracker. Large solar PV power plants with an output of 2–5 MW employed the tracker to great effect. Moreover, the developed system is modified and attached to a low concentration PV system. In order to track the electrical performance characteristics in terms of the I-V curve of a 5 kV solar PV facility, Valverde et al. (2016) designed a portable wireless device. The open circuit voltage can be measured up to 6 kV, while the short circuit current can be measured down to 100 mA. The suggested approach is designed for live, real-world characterization of large modules. The GUI system for reading sun irradiance, temperature, and humidity is implemented using an Arduino microcontroller system. The experimental setup of the developed model is shown in Figure 3.8.

Abdeen et al. (2017) focused on the study of power loss due to dust deposition on a solar PV surface. The study is carried out to optimize the tilt angle of the solar PV module. In this study, the authors used eight PV panels in the desert environment, fixed at 15°, 20°, 30°, and 45° tilt angles with the intent to investigate the performance of the cleaned and dusty solar PV panel. After ten months of study, it is observed that the power

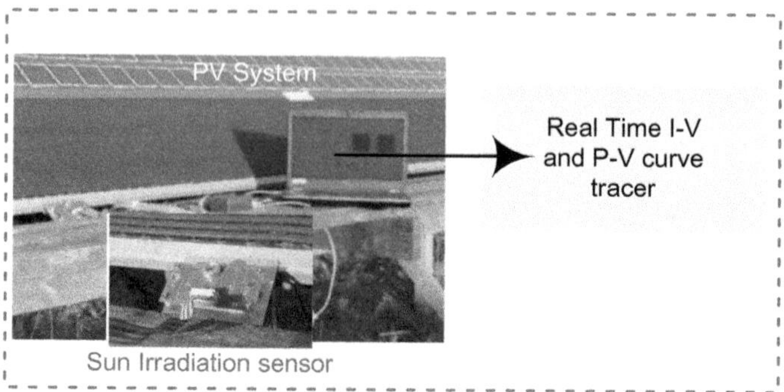

Figure 3.8 Arduino microcontroller-based sensor board for I-V measurements (Valverde et al., 2016.)

losses for dusty panels are 43%, 38%, 31%, and 25.5% for distinguish tilt angles of 15°, 20°, 30°, and 45°, respectively. A novel algorithm was developed by Ahmad et al. (2017) for tracking of the global maximum power point (GMPP) under non-uniform solar radiation and temperature conditions. The economical and efficient system is developed using Arduino Uno (ATmega 328 microcontroller), which shows good agreement with the theoretical and simulation results with the efficient tracking performance. The schematic implementation of the setup is shown in Figure 3.9.

An AVR microcontroller-assisted sun tracker with independent control over the azimuth and altitude axes has been built (Akbar et al., 2017). DC

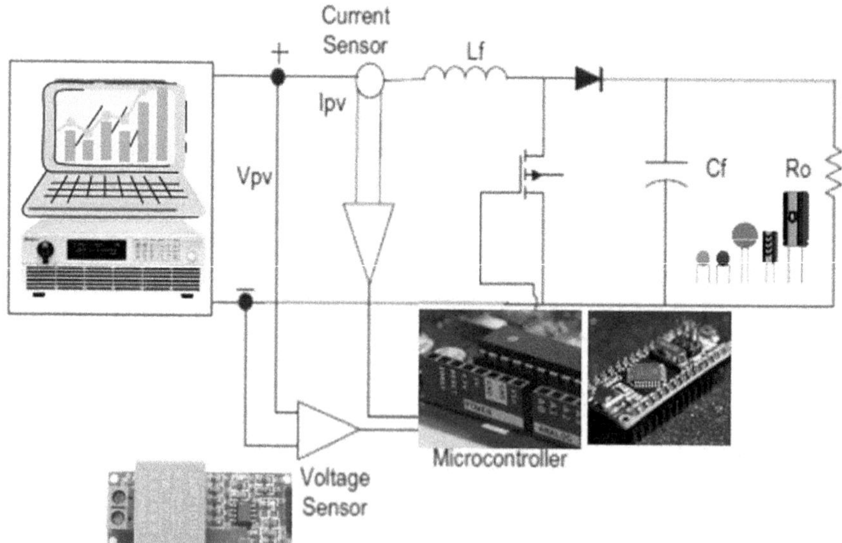

Figure 3.9 Schematic diagram of MPPT-assisted solar PV system (Abdeen et al., 2017).

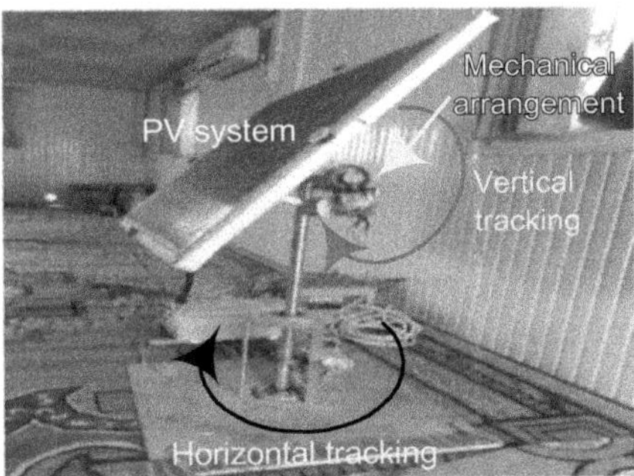

Figure 3.10 Experimental system of dual-axis solar tracker (Akbar et al., 2017).

motors, light sensors, and an electromechanical relay system are the key components of the system. The data demonstrates that the designed low-cost solar tracker increases the power production efficiency of a typical fixed solar PV panel by 25%–30% at maximum output. In Figure 3.10, we can see the created experimental system of the dual-axis solar tracker. Mondal et al. (2017) developed an Arduino-based solar tracking system for harvesting the maximum sun irradiation, which directly converts as output current. The primary components of an LDR and servomotor work as the sensor and actuator for the performance assessment of the developed system. Moreover, other developed dual-axis solar tracker systems have two DC motors for the movement of solar PV modules to cover directions of tilt and azimuth angles (Mustafa et al., 2017). The schematic diagram is shown in Figure 3.11.

A microcontroller-operated robot was developed to clean the dust from the PV module surface (Selvaganesh et al., 2017). In the initial testing phase,

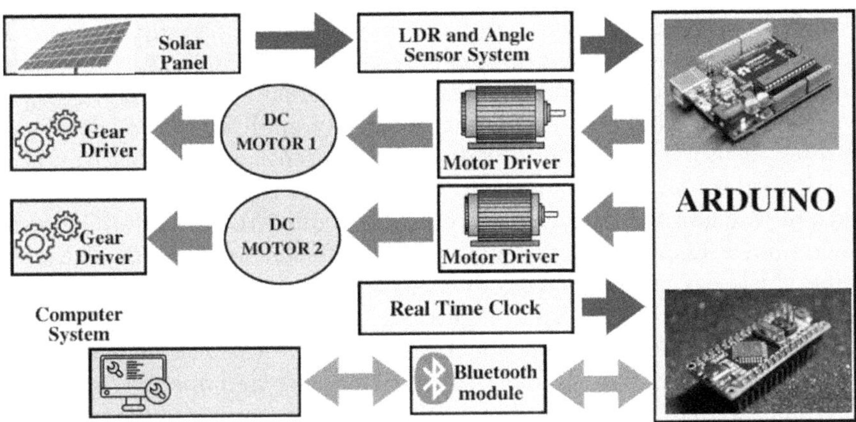

Figure 3.11 Schematic diagram of solar tracking system (Mustafa et al., 2017).

the cleaning robot showed extensive results and system adaptability due to improved panel efficiency. For accurate functioning of the developed model, the sensor and actuator are controlled by the Arduino platform. Chaieb and Sakly (2018) developed a novel MPPT technique based on the particle swarm optimization (PSO) algorithm. The required hardware system was comprised of an Arduino microcontroller (ATmega 328p), current-voltage sensors, DC-DC buck converter, and resistive load. A critical review based on the work done in the aforementioned literature is reported in Table 3.1.

3.3 STAND-ALONE ELECTRIFICATION

Various researchers have developed embedded-based systems for monitoring and controlling solar power generation for various applications such as water transporting systems and home energy management systems. A comprehensive study is carried out on the development of novel approaches using solar PV systems in literature resources.

Data logger systems built with data transmission modules may record a wide variety of information, including water flow rate, solar irradiation at the inclined and horizontal surface of the PV module, voltage, current of the PV field, and ambient temperature (Mahjoubi et al., 2012). An experimental study was carried out for up to ten months (from 6:00 a.m. to 8:00 p.m.) study. The block diagram of the system is shown in Figure 3.12.

In order to make the most efficient use of water for agricultural crops, Gutierrez et al. (2014) created an automated irrigation system. A distributed wireless sensor network was used to relay data from soil moisture and temperature sensors placed near plant roots. Based on an algorithm, threshold values of soil moisture and temperature were programmed into a microcontroller to control water flow. Moreover, the solar PV system was used for power assistance to the entire wireless sensor network and supported systems. The developed automated system was tested in a crop field for 136 days continuously and found water savings of up to 90% as compared to high water level consumption in traditional irrigation practices. A hardware system for lamp posts for the smart city application was developed and tested by Leccese et al. (2014). Locally, each lamp post uses an electronic card for on/off management. In addition to this, a ZigBee network is used to transmit the data to a central control unit. To perform this system operation, a Raspberry Pi system is used for efficient controlling. The street lighting isle has been realized and tested for some months in the field. The schematic diagram is shown in Figure 3.13.

An intelligent solar sterling water pump system was installed and a performance investigation was carried out by Saleh and Jokar (2016). The developed system is controlled by a microcontroller unit and the entire system is comprised of main components such as a temperature sensor, charge

Table 3.1 Comparison of embedded-based solar-powered systems for performance improvement

Reference	Major components of developed system	Controller/ technology	Parameters for measurement	Commercial/ laboratory-based experimental	Wired/wireless performance data transmission	Application
Koutroulis et al., 2001	• Voltage sensor • Current sensor	Microcontroller	• Power • Voltage • current	Experimental	Wired	A MPPT technique is developed to achieve global power from PV array.
Brunelli et al., 2009	• Voltage sensor • MPP regulator • DC-DC regulator: LTC3401	• MPP tracker • No need of any specific microcontroller	Voltage	Experimental	Wired	A batteryless solar energy harvesting system for low-power applications.
Al-Qubaisi et al., 2009	• LDR sensor • H-bridge • Buzzer • Motion detection sensor	Microcontroller: PIC18F442	• Voltage • Current • Power	Experimental	Wired	A microcontroller-based cleaning system designed to enhance solar PV performance.
Chow and Abiera, 2013	• Voltmeter • Data logger • Elevation and azimuth angle of solar irradiation sensor • Servo motor drive	• Arduino board • GUI display	Voltage	Experimental	Wired	An experimental system designed to harvest maximum solar power by tracking of sun position.

(Continued)

Table 3.1 (Continued) Comparison of embedded-based solar-powered systems for performance improvement

Reference	Major components of developed system	Controller/ technology	Parameters for measurement	Commercial/ laboratory-based experimental	Wired/wireless performance data transmission	Application
Chang et al., 2014	• Temperature sensor: SHT11 • Humidity sensor • Data logger	• Microcontroller: 8051 • Two USARTs with Support for SPI and UART	Voltage and current from hybrid solar and battery system	Experimental	• ZigBee sensor • RF Transceiver: 2.4 GHz	Wireless and sensor integrated-based system designed for transmitting and monitoring energy signal to receiver end.
Evans and Garcia, 2014	• Optical sensor • DAQ • Real-time clock • Relay circuit	Microcontroller: 8051	Voltage and current measured by sun tracker	Experimental	Wired	A sun tracking system developed to harvest maximum solar energy.
Barsoum et al., 2016	• Light dependence resistor (LDR) sensor • DC motor with driver circuit • H-Bridge IC: L293D	Arduino	• Voltage • Current • Irradiance	Experimental	Wired	A prototype model of dual-axis sun tracking is developed to harvest the maximum sun rays of the sun in four distinguished directions.
Zakariah et al., 2015	• LDR • DC motor • Motor driver: MD30C • Voltage and current sensor	Arduino Uno	Power	Experimental	Wired	A solar tracking system with fuzzy logic controller method aims to harvest the maximum solar radiance to generate maximum power from PV module.

(Continued)

Table 3.1 (Continued) Comparison of embedded-based solar-powered systems for performance improvement

Reference	Major components of developed system	Controller/ technology	Parameters for measurement	Commercial/ laboratory-based experimental	Wired/wireless performance data transmission	Application
Mondal and Bansal, 2015	• Wiper-assisted cleaning head • DC motor drive system • Air jet • Water sprinkler • Guide rail with T-beam shape	Microcontroller	Power	Prototype model	Wired	A robotic arm for automatic cleaning of the solar panel with four degrees of freedom to boost PV performance and reduce the effect of partial shading conditions.
Visconti et al., 2015	• DC motor drive • Proximity sensor • Voltage and current sensor	Master and slave electronic boards: C107	Voltage, current	Commercial	GPS	An electronic system developed for controlling and driving bi-axial sun trackers integrated with the PV plant and controlled by a PC with a GUI.
Babu et al., 2016	• LDR sensors (no. 02) • Stepper motor (PWM controlled) • DC motor • H-bridge motor driver	ATmega328 microcontroller	Voltage and current	Experimental	Wired	An embedded system-based single axis solar-tracking system.

(Continued)

Table 3.1 (Continued) Comparison of embedded-based solar-powered systems for performance improvement

Reference	Major components of developed system	Controller/ technology	Parameters for measurement	Commercial/ laboratory-based experimental	Wired/wireless performance data transmission	Application
Fathabadi, 2016	• Stepper motor • Stepper motor drive • PWM converter • Voltage and current sensor	Microcontroller: MC68HC11AB	• Power • Angle: azimuth and altitude	Experimental	Wired	A two-axis sun tracking system designed to track the sun position and harvest maximum energy.
Kaur et al., 2016	• Servo motor • LDR sensor	Arduino Uno LabVIEW for GUI	Angle: azimuth and altitude	Experimental	Wired	An Arduino-based active dual axis solar tracking system designed for system performance enhancement.
Mojallizadeh et al., 2016	• Oscilloscope • PV emulator • Programmable load • Boost converter	Atmega 2560 Microcontroller	Voltage	Experimental	Wired	A hardware implementation of maximum power point tracking method to achieve MPP of PV system.
Poulek et al., 2016	• Solar cells: mono and bi-facial • Measurement unit • Reflectors	Driving sensing unit	Voltage	Experimental	Wired	A design of polar axis solar tracker using bifacial cells are implemented and performance assessment is done.

(Continued)

Table 3.1 (Continued) Comparison of embedded-based solar-powered systems for performance improvement

Reference	Major components of developed system	Controller/ technology	Parameters for measurement	Commercial/ laboratory-based experimental	Wired/wireless performance data transmission	Application
Valverde et al., 2016	• Voltage and current sensor • DC-DC converter • HV capacitor and resistor bank	Microcontroller	• Voltage • Current • Power	Prototype	Bluetooth	A portable and wireless system is designed to plot the current-voltage performance curve of solar PV system.
Abdeen et al., 2017	• Solid-state relay • DC load • MPPT	Arduino and Lab-VIEW GUI	• Electrical performance • Temperature • Wind speed	Commercial	Wired	The presented study focused on identifying the tilt angle of the solar PV system to minimize power losses under duct accumulation conditions.
Ahmad et al., 2017	• Voltage and current sensor • Solar Emulator: E4360A • Rheostat load	Arduino	GMPP	Experimental	Wired	A hardware implementation of system is carried that is used to obtain the GMPP on P-V curve under shading conditions.

(Continued)

Table 3.1 (Continued) Comparison of embedded-based solar-powered systems for performance improvement

Reference	Major components of developed system	Controller/ technology	Parameters for measurement	Commercial/ laboratory-based experimental	Wired/wireless performance data transmission	Application
Akbar et al., 2017	• LDR • Relay (no. 04) • DC motor (no. 02)	AVR microcontroller	Power	Experimental	Wired	AVR microcontroller-based hardware implementation of dual-axis solar tracker to track the sun position in both azimuth and altitude axes.
Mandal and Singh, 2017	• LDR sensor • Servo motor	Arduino	Voltage	Experimental	LAN interfacing	A servo motor and LDR sensor-assisted real time data acquisition system is designed for solar performance storage purpose.
Mustafa et al., 2017	• DC motor (no. 2) and driver circuit • LDR • Angle sensor system • Gear driver	Arduino UNO controller	• Efficiency • Tracking angle	Experimental	Bluetooth mobile	An experimental system is used to track the sun position for performance enhancement of solar PV module.
Selvaganesh et al., 2017	• DC geared motor • Voltage regulator • Relay circuits • IR sensor • Cleaning brush	Arduino: Atmega 328	• Voltage • Current • Power	Experimental	Wired	A microcontroller based robotic system is proposed for solar PV module cleaning.

(Continued)

Table 3.1 (Continued) Comparison of embedded-based solar-powered systems for performance improvement

Reference	Major components of developed system	Controller/ technology	Parameters for measurement	Commercial/ laboratory-based experimental	Wired/wireless performance data transmission	Application
Chaieb and Sakly, 2018	• Current and voltage sensor • Resistive load • DC-DC Buck converter	Arduino	• Voltage • Current • Power	Experimental	Wired	A hardware implementation of MPPT technique to achieve MPP points under partial shading conditions.

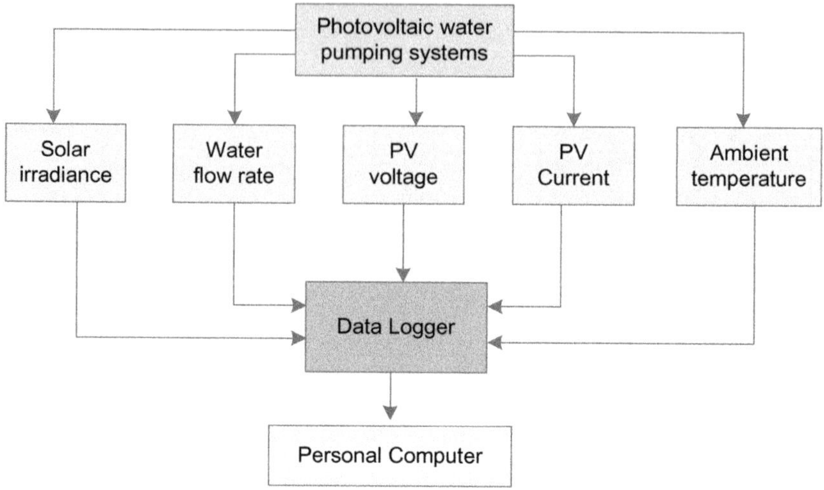

Figure 3.12 Data logger for field parameters of solar-assisted water pumping system (Mahjoubi et al., 2012).

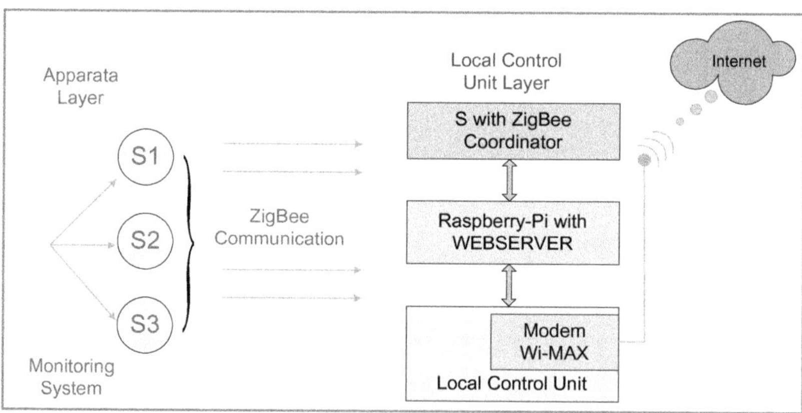

Figure 3.13 Block scheme of the system architecture (Leccese et al., 2014).

controller, and DC motor driver circuit. Jamaluddin et al. (2017) aim to evaluate the performance of lithium iron phosphate (LiFePO4) batteries as storage for the stand-alone PV street light system during the charging and discharging time on the Lab VIEW Interface. Using the wireless sensor network technology (WSNT), Kumar et al. (2017) integrated a solar photovoltaic system, an Arduino ATmega-2560 microcontroller, a soil moisture sensor, a mobile Bluetooth device, a water storage tank, and a pump to create an autonomous solar-powered drip irrigation system. The framework's minimal component count and three primary layers — an X-Bee-based circuit for processing and communication, a solar charge controller, and a solar battery load matching layer — led to its low power consumption. Using a sample frequency of 11–123 Hz, data on characteristics including voltage, current, and temperature was retrieved from a distant X-Bee node and sent in real time. By comparing test signals, we were able to confirm the accuracy of the prototype. A connected load's surge current, such as that of motor-based equipment upon startup, may be tracked by the system's measurement. In addition, a critical review based on work done in the aforementioned literature is reported in Table 3.2.

3.4 PROS AND CONS

Embedded systems play a crucial role in solar PV stand-alone applications, such as solar-powered street lights, remote monitoring systems, and off-grid power systems. Table 3.3 shows some pros and cons of using embedded systems in such applications. It is important to note that the pros and cons mentioned are general considerations and can vary based on specific system designs, implementation approaches, and the expertise of the individuals involved.

3.5 SUMMARY

In this chapter, a comprehensive study is carried out to identify the embedded applications in the area of solar PV systems, i.e., performance monitoring and stand-alone power generation systems. The salient features of the present study are concluded as follows:

- The solar PV plant has high installation costs, as the setup requires high-grade tools that increase the price. In addition, maintenance costs are always a matter. In this context, the designing of the analog sensors and microcontroller-controlled actuator systems is done for the dust cleaning of solar PV panels.
- A robotic arm-based wiper system is used for cleaning the dust from the solar PV surface and sun enhancement obstructions are removed. A PC program with an intuitive GUI controls and drives the bi-axial solar trackers of the PV plant.
- A MATLAB-based GUI system was developed to log the solar irradiation from horizontally and vertically installed sensors. Moreover, the energy harvesting unit functioning requires the ZigBee wireless sensor node. For a 5 kV solar PV plant, a portable wireless system for tracing the electrical performance parameters in the I-V curve was

Table 3.2 Comparison of embedded-based solar-powered systems for stand-alone electrification

Reference	Major components of developed system	Controller/ technology	Parameters for measurement	Commercial/ laboratory-based experimental	Wired/wireless performance data transmission	Application
Mahjoubi et al., 2012	• Voltage, current sensors • Temperature sensor • Flow sensor • Irradiation sensor	Microcontroller	• Voltage • Current • Temperature • Water flow • Irradiation	Prototype	GSM, GPRS	A prototype system is designed with the assistance of various sensors and transmission module to the DC motor-driven pumping system
Leccese et al., 2014	• Pressure sensor • LEDs	Raspberry Pi	Pressure of vehicle	Prototype	ZigBee	Remote controlled isle of lamp posts has been implemented and tested for satisfactory performance in the field.
Saleh and Jokar, 2016	• Charge controller • Voltage regulator • DC motor with driver • Battery system • Temperature sensor	Microcontroller unit with neural network algorithm	Water temperature control	Experimental	Wired	A mechatronic system is implemented experimentally and its performance demonstrated the feasibility of water pumping at variable operating conditions.

(Continued)

Table 3.2 (Continued) Comparison of embedded-based solar-powered systems for stand-alone electrification

Reference	Major components of developed system	Controller/ technology	Parameters for measurement	Commercial/ laboratory-based experimental	Wired/wireless performance data transmission	Application
Jamaluddin et al., 2017	• Relay circuit • Current sensor • Voltage sensor • Light sensor • Temperature sensor • Humidity sensor	• Microcontroller based on ATmega-328 and ATmega-2560 • Lab-VIEW for GUI	• Current • Voltage • Light • Temperature • Humidity	Experimental	Wired	The monitoring of battery-integrated solar PV system is carried out in terms of charging and discharging conditions for street lighting systems.
Kumar et al., 2017	• Soil moisture sensor • Water tank • Motor pump • Charge controller • LCD • Mobile	Arduino Mega 2560	• Soil moisture • Voltage	Experimental	GSM	In water pumping application, sensor-based smart water transporting system is developed to harvest maximum solar irradiance to run motor drive for a long duration.
Sabry et al., 2017	• Charge controller • MPPT • Current sensor • Voltage sensor • Temperature sensor	Microcontroller	• Current • Voltage • Temperature	Experimental	Wireless RF module: X-Bee	A solar-powered system is integrated with the battery system. The proposed system is used to monitor the performance data.

Table 3.3 Pros and cons of embedded system in solar power monitoring

Pros	Cons
• Intelligently manage power generation, storage, and distribution, ensuring that the energy generated by the solar panels is used effectively	• Any failure or malfunction in the embedded system can lead to system downtime
• Real-time monitoring and control	• Upgradability and compatibility
• Remote monitoring and control enable efficient maintenance, troubleshooting, and system configuration without the need for on-site personnel	• Restricts the flexibility and scalability of solar PV stand-alone applications
• Improving system reliability and reducing downtime	• This may involve periodic software updates, hardware checks, and troubleshooting
• Integration with other systems	• Adequate technical support and skilled personnel are essential for maintaining embedded systems effectively

developed. For solar PV plants like 2 to 5 MW, a self-powered solar tracker with bifacial solar cells has been developed and tested.

• For tracking the GMPP under non-uniform solar radiation and temperature conditions, a novel algorithm is used. The development of an automated irrigation system optimizes the water usage for agricultural crops. A hardware system for lamp posts for smart city applications requires the Raspberry Pi system for efficient controlling.

The present study will be helpful to beginners in the area of smart monitoring of solar PV system-based energy generation, especially in stand-alone applications.

REFERENCES

Abdeen, E., Orabi, M., Hasaneen, E. S., 2017. Optimum tilt angle for photovoltaic system in desert environment. *Solar Energy* 155; 267–280.

Ahmad, A., Khandelwal, A., Samuel, P., 2017. Golden band search for rapid global peak detection under partial shading condition in photovoltaic system. *Solar Energy* 157; 979–987.

Akbar, H. S., Siddiqi, A. I., Aziz, M. W., 2017. Microcontroller based dual axis sun tracking system for maximum solar energy generation. *American Journal of Energy Research* 5; 23–27.

Al-Qubaisi, E. M., Al-Ameri, M. A., Al-Obaidi, A. A., Rabia, M. F., El-Chaar, L., Lamont, L. A., 2009, November 10–12. Microcontroller based dust cleaning system for a standalone photovoltaic system. Proceedings IEEE Conference on Electric Power and Energy Conversion Systems at Sharjah, pp. 1–6.

Babu, M., Arasu, A. V., Ranjani, J. J., 2016. Arduino based sun positioning for linear Fresnel solar concentrating system with horizontal absorber and varying width reflectors. IJCTA 9; 377–388.

Barsoum, N., Nizam, R., Gerard, E., 2016. New Approach on development a dual axis solar tracking prototype. *Wireless Engineering and Technology* 7(1); 1–11.

Brunelli, D., Moser, C., Thiele, L., Benini, L., 2009. Design of a solar-harvesting circuit for batteryless embedded systems. *IEEE Transactions on Circuits and Systems* 56(11); 2519–2528.

Chaieb, H., Sakly, A., 2018. A novel MPPT method for photovoltaic application under partial shaded conditions. *Solar Energy* 159; 291–299.

Chang, C. C., Lin, C. B., Chan, C. B., 2014. ZigBee wireless sensor nodes with hybrid energy storage system based on Li-ion battery and solar energy supply. *International Journal of Industrial and Manufacturing Engineering* 8; 772–776.

Chow, L. S., Abiera, M., 2013. Optimization of solar panel with solar tracking and data logging. IEEE Conference, pp. 15–19.

Evans, R., Garcia, A. S., 2014. A novel open-source master-slave heliostat array for use as an experimental platform and solar power applications. *Energy Procedia* 57; 2080–2089.

Fathabadi, H., 2016. Comparative study between two novel sensorless and sensor based dual axis solar trackers. *Solar Energy* 138; 67–76.

Gutierrez, J., Medina, J. F. V., Garibay, A. V., Gandara, M. A. P.,2014. IEEE Transactions on Instrumentation and Measurement. Automated irrigation system using a wireless sensor network and GPRS module, pp. 166–176.

Jamaluddin, A., Ainia, A. N., Adhitamaa, E., Purwanto, A., 2017. Assessment of LiFePO4 battery performance in stand-alone photovoltaic street light system. *Procedia Engineering* 170; 503–508.

Kaur, T., Mahajan, S., Verma, S., Priyanka, G. J., 2016. Arduino based low-cost active dual axis solar tracker. IEEE International Conference on Power Electronics, Intelligent Control and Energy Systems (ICPEICES-2016), pp. 1–5.

Koutroulis, E., Kalaitzakis, K., Voulgaris, N. C., 2001. Development of a microcontroller-based, photovoltaic maximum power point tracking control system. *IEEE Transactions on Power Electronics* 16(1); 46–54.

Kumar, S., Sethuraman, C., Srinivas, K., 2017. Solar powered automatic drip irrigation system using wireless sensor network technology. *International Research Journal of Engineering and Technology* 04; 722–731.

Leccese, F., Cagnetti, M., Trinca,D., 2014. A smart city application: A fully controlled street lighting isle based on Raspberry-Pi card, a ZigBee sensor network and WiMAX. *Sensors* 14(12); 24408–24424.

Mahjoubi, A., Mechlouch, R. F., Brahim, A. B., 2012. Data acquisition system for photovoltaic water pumping system in the desert of Tunisia. *Procedia Engineering* 33; 268–277.

Mandal, S., Singh, D., 2017. Real time data acquisition of solar panel using Arduino and further recording voltage of the solar panel. *International Journal of Instrumentation and Control Systems* 7(3); 15–25.

Mojallizadeh, M. R., Badamchizadeh, M., Khanmohammadi, S., Sabahi, M., 2016. Designing a new robust sliding mode controller for maximum power point tracking of photovoltaic cells. *Solar Energy* 132; 538–546.

Mondal, A. K., Bansal, K., 2015. Structural analysis of solar panel cleaning robotic arm. *Current Science* 108; 1047–1052.

Munoz-Garc, M. M., Herreros, A. M., Balenzategui, J. L., Barrerio, P., 2014. Low-cost irradiance sensors for irradiation assessments inside tree canopies. *Solar Energy* 103; 143–153.

Mustafa, F. I., Al-Ammri, A. S., Ahmad, F. F., 2017. Direct and indirect sensing two-axis solar tracking system. The 8th International Renewable Energy Congress (I REC 2017), pp. 1–4.

Poulek, V., Khudysh, A., Libra, M., 2016. Self-powered solar tracker for low concentration PV (LCPV) systems. *Solar Energy* 127; 109–112.

Sabry, A. H., Hasan, W. Z. W., Kadir, M. Z. A. A., Radzi, M. A. M., Shafie, S., 2017. DC based smart PV powered home energy management system based on voltage matching and RF module. *PLoS One* 12(9); 1–22.

Saleh, A. R. T., Jokar, H., 2016. Neural network-based control of an intelligent solar Stirling pump. *Energy* 94; 508–523.

Selvaganesh, V., Manoharan, P. S., Seetharaman, V., 2017. Cleaning solar panels using portable robot system, *IJCTA* 10(02); 195–203.

Valverde, R., Almagro, S. C., Corazza, M., Espinosa, N., Hosel, M., Sondergaard, R. R., Jorgensen, M., Villarejo, J. A., Krebs, F. C., 2016. Portable and wireless I-V curve tracer for 45kV organic photovoltaic modules. *Solar Energy Materials and Solar Cells* 15; 60–65.

Visconti, P., Costantini, P., Orlando, C., Ekuakille, A. L., Cavalera, G., 2015. Software solution implemented on hardware system to manage and drive multiple bi-axial solar trackers by PC in photovoltaic solar plants. *Measurement* 76; 80–92.

Zakariah, A., Jamian, J. J., Yunus, M. M., 2015. Dual Axis solar tracking system based on fuzzy logic control and light dependent resistors as feedback path elements. IEEE Student Conference on Research and Development, pp. 139–144.

Chapter 4

PID embedded control and DC motor behavior for FPGA-based simulation

Pankaj Singh Panwar, Monika Gupta, Surajit Mondal, and Adesh Kumar

4.1 INTRODUCTION

The proportional–integral–derivative (PID) control technique is widely used in practical engineering, but it faces challenges in terms of complex parameter adjustment and limited ability to respond to nonlinearity when applied to varied equipment. Hence, it is becoming progressively evident that a conventional PID exhibits constraints in engineering applications (Kumar et al. 2021). An increased emphasis on neural network regulation is a result of the growth in artificial intelligence (Rawat et al. 2018). The traits of neural networks include excellent robustness, self-adaptation, and self-learning. The real requirement of the reply time and balance in the work process can be met by combining a PID controller and neural network. The innovation involved customized switching PID markers by a system with online support that is a melancholic alternative to distinctive attributes of distinctive controlled equipment, which is easily accessible in several varieties and alternatives (Gupta et al. 2017). It is more improvised than the PID controller used in the real mechanism production process in industry. However, standard microcontroller units (MCUs) cannot handle high-reliability application scenarios. This solution is improving all the time. It uses a backpropagation neural network (BPNN) PID controller to aid in aiding movement. The controller is a self-adjusting system component. This is created using Xilinx FPGA (Kumar et al. 2017). The controller creates numerous sub-modules, as shown in the suggested modular schematic. After finishing the further broadcasting process, the output layer is formed from the input layer, and the further broadcasting sample is used.

The BPNN-based PID controller serves as the fundamental component of a closed-loop motion control system. This system utilizes a solution implemented on a Xilinx FPGA, which is specifically designed to meet the demanding requirements of real-time performance and exceptional reliability. The suggested system has two fundamentally distinct components: the sketch of the control system's closing-loop peripheral module and the sketch of the BPNN PID control algorithm. The first component now includes the further broadcasting module, primary state machine module, PID module, failure backpropagation module, and weight update module. A group of nerve cells clump together to form the forward propagation module. This module allows you to generate the output layer from the inside layer. The increased methodology of PID mapping, register transfer level (RTL) is implemented by the PID module, which is responsible for determining the outcome of the handled amount. Its sub-modules are organized into three sections and were uniquely sketched following the arithmetic structure. Each module's enable control signals are generated by the major condition of the machine module, which is also responsible for directing each thread module following the algorithm's execution sequence. The four-point theory is based on fallacy backpropagation and weight module updates, which are required to keep the weights of each layer of the neural network up to date. The auxiliary region is divided into two segments: design of the PWM signal generation module and estimation of the motor speed module. The encoder recurrence multiplying innovation was used to construct a fourfold recurrence interface circuit for a motor speed measurement module, and a recurrence method was used to estimate the true speed of the motor, as depicted by the module, which is based on PWM signal production. The frequency's four times greater circuit's topics interaction features a motor speed measurement module that was created using the encoder's frequency doubling technology, and the motor's actual speed was measured using the frequency practice method. PID control is often denoted by the abbreviations K_p for proportional, K_i for integral, and K_d for derivative. It is an efficient and powerful control method that is commonly used in industrial control systems. Its essential preface is that the controlled object is finally scrutinized and figured out when the system's inaccuracy e(k) between the appealed yield r(k) and the actual yielded value y(k) is returned to the objects that are dealt with by a subsequent combination of proportional, integral, and differential functions (Chhaya et al. 2017). Because of the implications of the complicated monitored environment and the convoluted contrast of time parameters of the handled object, the traditional PID is unable to demonstrate adaptive criteria of the handled object within a real motion-monitored technique. In theory, BPNN may imitate any nonlinear function and achieve identical outcomes, as well as dynamic behavior such as self-learning and adaptation (Yadav et al. 2021). As a result, the optimum motion control effects can be designed to achieve live autotuning of PID control parameters when comparing BPNN with PID control. Figure 4.1 presents a PID controller based on the BPNN.

The PID sample is responsible for finishing the control amount output, after which the PID arithmetic is transformed into RTL. The main state machine module generates an enable signal, which controls the sub-module that is executed in chronological order. The combination of the weight update module and backpropagation completes the network's weight update of all separated layers. The control system's peripheral modules are divided

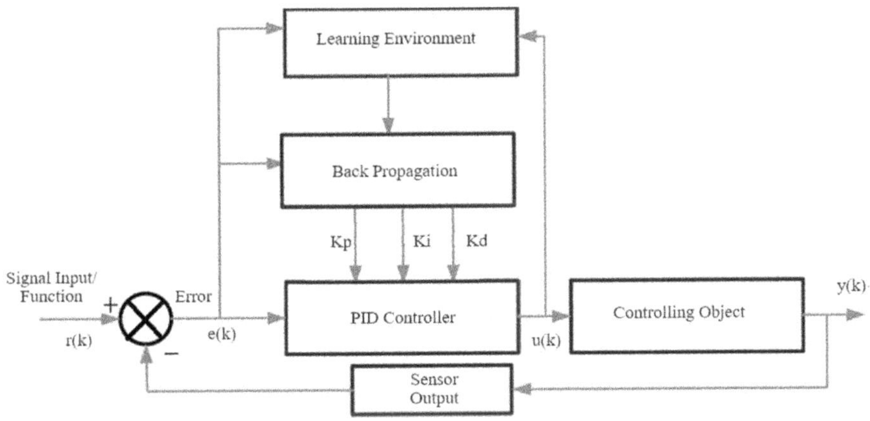

Figure 4.1 PID controller based on BPNN.

into two groups. The speed module uses an encoder and motor signals to measure how fast something is moving. To control the motor's rotation speed, the signal generation module generates PWM waves with varying duty cycles (Gehlot et al. 2021). The system is verified and simulated using a co-simulation of ModelSim and Simulink, as well as an experiment on the development platform. The final output demonstrates that the existing system can automatically ramify PID control and has features such as synchronized display, consistent effective presentation, and is unaffected by other variables (Pandey et al. 2023). The MCU can converge significantly faster than other systems, which makes it superior. The ModelSim simulator, which is based on MATLAB, was used in previous research articles. Xilinx ISE 14.7 software is used in design and does not include any MATLAB simulations. In the previous design, the constant values of the PID controller such as Kp, K_i, and K_d values were obtained via MATLAB. In the suggested architecture, the data is saved using the numerical controller oscillator (NCO), and the corresponding data is stored in read-only memory (ROM). The previous designs support 50 MHz. Moreover, the suggested design can handle 230.00 MHz frequency compared to the previous design, and the system runs faster.

4.2 RELATED WORK

Chan et al. (2007) used FPGA technology to program integrated circuits and better comprehend the modular design of integrated feedback controllers. To accomplish this, a controller technique is proposed, namely, a novel distributed-arithmetic (DA)-based PID, which is integrated into a digital feedback control system. A comparison with the multiplier-based scheme demonstrates the DA-based PID controller's 80% hardware utilization savings and 40% power savings. Furthermore, good closed-loop performance is expected when it is used with a lesser number of resources, as it reduces costs and produces rapid speeds and low power consumption, all of which are desirable in embedded control applications. For fast prototyping applications, the reusable modules can be integrated into MATLAB/Simulink as Simulink blocks for hardware/software co-stimulation applications or consolidated into a larger layout in the MATLAB/Simulink environment. Wang et al. (2022) introduced a concept for a closed-loop motion system that utilizes a BPNN-based PID controller implemented through a BPA FPGA setup from Xilinx. The controller's design incorporates the concept of measured quantity by dividing it into several smaller components. The forward propagation module encapsulates the process of propagating information from the input layer to the output layer. The PID module performs the translation of the PID arithmetic RTL into a configuration that produces the overall sum. Enable signals are produced via the principal state machine module, which generates enable signals that are used to facilitate the sequential execution of each sub-module. The error propagates the weight modifications for each network layer. The auxiliary modules of the control system are divided into two major types. The speed measurement module encompasses the acquisition of the encoder's output pulse signal and the calculation of the motor speed. The motor's rotational speed is controlled using pulse width modulation (PWM) waves with several duty cycles, created by a PWM signal-generating module. The framework is reconstructed and validated by employing ModelSim and Simulink in conjunction with a test analysis on the development platform. Zhao et al. (2022) provided a comprehensive description of the system requirements, including the need for a multilayer DC voltage supply. Converters are commonly employed in smart grid applications. The current converters encounter some notable issues, such as suboptimal efficiency, extended response time, a substantial circuit size due to multiple switches, and inadequate PWM signal quality. Furthermore, each source employed in the circuit necessitates its converter. This work introduces and evaluates a very efficient hardware co-simulation for a DC-DC buck-boost converter with a PID controller. The control of this converter design is accomplished by employing the Virtex5 and Virtex7 FPGA kits. The efficacy of both kits is also assessed and contrasted. The recommended converter has high efficiency, fast response time, and compact size due to its reduced number of switches compared to standard converter topologies. Qiu et al. (2015) initiated the examination and creation of fuzzy-model-based nonlinear networked control systems (NCSs), which have garnered attention from both the scientific and industrial worlds. Several significant discoveries have been proposed as a result. This study focuses on evaluating the assessment and design of fuzzy-model-based nonlinear NCSs with different constraints caused by network issues, such as bundle dropouts, downtime, and flag quantization. An in-depth analysis is conducted on the progress made in addressing various command and control issues caused by these networks, and numerous significant technological problems are identified. Additionally, this paper provides a summary of subsequent discoveries about the strategy of event-triggered control and filtering for nonlinear fuzzy NCSs. Subsequently, some conclusions are drawn and other potential areas for future investigation are emphasized.

Verma and Padhy (2019) examined the basic plan strategy of the PID controller, specifically focusing on the internal model controller (IMC) PID design method, the filter that resolves the three-inconsistent PID layout issue, hence forming a single variable design problem in the IMC-PID layout. The IMC filter is utilized to control process robustness and transient performance. To maintain the transient responsiveness, the stability of the process needs to be compromised due to the presence of an additional phase lag caused by the IMC filter. An advanced approach is proposed for the indirect layout plan of an IMC-PID controller, which involves the use of a first-order filter with dead time (FOPDT) and a second-order filter with dead time (SOPDT) processes. The importance of control systems in engineering has significantly increased due to technological improvements, as outlined by Yalçin et al. (2022). The primary applications of control theory involve the examination and creation of controllers and controller circuits, which enable systems to operate following their desired characteristics. The Z-source inverter (ZSI) circuit in this study experiences an error caused by the discrepancy between the input flag and the yield flag. It is recommended to address this issue by utilizing the PID controller for correction. The differential evolution algorithm, a heuristic optimization algorithm, is employed to determine PID settings. Hence, the primary objective is to achieve the most seamless functioning of the system. The model was estimated using support vector regression (SVR) by employing a numerical cost function that can be executed on an FPGA. Lins and Krishnakumar (2022) examined the extensive use of brushless DC (BLDC) motors in various applications such as autos and water pumping systems. These applications heavily depend on solar energy as their major source of power, supplemented by a battery. The wide speed range capability of BLDC motors is essential for electric vehicle applications. In this experiment, the Zeta converter is utilized to track the highest power generated by the PV panel. To get optimal performance, the PV should be programmed at its maximum control point. To achieve optimal control of the PV panel, this task entails adjusting the duty cycle of the Zeta converter. It also utilizes optimization techniques such as teaching learning-based optimization and particle swarm optimization. A battery is linked in parallel in both directions to account for the intermittent nature of solar generation. Phan et al. (2022) conducted research intending to develop a torque position-based PID controller for a 6-DOF HIWIN RA605 articulated robot, based on its motion geometry, and implement it on a high-speed FPGA. This direct torque controller is appropriate for collaborating with individuals on tasks such as grinding, winding, or any other processes that demand precise torque control. This application effectively mitigates the harm caused by both human–robot and robot–robot attacks. A digital algorithm for a 6-DOF controller is developed using Verilog, a hardware descriptive language, to incorporate various functionalities such as encoder counters, digital filters, analog generators, PID controllers, and AC servo motor drives. The algorithm is then verified using an FPGA. To reduce the imprecision of the encoder signals, an essential technique known as digital filtering is employed within the FPGA. Wu et al. (2022) researched optimizing computing correctness in the execution of hardware programmers. The study focused on approximating the estimated value, highlighting the importance of this research. Conventional approaches are based on either dynamic or static analysis of arithmetic mistakes. The examination of mathematical fallacies, whether conducted on a mobile or stationary platform, serves as the foundation for traditional methodologies. However, only a stationary review may be employed to definitively confirm the necessary worst-case accuracy. This paper presents a computerized approach to calculate the mathematical binary representations and assess the computational sensitivity of one-dimensional feedback-loop algorithms. It enables the estimate of both affine active points and stationary points. A novel automated technique has been developed to accurately determine the exponent and mantissa of floating-point signals, as well as the integer and fractional components of static-point arrays, by employing universal criteria for iterative PID control. This reduces the circuit area and power consumption of the FPGA implementation. Wiegand et al. (2022) developed a versatile and user-friendly device to measure the changing frequency of laser drift over time. The software is primarily designed for spectrum locking, but it can also be used for other comparable interactive methods including basic PID operation. Linien has proposed two distinct methods for selecting an automated lock point. The first method is time-dependent and exclusively operates on FPGA. It also provides support for sinusoidal modulation (up to 50 MHz), triangle ramp scanning, IQ demodulation (1 f to 5 f), and digital filters that utilize IIR (infinite impulse response (IIR) and PID techniques. Linien could calculate the density of false alerts in a power spectrum and automatically enhance spectroscopic parameters using machine learning. The software's modular structure provides both a graphical user interface and a Python programming interface. The Red Pitaya STEM Lab serves as the basis for its establishment. Fan et al. (2022) examined the remarkable precision of deep convolutional neural networks (CNNs) in a wide range of cognitive processes. Scientists have developed a strong interest in this phenomenon. FPGAs are highly sought-after for accelerating convolutional neural networks due to their exceptional performance, adaptability, and energy efficiency. Several accelerators, including the Australian Security & Investment Commission, FPGAs, and graphics processing units (GPUs), have been employed to enhance the performance of CNNs. FPGAs are commonly employed to enhance the performance of deep learning networks owing to their exceptional power efficiency and ability to optimize parallelism. The works were categorized to facilitate comparisons and distinctions about the various techniques employed to create and optimize CNN algorithms on FPGA. This paper presents the most recent developments in FPGA-based accelerators designed for deep learning networks. This assessment is expected to enhance the effectiveness of deep learning research by offering guidance for future hardware advances. Shah and Agashe (2016) conducted an extensive study on fractional calculus, which has been widely employed in both science and technology for over three centuries. They discussed the initial implementation of control systems, specifically focusing on the fractional PID method in 1993. Subsequently, it introduces an innovative fractional PID controller that integrates the latest developments in this field. This text focuses on the progress made in the field of fractional PID controllers, explores their various versions, and discusses their design and calibration. The utilization of software tools in the development of fractional PID controllers is also addressed. Lima et al. (2006) investigated the use of fixed-point numerical representation for the development of PID controls in FPGAs. The MATLAB/Simulink environment is utilized to create, simulate, and evaluate the performance of various fixed-point representations while employing a specific control technique. The controller system employs a static bit-width analyzer to produce a precise fixed-point representation for every operand/operator. After conducting bit-width analysis, the system is implemented using VHDL. The findings indicate that the utilization of tailored fixed-point representations reduces the duration of design cycles while conserving substantial resources. Chander et al. (2010) examined the PID controller as the controller of choice in enterprises when a comprehensive analytical description of the system being handled is not necessary. This paper presents the development and execution of a PID controller using FPGA technology for a low-voltage synchronous buck converter. The MATLAB/Simulink environment is utilized for the design of PID controllers to generate a set of coefficients that correspond to the desired attributes of the controller. Subsequently, these controller coefficients are integrated into VHDL, a hardware description language, to execute the PID controller on the FPGA. The two PID controller architectures are assessed, together with their device utilization and power dissipation statistics, to showcase the resource usage and power dissipation of the chosen FPGA. The designs are executed using an FPGA chip called Virtex-5 (ML505) XC5VLX50T-1FF1136 (-1 speed grade).

Kocur et al. (2014) discussed that FPGA enables efficient implementation of digital PID control algorithms. PID control techniques and algorithms are widely used feedback controllers in various industrial processes for automated control. PID controllers are extensively utilized in diverse domains such as power systems, drive control, automotive mechatronics, aerospace, process control, and robotics. PID controllers are widely used in several fields, including drive control, robotics, automotive mechatronics, power systems, aerospace, and process control. PID control algorithms progress through multiple stages, starting from initial mechanical, electrical, and pneumatic designs and evolving into microprocessor-based systems. Another method for determining digital control algorithm systems involves utilizing FPGAs, previously utilized by general-purpose microprocessor systems. Compared to typical PID controller implementation, utilizing an FPGA-based controller offers several advantages, including high speed, advanced functionality, and low power consumption. Supervising concurrent activities is a notable feature of FPGA-based platforms, allowing for the simultaneous design of well-defined digital controller systems through parallel structural planning. This study elucidates the implementation of a digital PID control approach, which involves the integration of hardware and software components, to achieve quick dynamics in a specific application. The implementation of the FPGA PID control algorithm for high-speed DC motors involves a series of construction, validation, and exploration. This process utilizes FPGA technology, specifically the Xilinx Spartan-6 FPGA Family, which helps to reduce the risks, costs, and power requirements associated with this application. Dhanabalan et al. (2022) discussed the ladder diagram (LD), which utilizes input and output segments within PLC. An LD also includes function blocks for PID controllers. In addition, an LD has function blocks for PID control. It is also effective in managing several PID function blocks for a range of process parameters. The time of the PLC examination is influenced by an increase in the number of process specifications. Analog indicators regulate the process specifications. The analog indications or signals are transformed into digital signals by the analog input specification of the PLC (Dhanabalan & Selvi 2015), which subsequently transmits them as inputs to the PID controller. The study demonstrates the reduction in PLC scan time by implementing an FPGA-based multiple PID controller. It was confirmed that multiple PID controllers can be constructed concurrently by assigning distinct FPGA hardware resources to each one. The PID controller makes use of an analog-to-digital converter (ADC), of which FPGA, comparator, and digital-to-analog (DAC) modules are required for this process. In a closed-loop control system, an ADC and dedicated PID controller circuitry on an FPGA can be used to run many PID controllers at once. The specified time limit for conducting two closed-loop controls has been determined to be 18.96000004 ms. The utilization of the layout for this purpose can be achieved with or without a PLC. The role of PLC is a critical component of a computerized system. The closed-loop system is controlled by PLC and performs several iterations of combinational or sequential logic. The ladder diagram is also utilized in this process. The accumulation of information that is referred to as a load is the result of a significant quantity of analog and digital signals being broadcast through it, as well as additional signals being added to it simultaneously. A PID functional block, an analog input module (AIM), and an analog output module (AOM) are all components that must be present in the logic device to complete the establishment of a closed-loop control system. The combination of an ADC and dedicated PID controller circuitry on FPGA makes it possible for any closed-loop control system to execute several PID controllers simultaneously. These components are essential for the implementation of a PLC. This ensures that the closed-loop control system's output consistently achieves the desired value. The PLC examines the PID function block in its ladder diagram and provides a reference to AIM to monitor the corresponding analog value. Upon the completion of the execution of the PID function block, it generates a computerized output and sends it to the AOM. The primary function of AOM-DAC is to transform the computerized output or digital output into an analog signal and activate the final control element. By adding more PID function blocks in a stack, the execution of the remaining rungs in the ladder program is slowed down and the scanning time of the PLC is increased. The main reason for this is that the processor is shared by multiple entities such as PLC, AIM, and AOM, resulting in delays in processing their tasks. A monitored activity is established by integrating a genetic algorithm with a neural network, while the computer functions as the control unit. This system utilizes PLC and an industrial Wide Area Network (WAN) to monitor the water level in the tank. The exploration of a parallelized multi-PID controller design is discussed. An analysis is conducted on the design of a parallelized multi-PID controller, which utilizes an FPGA-based multiprocessor. This also incorporates an enhanced Pico Blaze microcontroller (EPM). Four PID controllers were designed to be simultaneously tested to evaluate their performance. However, this research lacked a comprehensive analysis of the functioning of analog signals by this multiple PID controller. An illustration presented on paper does not necessarily require the conversion of analog signals to digital signals. It is uncertain whether this design can effectively run several PID controllers simultaneously under a certain situation to maintain an analog signal at a predetermined point. Therefore, this feedback proposes the use of a PLC computer, DSP, FPGA, or a combination of these hardware components, to build one or more closed-loop control systems. The FPGA has been identified as a superior alternative for executing CPU functions with greater efficiency. An instance of the FPGA-based floating-point processor layout is provided as an illustration. FPGAs can also be utilized for the implementation of PID controllers. Zhang and Guo (2019) discussed why linear PID controllers may efficiently regulate nonlinear uncertain dynamical systems. Additionally, Zhang and Guo introduced a method to get an accurate design equation for the PID parameters, which is still to be developed. This is the catalyst for our latest inquiry into the theoretical foundations of PID control. The main goals of this chapter are to extend the 1-D findings to nonlinear systems in higher dimensions and to significantly improve the results by employing a more advanced methodology. Moreover, it considers a complex system including several agents, where each agent is controlled by a PID controller that utilizes its governing error. Moreover, an illustration demonstrates the unambiguous construction of a parameter manifold in multi-agent frameworks that exhibit universal stability. The PID specifications are selected from this manifold to ensure that tracking errors approach zero exponentially quickly.

4.3 SYSTEM DESIGN

Three fundamental modes – the proportional, integral, and derivative – make up the PID algorithm. It is important to select the mode (P, I, or D) to be used before specifying the parameters for each mode before applying the algorithm. There are typically three fundamental algorithms utilized: P, PID, or PI. Figure 4.2 displays the architecture block diagram of a PID-controlled DC motor. The DC motor plant's chosen speed in this is a set point or reference signal. The detector receives this reference signal. To compare the value of the feedback element with the selected value input, a detector is needed. At the final stage, it aids in obtaining an accurate estimate or an approximation number that is substantially comparable to the chosen value. The PLL input receives the error detector's output. The clock signal creation in this case uses a PLL (Kumar et al. 2017).

The numerically controlled oscillator is the primary component of the PLL. The PID controller receives the error detector's output as input. The discrepancy between the reference signal and the feedback element, which activates the regulator components, is the error detector's output. The control components alter the environment in the plant to lessen the initial

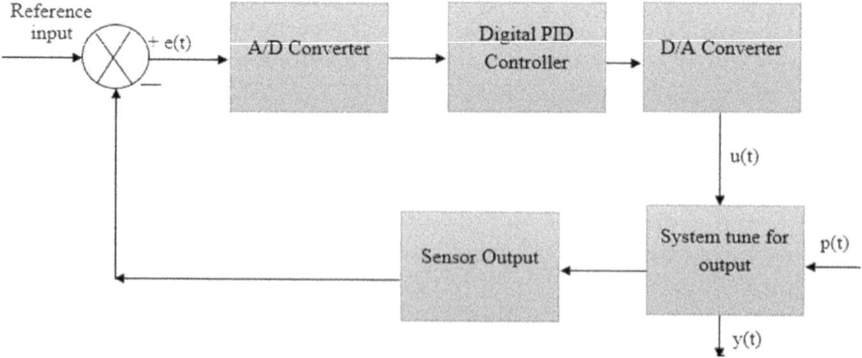

Figure 4.2 Blocks of motion control system.

inaccuracy. The controller uses a PWM generator to create the PWM waveform that will be the system's input, in this case, a DC motor. The FPGA control-based motion control level module architecture is shown in Figure 4.3, which consists of the neural network-based module. The Virtex-7 FPGA is used for the synthesis of the different modules of the motion control-based system (Bhatia et al. 2022).

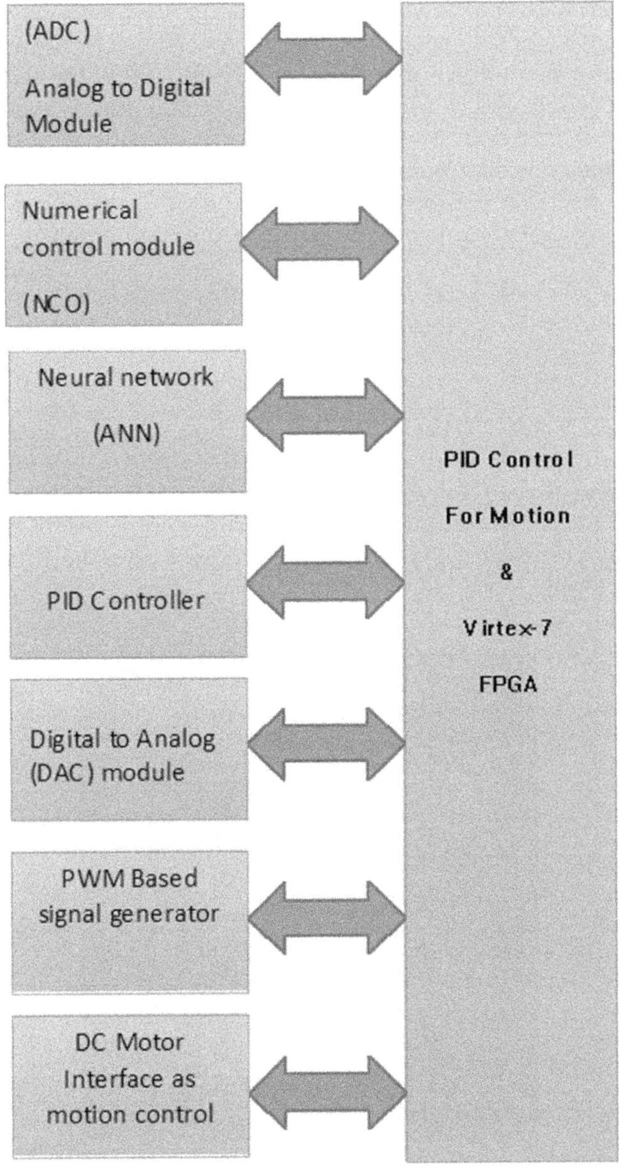

Figure 4.3 Modules of motion control modules integrated with PID and FPGA.

The ADC interface, numerical controlled oscillator, PID controller, DAC interface, DC motor, and PWM generator are the key elements of the system block diagram. The DC motor and error signal are analog, necessitating the use of ADC and DAC to transform them into digital. Only digital signals are accepted by the FPGA and digital PID chip. Following PID tuning and constant processing of the output signal, an analog conversion by an FPGA and PID controller is required. A DAC is therefore necessary (Mishra and Kumar 2022). With the PWM signal adjusted, the DC motor may also need to revolve. The clock is used to generate the PWM signal. The variation between an intended set point and a measured process variable is the error value e(t), which a PID controller continuously calculates and uses to apply a proportional, integral, and derivative term-based correction. To reduce time-dependent error, the controller balances a control variable u(t) to a weighted sum value. This affects the position of a damper, control valve, or heating element.

$$u(t) = K_p e(t) + K_i \int_0^t e(t)\,dt + K_d \frac{de(t)}{dt} \tag{4.1}$$

4.4 RESULTS AND DISCUSSION

The RTL view of the PID control module is shown in Figure 4.4. The RTL describes the input/output of the PID controller components. The internal schematic diagram of the PID controller obtained from the Xilinx software is shown in Figure 4.5.

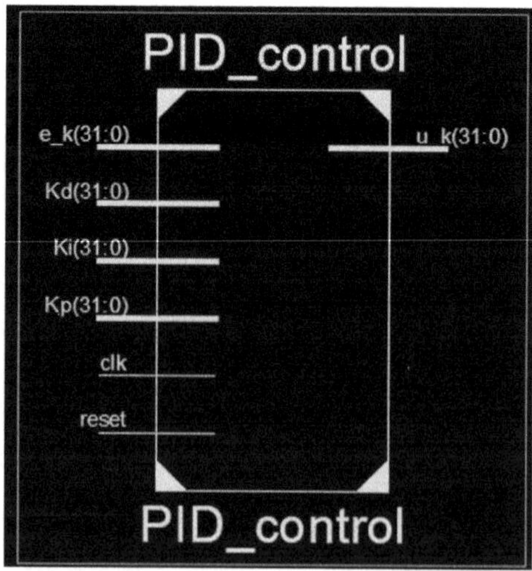

Figure 4.4 PID control module presentation as RTL.

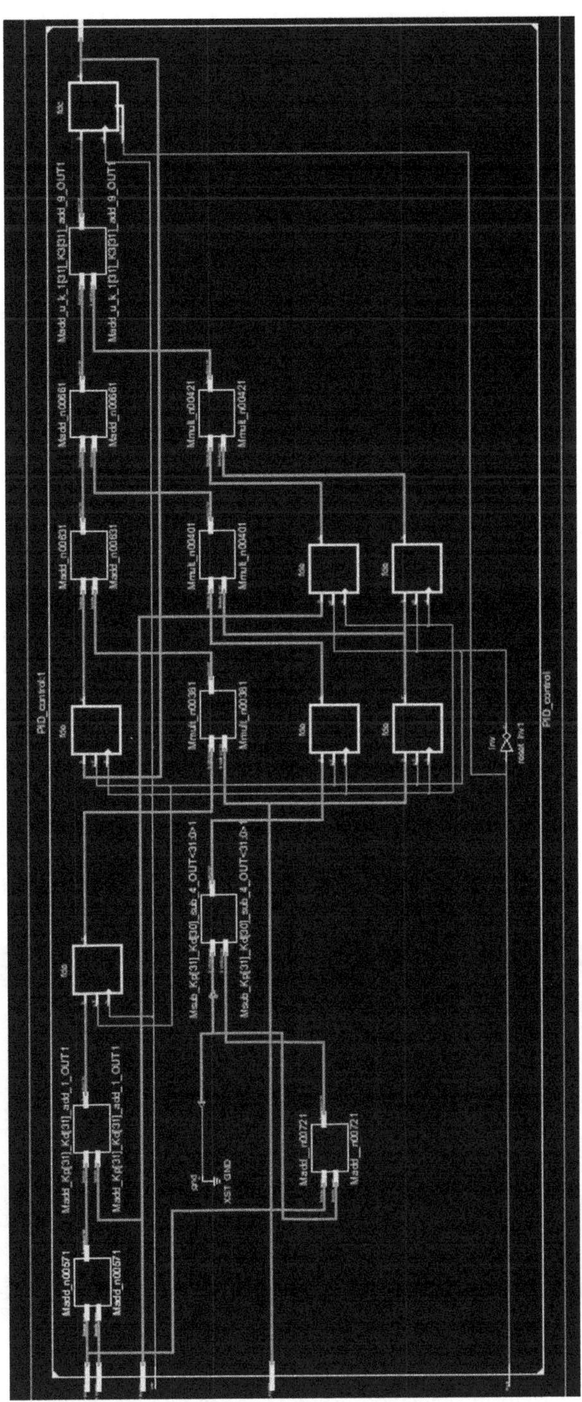

Figure 4.5 Internal schematics presentation of PID in Xilinx.

The main inputs of the PID control module are clock input (CLK) and reset input (reset) (Ompal et al. 2021). The other inputs are e(k), K_p, K_i, and K_d which are given in the form of the decimal value. The output of this model is given as u(k). The behavior of the PID control is simulated in the Xilinx ISE waveform simulation platform. The simulation is based on the following inputs.

Case 1: Assign the clock signal to clock input and reset is kept high in the initial stage, also error, reset = 1. The positive or the rising edge of the clock signal is given to the functional block.

Case 2: Keep the same edge of the clock signal and reset = 0, e(k) = 0. Assign the values of the control parameters or the PID such as K_p = 90, K_d = 5, and K_i = 98 as integer values, and the pwm_out is obtained with the different increasing sequences. Figure 4.6 presents the hardware simulation in ISIM waveform simulation presentation of the PID control in Xilinx for the parameters for the PID module taken from the hardware report summary in Xilinx after pre-synthesis. Figure 4.7 presents the ISIM waveform simulation presentation of PID control in Xilinx with increasing values of DC motor speed. Figure 4.8 presents the PID control module's estimated hardware parameters.

The RTL view of the DC motion control module is shown in Figure 4.9. The RTL describes the input/output of the motion control component. The internal schematic view of the same PID is depicted in Figure 4.10.

In the RTL, the term e(k) denotes the error signal. The clock pulse is supplied using the clock signal, which works on the 50% duty signal, and the reset is applied to work and keep the outcome exactly zero which are working together. The components of the control module of the PID controller are K_p, K_i, and K_d, which are given based on the simulation inputs and the load are the signals applicable with the numerically controlled oscillator to

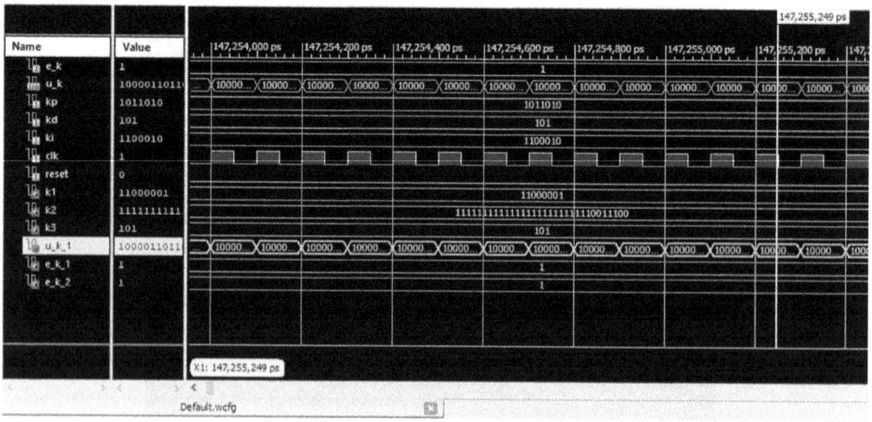

Figure 4.6 ISIM waveform simulation presentation of PID controller in Xilinx with the constant value.

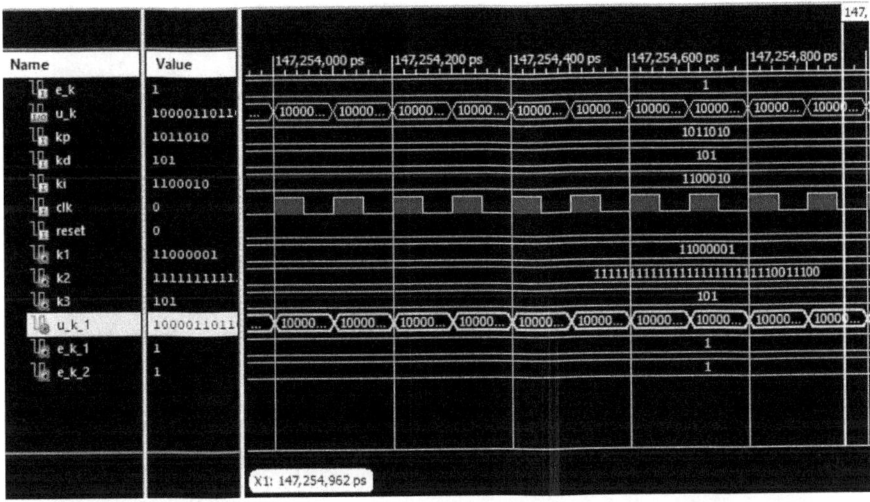

Figure 4.7 ISIM waveform simulation presentation of PID controller in Xilinx with increasing values.

PID_control Project Status (11/04/2022 - 12:43:59)			
Project File:	PID_Control.xise	Parser Errors:	No Errors
Module Name:	PID_control	Implementation State:	Synthesized
Target Device:	xc7vx330t-3ffg1157	• Errors:	
Product Version:	ISE 14.7	• Warnings:	
Design Goal:	Balanced	• Routing Results:	
Design Strategy:	Xilinx Default (unlocked)	• Timing Constraints:	
Environment:	System Settings	• Final Timing Score:	

Device Utilization Summary (estimated values)				[-]
Logic Utilization	Used	Available	Utilization	
Number of Slice Registers	128	408000	0%	
Number of Slice LUTs	224	204000	0%	
Number of fully used LUT-FF pairs	55	297	18%	
Number of bonded IOBs	162	600	27%	
Number of BUFG/BUFGCTRLs	1	32	3%	
Number of DSP48E1s	9	1120	0%	

Figure 4.8 PID control module estimated hardware parameters.

count the first load values to the PWM and notice the sequence in the increment order based on the motion load value to the DC motor controller. The input for the PWM control model is given using the Motor_PWM_in (31:0) and the corresponding outcome is taken using Motor_PWM_out (31:0). Figure 4.11 presents the device utilization report extracted from the software to estimate the hardware parameters' utilization.

Figure 4.12 presents the waveform simulation for motion control with the increasing speed of the motor and Figure 4.13 presents the waveform simulation for motion control with other data inputs in the ModelSim. All the designed modules in the VHDL are synthesized in the Virtex-7 FPGA with

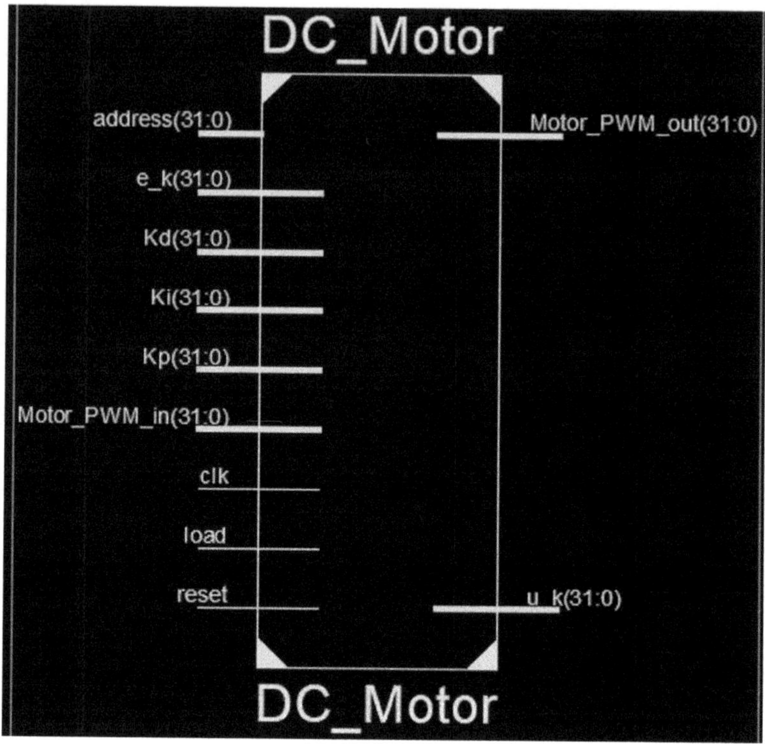

Figure 4.9 RTL view of the motion control module.

the device number-xc7vx330t-3ffg1157. Table 4.1 shows the memory utilization and input/output block utilization summary obtained by the synthesis of the different blocks and modules. Table 4.2 lists the delay estimation for different modules based on simulation. The corresponding graph that shows the input-output blocks (IoBs) usage is depicted in Figure 4.14. In the graph pictorial, the X-axis denotes the module name and the Y-axis shows the value of the IoBs used in the design. Figure 4.15 presents the memory (kB) Utilization for Virtex-7 FPGA in all modules. Table 4.3 lists the values of the total timing as a delay (ns) parameter. Figure 4.16 presents the delay (ns) summary on Virtex-7 FPGA in all modules.

In comparison to previous work, the designed chip supports 230.00 MHz frequency for motion and PID control, whereas the Wang et al. (2022) design supports 50.0 MHz frequency, indicating that the designed chip is quicker. The chip supports a higher frequency, which means faster switching. The chip design supports a delay of 9.203 ns, which is less than the other 10.00 ns. The work is performed using a Virtex-7 FPGA.

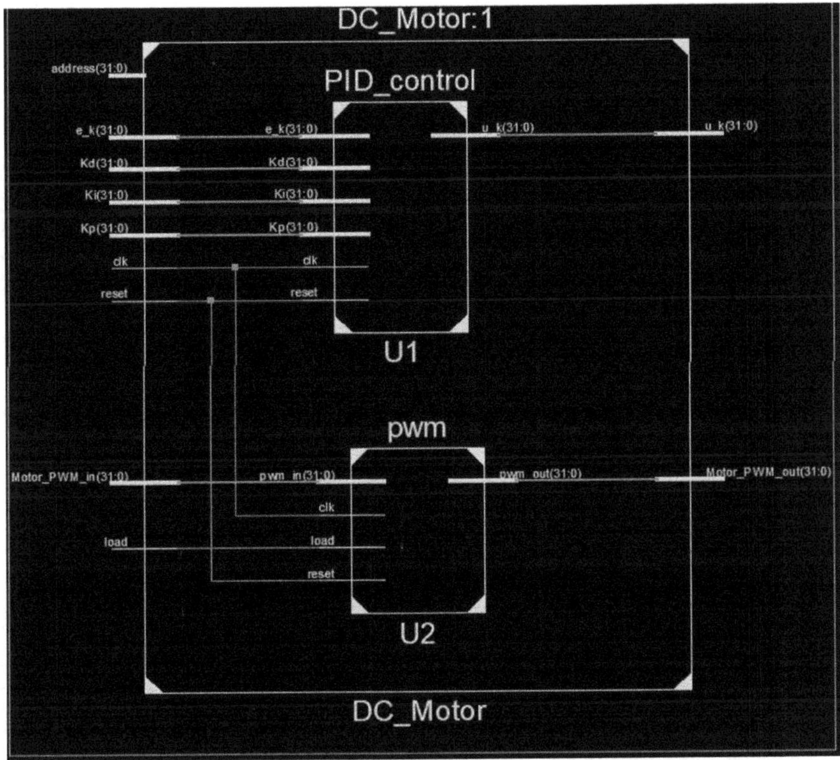

Figure 4.10 Internal view of the motion control module in RTL.

PID_control Project Status (11/13/2022 - 12:13:31)			
Project File:	DC_Motor.xise	Parser Errors:	No Errors
Module Name:	DC_Motor	Implementation State:	Synthesized
Target Device:	xc7vx330t-3ffg1157	• Errors:	No Errors
Product Version:	ISE 14.7	• Warnings:	10 Warnings (10 new)
Design Goal:	Balanced	• Routing Results:	
Design Strategy:	Xilinx Default (unlocked)	• Timing Constraints:	
Environment:	System Settings	• Final Timing Score:	

Device Utilization Summary (estimated values)				[-]
Logic Utilization	Used	Available	Utilization	
Number of Slice Registers	224	408000	0%	
Number of Slice LUTs	391	204000	0%	
Number of fully used LUT-FF pairs	88	527	16%	
Number of bonded IOBs	227	600	37%	
Number of BUFG/BUFGCTRLs	1	32	3%	
Number of DSP48E1s	9	1120	0%	

Figure 4.11 Xilinx response utilization summary.

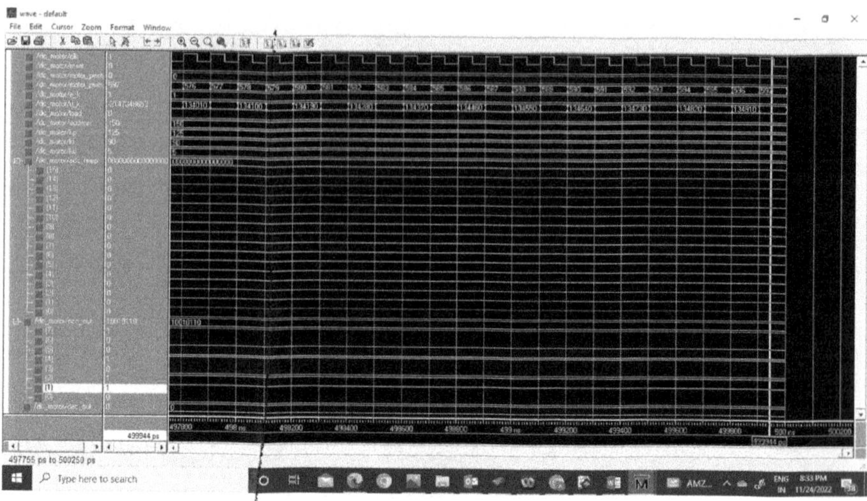

Figure 4.12 Waveform simulation for motion control.

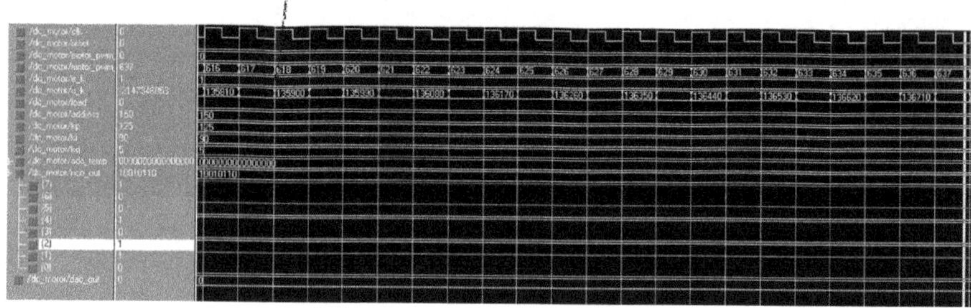

Figure 4.13 Waveform simulation for motion control with a functional simulation.

Table 4.1 Hardware parameter utilization of different modules

	ADC	DAC	NCO	PWM	PID control	Main DC motor
Bonded IoBs	32	48	42	96	162	227
Memory (kB)	4690032	4676848	4692400	4692592	4695408	4704880

Table 4.2 Dealy estimation for different modules based on simulation

Parameter	ADC	DAC	NCO	PWM	PID control	Main DC module
Delay (ns)	0.279	0.279	0.511	1.086	7.048	9.203

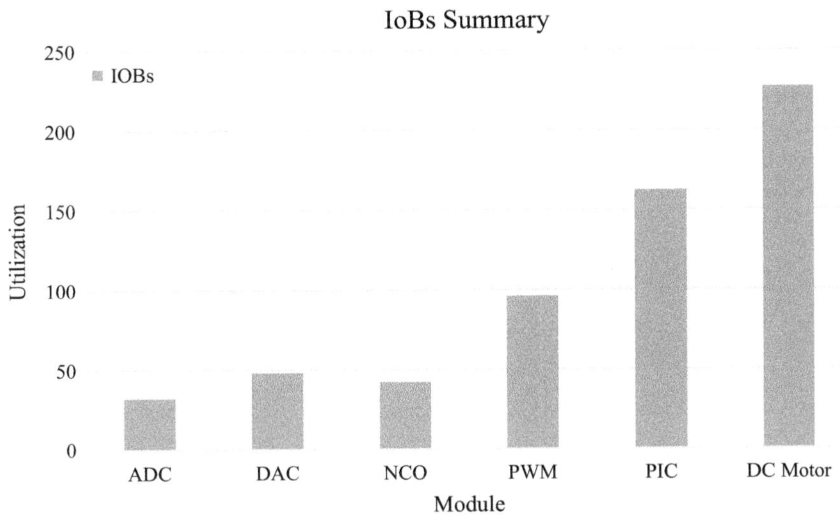

Figure 4.14 Utilization of IoBs for the different modules of DC motor control.

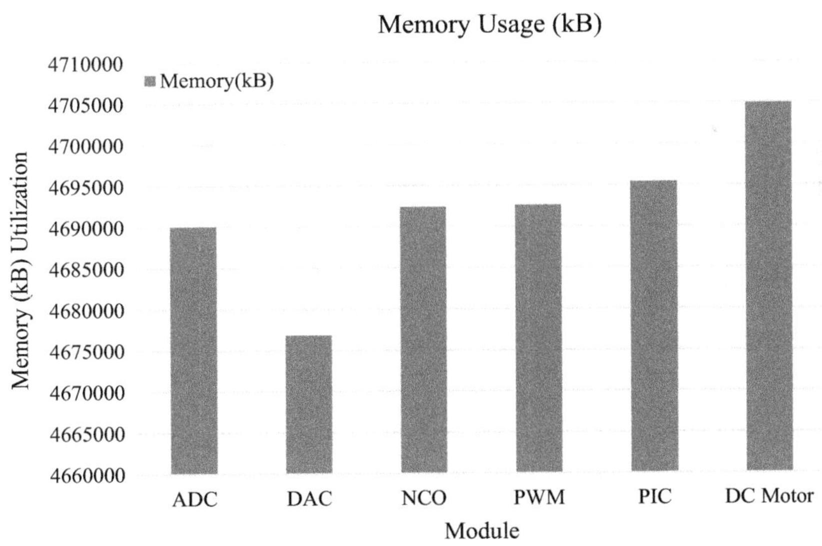

Figure 4.15 Utilization of memory for all modules in Virtex-7 FPGA.

Table 4.3 Comparison with the existing work

Parameter	Wang et al. (2022)	Proposed work
Frequency	50.00 MHz	230.00 MHz
Delay	10.00 ns	9.203 ns
Technology	Artix-7 Xc7a200tfbg484-2, 28 nm	Virtex-7, 28 nm

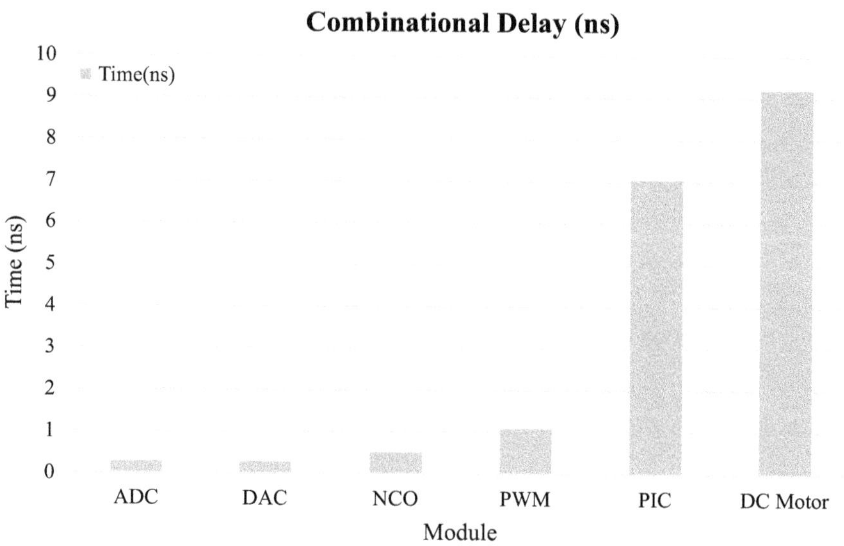

Figure 4.16 Combinational delay (ns) for all modules in Virtex-7 FPGA.

4.5 SUMMARY

The hardware chip design for the digital PID controller has been successfully implemented to effectively manage the behavior of a DC motor. This design allows for the adaptation of various attributes of different controlled objects and enables the adaptive tuning of PID parameters. The unique microcontroller unit cannot function. Consequently, it is necessary to manage the system in real-time and high-reliability application situations. Therefore, this research aims to develop a closed-loop motion control system using a Xilinx FPGA solution that utilizes a BPNN-based PID controller. The PID module integrates the output of the control values and applies it to the application, which includes the conversion of PID calculations to the RTL. The main state machine module is responsible for implementing each subsequent module and generating enable signals. The process of updating the weights of each layer in a network involves the use of error backpropagation and a weight update module. The control system's auxiliary parts are divided into two fundamental kinds. The motion control speed is regulated by utilizing a clock signal to generate PWM waves with specific duty cycles, which is accomplished by the PWM signal production module. The system is simulated and validated using ModelSim and Xilinx, together with a test analysis conducted on the development platform. The comprehensive block diagram comprises an ADC, DAC, PWM, NCO, PID control, and a final motion controller integrated module. The RTL design of the ADC, DAC, PWM, NCO, and PID has been completed using Xilinx. The functionality has been verified by testing the inputs in the Xilinx ISIM simulator. The RTL design of the PID controller was completed in Xilinx. The functionality has been tested using test inputs in the Xilinx ISIM simulator, considering various values of K_p, K_i, and K_d, as well as the PID parameters. The RTL for the entire chip, which is based on motion control, is derived from the Xilinx ISE software. It is then tested with the ISIM simulator to ensure that the waveforms are accurate.

The expansion of PID control design approaches appears to be nearing a threshold of diminishing returns in the realm of research. It is still necessary to resolve certain difficult issues. For example, the processes that are enhanced by PID control have not been identified. The identical modules can subsequently be generated using the more advanced FPGA, which provides reduced latency and memory concerning the outcomes. Therefore, for certain robotics tasks, the integrated chip may be the best option. After that, we can evaluate the controller's execution on the FPGA and see how much power it consumes.

REFERENCES

Bhatia, A., Kumar, A., Jain, A., Kumar, A., Verma, C., Illes, Z., & Raboaca, M. S. (2022). Networked control system with MANET communication and AODV routing. *Heliyon*, 8(11), doi: 10.1016/j.heliyon.2022.e11678.

Chan, Y. F., Moallem, M., & Wang, W. (2007). Design and implementation of modular FPGA-based PID controllers. *IEEE Transactions on Industrial Electronics*, 54(4), 1898–1906, doi: 10.1109/TIE.2007.898283.

Chander, S., Agarwal, P., & Gupta, I. (2010, December). FPGA-based PID controller for DC-DC converter. In: 2010 I Joint International Conference on Power Electronics, Drives and Energy Systems & 2010 Power India (pp. 1–6). IEEE, doi: 10.1109/PEDES.2010.5712454.

Chhaya, L., Sharma, P., Bhagwatikar, G., & Kumar, A. (2017). Wireless sensor network based smart grid communications: Cyber attacks, intrusion detection system, and topology control. *Electronics*, 6(1), 5, doi: 10.3390/electronics6010005.

Dhanabalan, G., & Selvi, S. T. (2015). Design of parallel conversion multichannel analog to digital converter for scan time reduction of programmable logic controller using FPGA. *Computer Standards and Interfaces*, 39, 12–21, doi: 10.1016/j.csi.2014.12.002.

Dhanabalan, G., Tamil Selvi, S., & Mahdal, M. (2022). Scan time reduction of PLCs by dedicated parallel-execution multiple PID controllers using an FPGA. *Sensors*, 22(12), 4584, doi: 10.3390/s22124584.

Fan, H., Ferianc, M., Que, Z., Liu, S., Niu, X., Rodrigues, M. R., & Luk, W. (2022). FPGA-based acceleration for Bayesian convolutional neural networks. *IEEE Transactions on Computer-Aided Design of Integrated Circuits and Systems*, 41(12), 5343–5356, doi: 10.1109/TCAD.2022.3160948.

Gehlot, A., Singh, R., Kuchhal, P., Kumar, A., Singh, A., Alsubhi, K., … Brenosa, J. (2021). WPAN and IOT enabled automation to authenticate ignition of vehicle in perspective of smart cities. *Sensors*, 21(21), 7031, doi: 10.3390/s21217031.

Gupta, A., Verma, V., Kumar, A., Sharma, P., Gupta, M. K., & Meera, C. S. (2017). Stabilization of underactuated mechanical system using LQR technique. In: Proceeding of International Conference on Intelligent Communication, Control and Devices: ICICCD 2016 (pp. 601–608). Springer, doi: 10.1007/978-981-10-1708-7_68.

Kocur, M., Kozak, S., & Dvorscak, B. (2014, May). Design and implementation of FPGA-digital based PID controller. In: Proceedings of the 2014 15th International Carpathian Control Conference (ICCC) (pp. 233–236). IEEE, doi: 10.1109/CarpathianCC.2014.6843603.

Kumar, A., Bansal, K., Kumar, D., Devrari, A., Kumar, R., & Mani, P. (2021). FPGA application for wireless monitoring in power plant. *Nuclear Engineering and Technology*, 53(4), 1167–1175, doi: 10.1016/j.net.2020.09.003.

Kumar, A., Verma, G., & Gupta, M. K. (2017). FM receiver design using programmable PLL. *Wireless Personal Communications*, 97(1), 773–787, doi: 10.1007/s11277-017-4536-1.

Kumar, R., Divyanshu, & Kumar, A. (2021). Nature based self-learning mechanism and simulation of automatic control smart hybrid antilock braking system. *Wireless Personal Communications*, 116(4), 3291–3308, doi: 10.1007/s11277-020-07853-7.

Lima, J., Menotti, R., Cardoso, J. M., & Marques, E. (2006, October). A methodology to design FPGA-based PID controllers. In: 2006 3 IEEE International Conference on Systems, Man and Cybernetics (Vol. 3, pp. 2577–2583). IEEE, doi: 10.1109/ICSMC.2006.385252.

Lins, A. W., & Krishnakumar, R. (2022). Tuning of PID controller for a PV-fed BLDC motor using PSO and TLBO algorithm. *Applied Nanoscience*, 1–24, doi: 10.1007/s13204-021-02272-x.

Mishra, V. M., & Kumar, A. (2022). FPGA integrated IEEE 802.15. 4 ZigBee wireless sensor nodes performance for industrial plant monitoring and automation. *Nuclear Engineering and Technology*, 54(7), 2444–2452, doi: 10.1016/j.net.2022.01.011.

Ompal, Mishra, V. M., & Kumar, A. (2021). Zigbee internode communication and FPGA synthesis using mesh, star, and cluster tree topological chip. *Wireless Personal Communications*, 119(2), 1321–1339, doi: 10.1007/s11277-021-08282-w.

Pandey, U., Pathak, A., Kumar, A., & Mondal, S. (2023). Applications of artificial intelligence in power system operation, control and planning: A review. *Clean Energy*, 7(6), 1199–1218, doi: 10.1093/ce/zkad061.

Phan, T. P., Chao, P. C. P., & Huang, Z. W. (2022). Design and implementation of a new torque controller via FPGA for 6-DOF articulated robots. *Microsystem Technologies*, 1–18, doi: 10.1007/s00542-022-05292-x.

Qiu, J., Gao, H., & Ding, S. X. (2015). Recent advances on fuzzy-model-based nonlinear networked control systems: A survey. *IEEE Transactions on Industrial Electronics*, 63(2), 1207–1217, doi: 10.1109/TIE.2015.2504351.

Rawat, A. S., Rana, A., Kumar, A., & Bagwari, A. (2018). Application of multilayer artificial neural network in the diagnosis system: A systematic review. *IAES International Journal of Artificial Intelligence*, 7(3), 138, doi: 10.11591/ijai.v7.i3.pp138-142.

Shah, P., & Agashe, S. (2016). Review of fractional PID controller. *Mechatronics*, 38, 29–41, doi: 10.1016/j.mechatronics.2016.06.005.

Verma, B., & Padhy, P. K. (2019). Indirect IMC-PID controller design. *IET Control Theory and Applications*, 13(2), 297–305, doi: 10.1049/iet-cta.2018.5454.

Wang, J., Li, M., Jiang, W., Huang, Y., & Lin, R. (2022). A design of FPGA-based neural network PID controller for motion control system. *Sensors*, 22(3), 889, doi: 10.3390/s22030889.

Wiegand, B., Leykauf, B., Jördens, R., & Krutzik, M. (2022). Linien: A versatile, user-friendly, open-source FPGA-based tool for frequency stabilization and spectroscopy parameter optimization. *Review of Scientific Instruments*, 93(6), doi: 10.1063/5.0090384.

Wu, Y., Zhang, Y., Hamadouche, A., Mota, J. F., & Wallace, A. M. (2022, September). Automatic approximation for 1-dimensional feedback-loop computations: A PID benchmark. In: 2022 Sensor Signal Processing for Defence Conference (SSPD) (pp. 1–5). IEEE, doi: 10.1109/SSPD54131.2022.9896191.

Yadav, J., Kurre, S. K., Kumar, A., & Kumar, R. (2021). Nonlinear dynamics of controlled release mechanism under boundary friction. *Results in Engineering*, 11, 100265, doi: 10.1016/j.rineng.2021.100265.

Yalçin, O., Canli, A., Yilmaz, A. R., & Erkmen, B. (2022, March). Robust tuning of PID controller using differential evolution algorithm based on FPGA. In: 2022 9th International Conference on Electrical and Electronics Engineering (ICEEE) (pp. 180–184). IEEE, doi: 10.1109/ICEEE55327.2022.9772529.

Zhang, J., & Guo, L. (2019). Theory and design of PID controller for nonlinear uncertain systems. *IEEE Control Systems Letters*, 3(3), 643–648, doi: 10.1109/LCSYS.2019.2915306.

Zhao, Z., Cao, R., K. F., Yu, W. H., Mak, P. I., & Martins, R. P. (2022). An FPGA-based transformer accelerator using output block stationary dataflow for object recognition applications. *IEEE Transactions on Circuits and Systems. Part II: Express Briefs*, 70(1), 281–285, doi: 10.1109/TCSII.2022.3196055.

Chapter 5

Design of rectangular microstrip patch array antenna at sub-6 GHz

*F. B. Shiddanagouda, N. Dinesh Kumar,
M. Sujeeth Reddy, T. Anirudh, and V. Lokesh*

5.1 INTRODUCTION

The current wireless world relies heavily on high data rate transmission. Even after the COVID-19 pandemic, internet usage has doubled compared to previous years, as many sectors such as corporate, industrial, educational, and government institutions depend on the internet for smooth remote work. As digital devices become more interconnected through the Internet of Things, bandwidth and data rate issues arise to support customers' needs [1, 2]. To address these issues, antennas are crucial components in wireless devices. The microstrip antenna is commonly used and developed for wireless communication, particularly for spacecraft and low-profile antenna applications (Balanis (1982)) et al. This antenna consists of a substrate and a thin metallic layer that acts as a radiating element or a ground. The microstrip antenna patch can be made in different shapes, including rectangular, triangular, and circular. This antenna has many advantages, such as low cost, lightweight, and ease of manufacture. The gain, directivity, and bandwidth of the antenna system can be improved by using an array of microstrip patch antennas, as reported in many studies [3–7]. In this work, a rectangular microstrip patch antenna is considered due to its simple construction, consisting of a metal patch on top of a dielectric substrate, with a ground plane at the bottom. These antennas can be arranged by placing numerous patches on a single substrate and feeding each patch with its own feed line. The main goal is to create a rectangular microstrip patch array antenna at 2.4 GHz frequency utilizing the array principle. Therefore, this chapter is organized as follows: the second section presents the design of microstrip patch array antenna for sub 6GHz applications, and the results and their discussions are introduced in the third section, followed by a conclusion in the fourth section.

5.2 DESIGN

The design process for a microstrip patch antenna involves four main steps:

DOI: 10.1201/9781003510420-5

Table 5.1 Specifications of the substrate materials

Material	FR-4
Relative permittivity	4.4
Loss factor	0.02
Thickness	1.6 mm

- Selecting the resonating frequency and dielectric substrate material for the antenna.
- Calculating the dimensions of the antenna for the desired resonating frequency.
- Simulating the dimensions using an electromagnetic (EM) simulator.
- Continuing the parametric study using the EM simulator until the desired results are achieved.

Initially, a resonating frequency of 2.4 GHz was chosen for the single microstrip patch antenna. The antenna was designed using FR-4 substrate material, and its specifications are shown in Table 5.1. The single microstrip patch antenna (SMPA) structures were designed using an HFSS EM simulator and fed with a 50 Ω microstrip line as shown in Figure 5.1. The dimensions were calculated from equations [6–9] and summarized in Table 5.2.

$$W_p = \frac{c}{2f_c}\sqrt{\frac{2}{2\mu_e+1}} \tag{5.1}$$

$$\varepsilon_{ref} = \frac{\varepsilon_r+1}{2} + \frac{\varepsilon_r-1}{2\sqrt{1+12\dfrac{h}{W}}} \tag{5.2}$$

$$L_p = \frac{c}{2f_c}\sqrt{\frac{2}{\varepsilon_{ref}}} \tag{5.3}$$

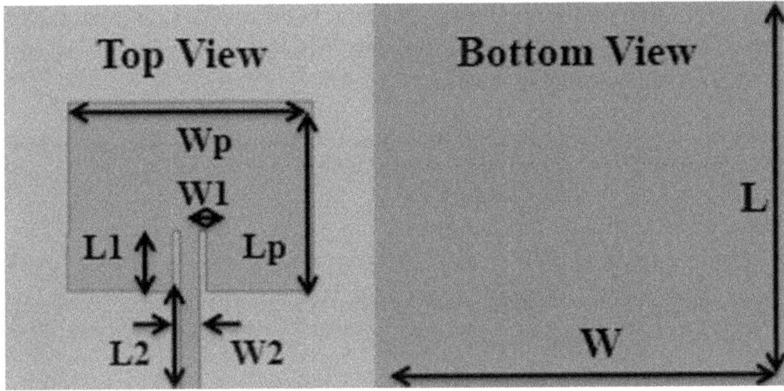

Figure 5.1 Structure of SMPA.

Table 5.2 Dimensions of SMPA

Variable	L	W	Lp	Wp	L1	W1	L2	W2
Dimension (mm)	60	60	29.4	38	9.5	2	20	5

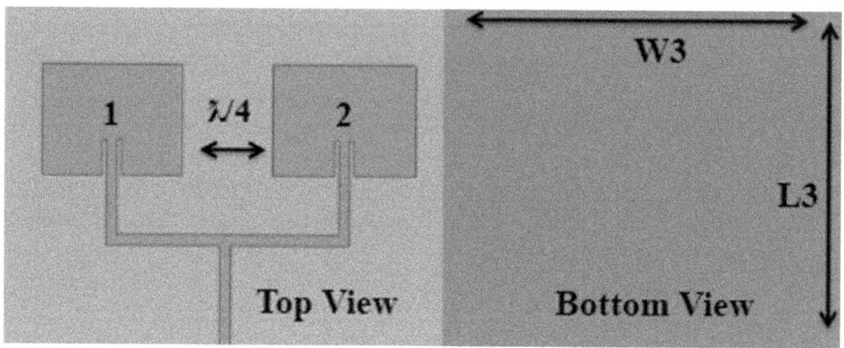

Figure 5.2 Structure of TEMPAA.

The SMPA has been upgraded to a two-element microstrip patch array antenna configuration. This setup includes two identical microstrip patch antennas with a distance of $\lambda/4$ (where λ is the free space wavelength) to prevent poor diversity and coupling effects between the antenna elements. To enhance efficiency, the ground plane size is set to 90 mm × 120 mm. The final two-element microstrip patch array antenna (TEMPAA) structure is depicted in Figure 5.2.

To further enhance the antenna parameters, the two-element microstrip patch array antenna configuration has been extended to a four-element microstrip patch array antenna configuration. This new configuration consists of four identical microstrip patch antennas with a distance of $\lambda/4$ (where λ is the free space wavelength) maintained to prevent poor diversity and coupling effects between the antenna elements. In order to improve efficiency, the size of the ground plane has been set to 120 mm × 280 mm. The final structure of the four-element microstrip patch array antenna (FEMPAA) is shown in Figure 5.3.

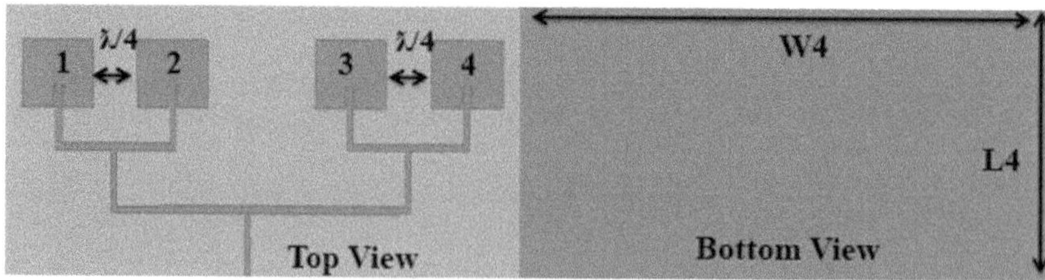

Figure 5.3 Structure of FEMPAA.

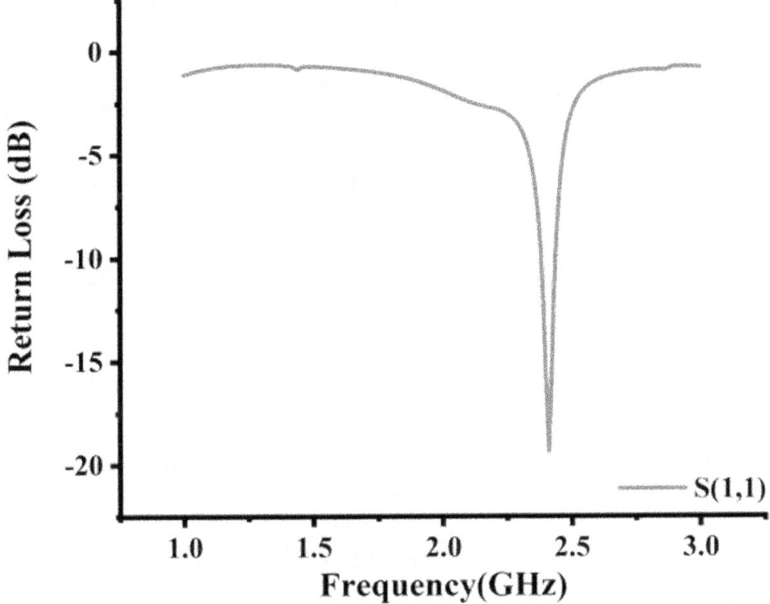

Figure 5.4 Return loss characteristics of SMPA.

5.3 RESULTS AND DISCUSSION

5.3.1 Return loss

Initially, the return loss characteristics of the SMPA are computed and shown in Figure 5.4. The SMPA resonates at 2.4 GHz with a return loss of –20 dB and a bandwidth of 70 MHz. This refers to the range of frequencies over which the antenna resonates at 2.4 GHz.

The return loss characteristics of the TEMPAA are depicted in Figure 5.5. This antenna resonates at 2.4 GHz with a return loss of –20dB and a wide bandwidth of 210 MHz due to its array configuration.

The return loss characteristics of the FEMPAA are depicted in Figure 5.6. This antenna resonates at 2.4 GHz with a remarkably low return loss of –27.16dB, thanks to its four-element array configuration. Additionally, the array configuration allows for a wide bandwidth of 250 MHz.

5.3.2 Total peak gain

The total peak gain of an antenna measures its ability to radiate electro-magnetic waves toward the far-field region. An SMPA has a total peak gain of 2.91 dB, as shown in Figure 5.7. A TEMPAA has a total peak gain of 5.56 dB, as shown in Figure 5.8. Additionally, an FEMPAA has a total peak

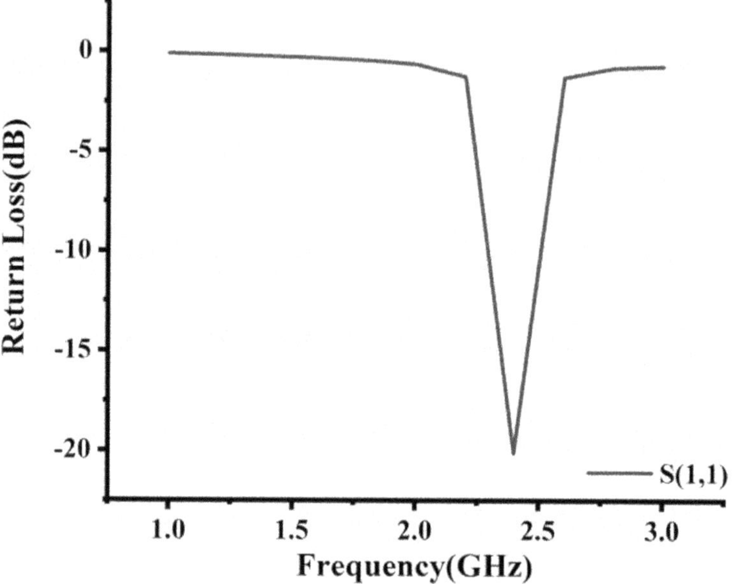

Figure 5.5 Return loss characteristics of TEMPAA.

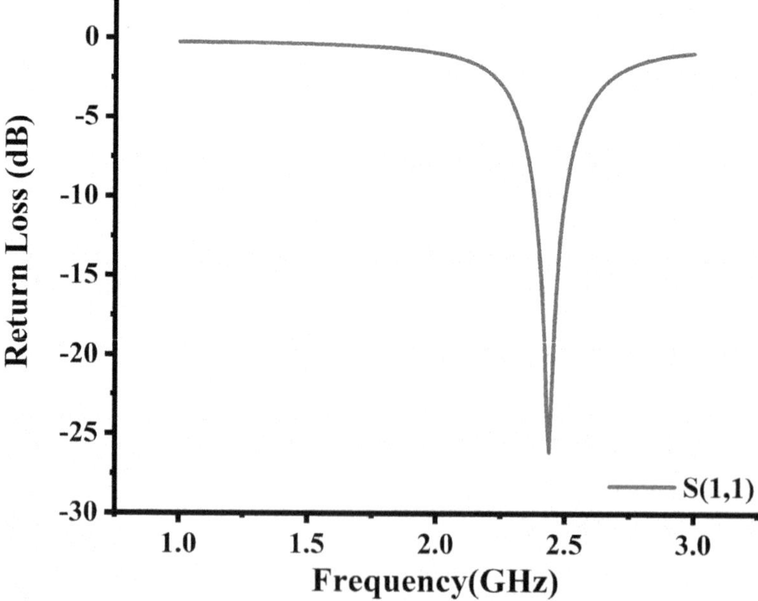

Figure 5.6 Return loss characteristics of FEMPAA.

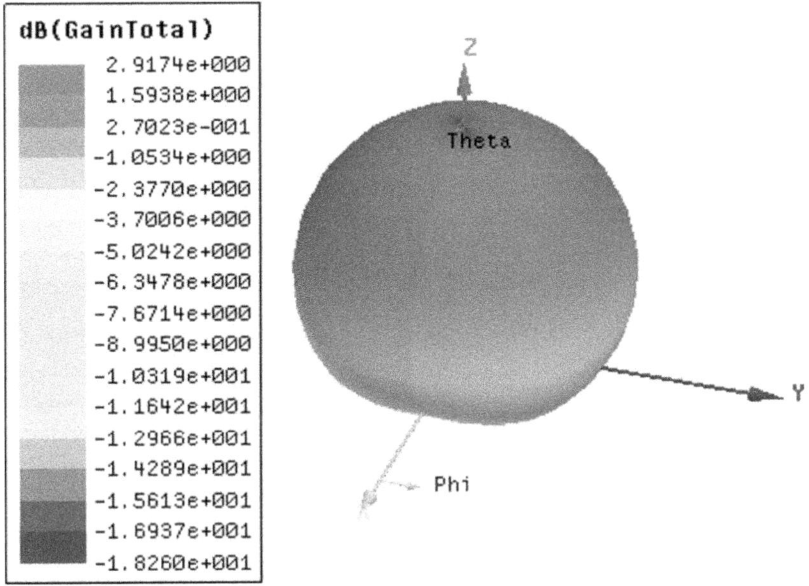

Figure 5.7 Total peak gain of SMPA.

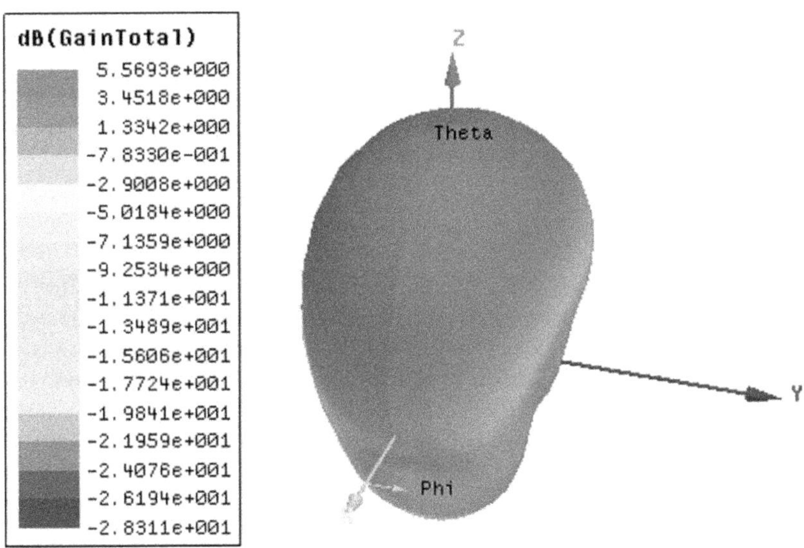

Figure 5.8 Total peak gain of TEMPAA.

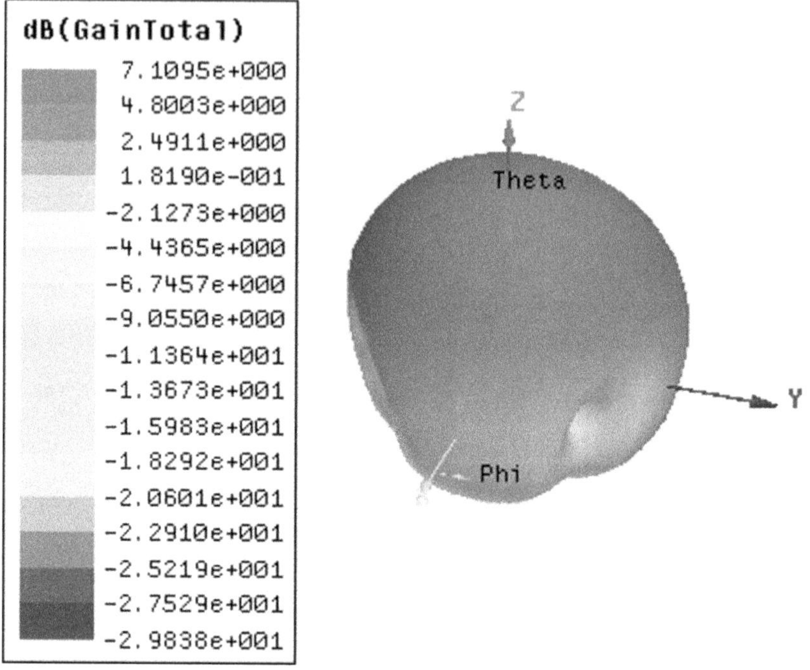

Figure 5.9 Total peak gain of FEMPAA.

gain of 7.10 dB, as shown in Figure 5.9. These results indicate that the four-element array configuration on the same ground plane enhances the total peak gain compared to the SMPA and the TEMPAA.

5.3.3 VSWR

The next antenna parameter to be considered is the voltage standing wave ratio (VSWR), which measures the efficiency of the antenna by considering the amount of radio frequency power transmitted from the antenna toward the load. A smaller VSWR indicates better efficiency of the antenna. Figure 5.10 shows the VSWR of SMPA as 1.20, Figure 5.11 shows the VSWR of TEMPAA as 1.17, and Figure 5.12 shows the VSWR of FEMPAA as 1.15, respectively.

5.3.4 Radiation pattern

The radiation pattern demonstrates the propagation of electromagnetic waves toward the far-field region. In Figure 5.13, the radiation pattern of the single microstrip patch antenna displays broadside radiation characteristics. Additionally, Figure 5.14 illustrates the radiation characteristics of a TEMPAA, while Figure 5.15 depicts the radiation characteristics of

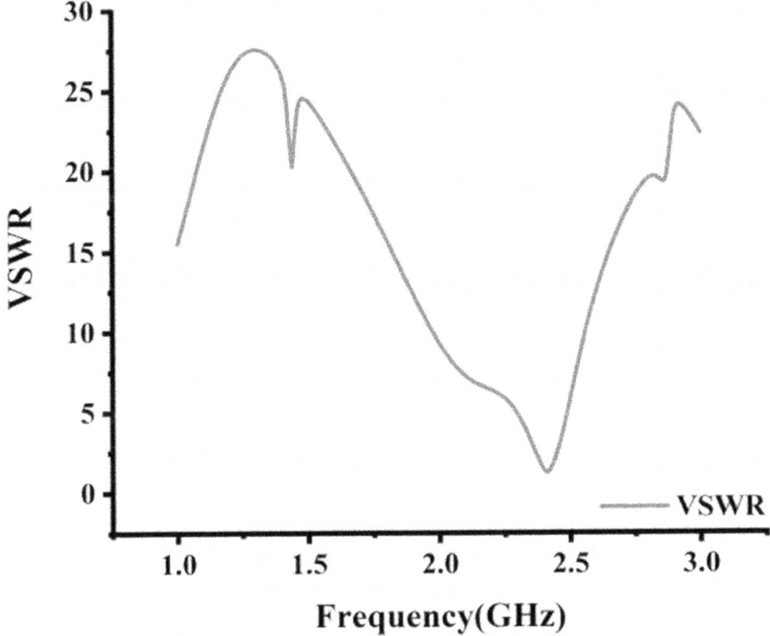

Figure 5.10 VSWR of SMPA.

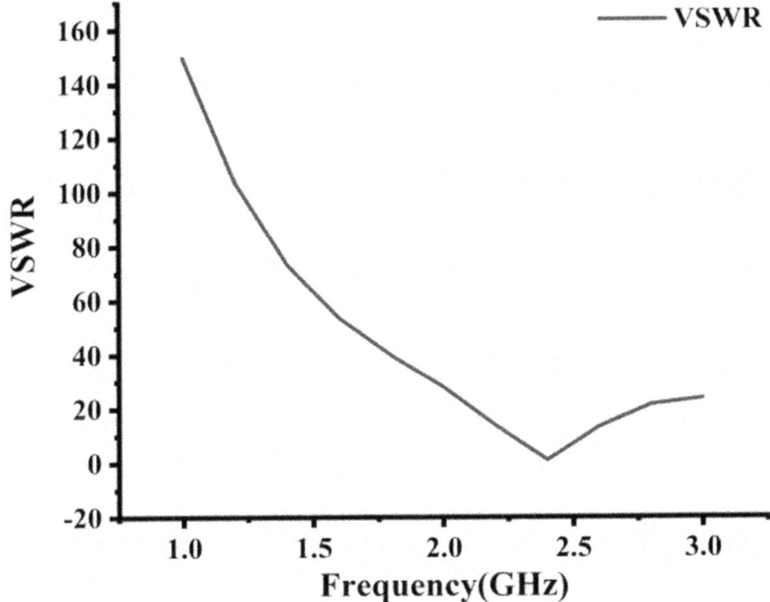

Figure 5.11 VSWR of TEMPAA.

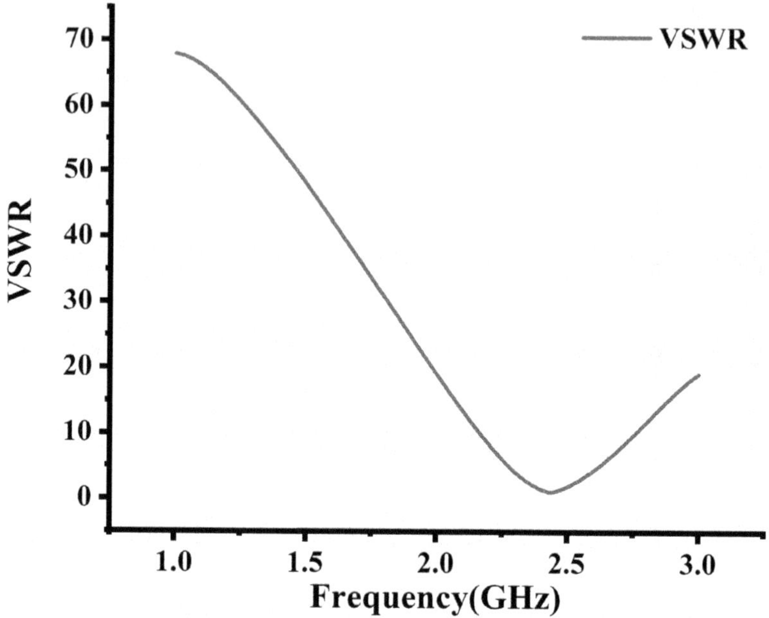

Figure 5.12 VSWR of FEMPAA.

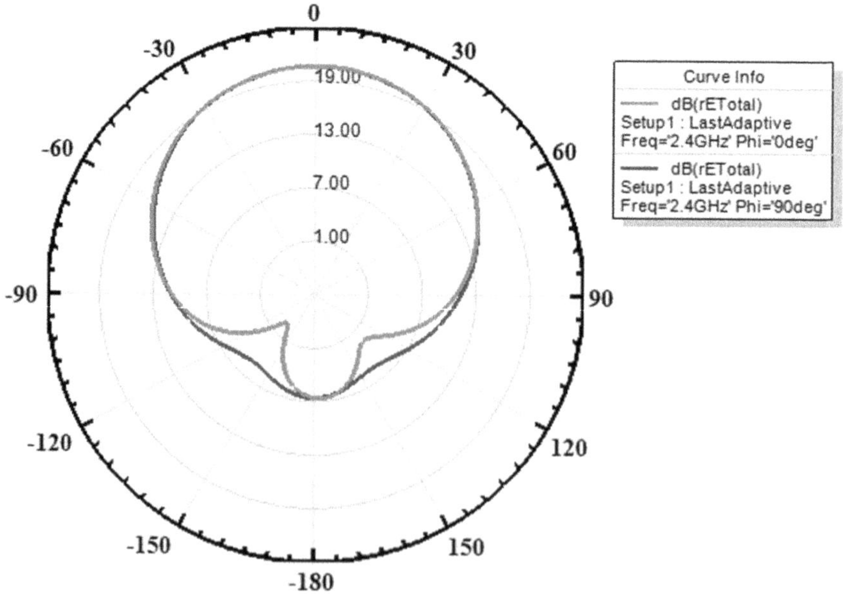

Figure 5.13 Radiation pattern of SMPA.

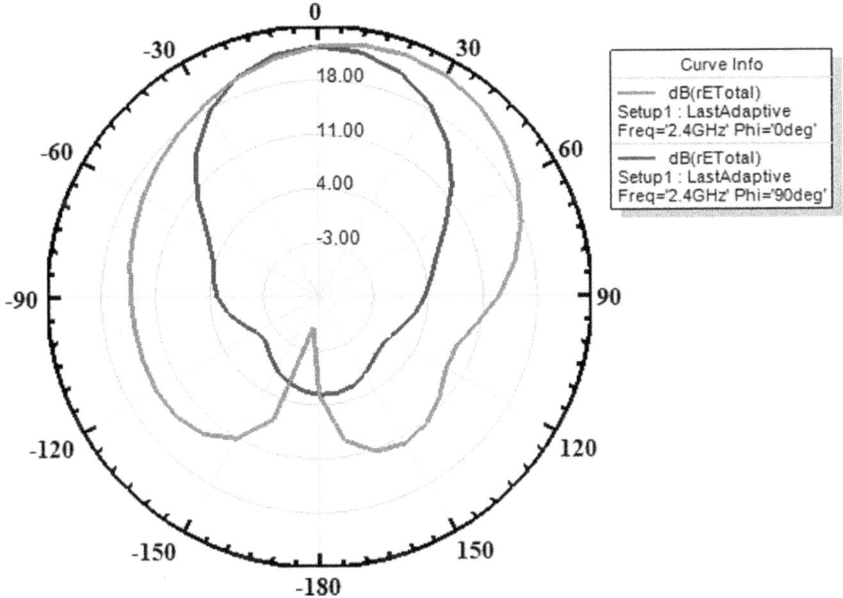

Figure 5.14 Radiation pattern of TEMPAA.

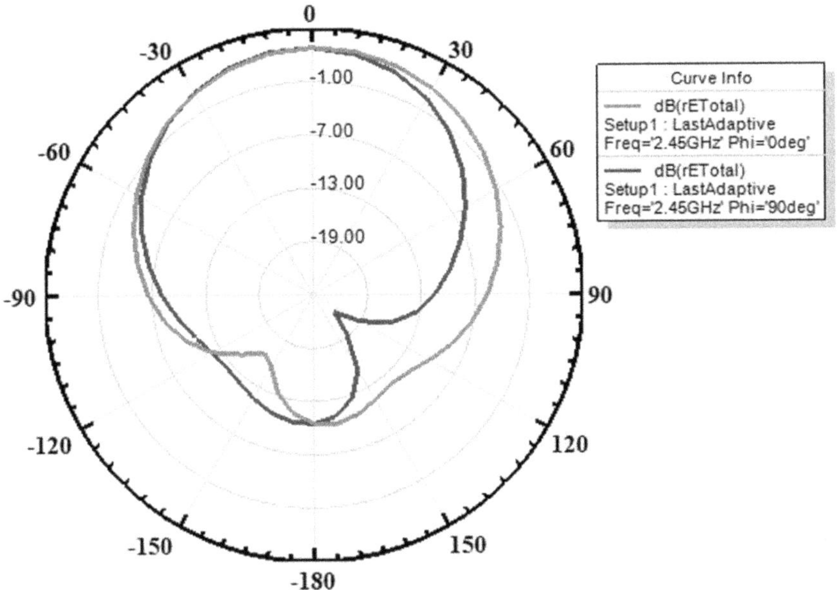

Figure 5.15 Radiation pattern of FEMPAA.

Table 5.3 Comparisons of results of previously published and proposed antenna results

Reference	Frequency (GHz)	Return loss (dB)	BW (MHz)	Gain (dB)	VSWR
[2]	2.4	−27	170	2	1.8
[3]	2.4	−40	150	1	1.14
[4]	2.4	−23	90	6.7	1.9
[5]	2.4	−40	185	4.7	1.95
[6]	2.4	−40	167	4.2	1.8
[7]	2.4	−16	195	5.8	1.9
SMPA	2.4	−20	70	2.94	1.2
TEMPAA	2.4	−20	210	5.56	1.17
FEMPAA	2.4	−27.16	250	7.1	1.15

an FEMPAA, both showing broadside radiation at resonating frequency points.

It has been observed that the FEMPAA resonates at 2.4 GHz with a return loss of −27.16 dB and a wide bandwidth of 250 MHz. This antenna also provides a total peak gain of 7.10 dB. Therefore, the FEMPAA represents a novel design compared to SMPA, TEMPAA, and other microstrip patch array antennas available in the literature. Table 5.3 compares the results of previously published antennas with the proposed antenna results.

5.4 SUMMARY

The antennas were designed using an HFSS electromagnetic simulator. The designed two-element and four-element microstrip patch array antennas effectively improved the performance in terms of gain and bandwidth compared to a single microstrip patch antenna. Both antennas were resonated at 2.4 GHz with a return loss of −20 dB and −27.16 dB, respectively. The TEMPAA antenna has a bandwidth of 210 MHz and a total peak gain of 5.56 dB, and the FEMPAA antenna has a wide bandwidth of 250 MHz and a total peak gain of 7.10 dB, respectively. The FEMPAA has a decent scope and provides satisfying results such as wide bandwidth, gain, and VSWR compared to the SMPA and TEMPAA, making it well-suited for sub-6 GHz applications.

REFERENCE

C. A. Balanis. *Antenna Theory: Analysis and Design.* 2nd ed., John Wiley & Sons, 1982, 0471592684.ch, 14

Chapter 6

Weather prediction model for Telangana state using ANNs

N. Dinesh Kumar, B. Vinay Kumar,
and M. Rohith Reddy

6.1 INTRODUCTION

The task of predicting the atmosphere at a future time and a specific area is known as weather prediction. In the beginning, when the atmosphere was still thought of as a fluid, this was accomplished using physical equations. By numerically resolving those equations, the state of the environment is examined, and the future state is anticipated, however, the process is unable to predict the weather with great accuracy for periods longer than 10 days. Science and technology can improve this situation.

Instant comparisons between past weather forecasts and observations can be processed using machine learning. With the aid of machine learning, weather models may make predictions that are more accurate by better accounting for prediction errors like overestimated rainfall. The ability to predict temperature is crucial in a variety of fields, including climate research, energy, agriculture, and medicine.

6.2 PROBLEM STATEMENT

The problem statement for weather prediction using machine learning methods is to implement a model that can accurately forecast future weather conditions based on historical meteorological data and other essential factors like location, season, and atmospheric pressure. To generate forecasts in real time, the model must be able to process a large amount of data from many sources, including radar, weather stations, and satellite photos. The key problem is to develop a model that can adequately describe the intricate interactions between numerous weather-related components and produce precise predictions. The model must be adaptable enough to consider changing conditions and modify its predictions because forecasting the weather is a challenging and dynamic operation that incorporates multiple variables that can change quickly.

Another difficulty is ensuring that the model is scalable and capable of handling enormous amounts of data in real time.

DOI: 10.1201/9781003510420-6

6.3 OBJECTIVES

The project's main goal is to improve the accuracy of weather predictions. This can be done by developing models that can produce precise forecasts and better depict the intricate relationships between a variety of weather-related phenomena. Additionally, to make predictions for diverse climates and geographical regions, the model must be scalable and capable of managing massive amounts of data in real time.

The project's secondary objective is to develop models that can provide tailored weather predictions based on an individual's preferences and location.

Another objective is to help people, organizations, and governments make decisions that can save lives, property, and money, projects using machine learning models for weather prediction.

See Figure 6.1 for a flow chart of the process.

6.4 IMPLEMENTATION

6.4.1 Import libraries

The following libraries are used:

Pandas: Pandas is a Python library designed to efficiently handle and analyze structured data, providing various tools for data manipulation.

NumPy: NumPy is a fundamental Python library that enables numerical computing, supporting large arrays and matrices, and offering mathematical functions for data manipulation.

Scikit-learn: Scikit-learn is a popular Python library for machine learning tasks, offering a user-friendly interface and supporting various machine learning algorithms.

Keras and TensorFlow: Keras, running on TensorFlow, is a high-level API for building and training deep learning models, simplifying the process of neural network implementation.

Matplotlib: Matplotlib is a Python 2D plotting library, used to create high-quality visualizations and graphs for data analysis.

Urlib library: The Urlib library in Python is used for fetching and handling data from URLs, making it useful for web-based data access.

Geocoding library: Geocoding is a specialized library that converts location names or addresses into geographic coordinates, facilitating mapping and location-based analysis.

Min-Max Scalar: Min-Max Scalar is a data preprocessing technique that scales numeric features to a specific range, often between 0 and 1, ensuring uniformity in feature importance during model training.

K-fold cross-validation: K-fold cross-validation is a model evaluation method that involves partitioning data into K subsets, using K-1

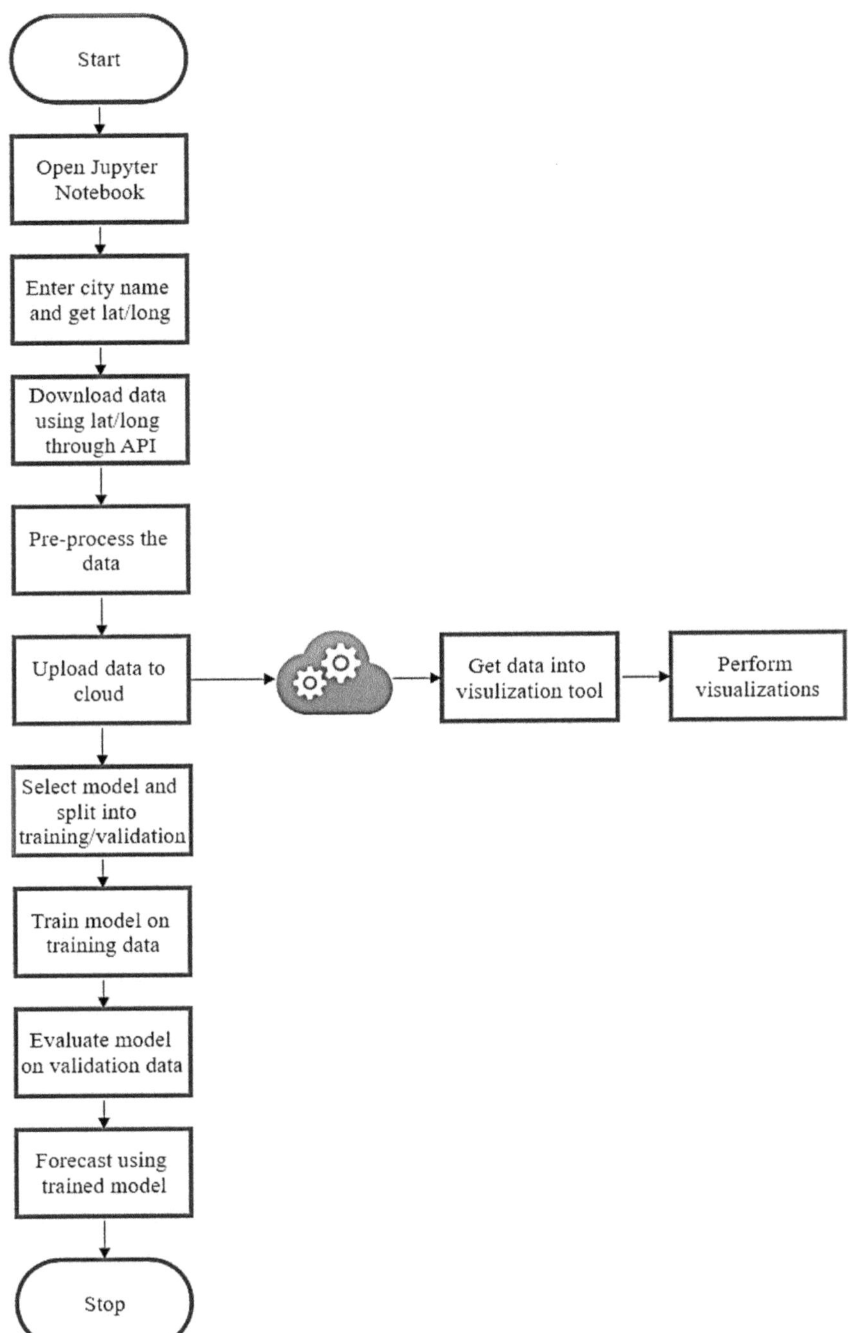

Figure 6.1 **Flow chart.**

subsets for training and one subset for testing in each iteration, leading to robust model performance assessment.

6.4.2 Importing data from NASA POWER data access viewer

Prediction of Worldwide Energy Resource, or NASA POWER, is a project that aims to make a variety of climatic data and weather forecasts available for use in renewable energy applications. Developers can access and retrieve data from the project's various data products, which include information on temperature, wind speed, solar radiation, and more, using an available API. Anyone working on renewable energy applications has access to and can use climatic data through the NASA POWER API to maximize the performance of their systems (https://power.larc.nasa.gov/data-access-viewer/) [1]. (also see Figure 6.2).

1.Importing necessary libraries.

```
import urllib.request
from geopy. geocoders import Nominatim
pd.set_option('display.max_columns',None)
```

2.Selecting start date and end date.

```
start_date = input("Enter start date in yyyy/mm/dd format:")

start_date = start_date.replace('/',")

end_date = input("Enter end date in yyyy/mm/dd format:")

end_date = end_date.replace('/',")
```

3.Selecting Location for weather prediction.

```
address = input("Enter City Name:")

geolocator = Nominatim(user_agent = "Vinay")

location = geolocator.geocode(address)

print(location.address)

print((location.latitude,location.longitude))

latitude = location.latitude

longitude = location.longitude
```

4.NASA POWER API:

```
Nasa Power API URL link
```

5.Requesting Data through api:

```
urllib.request.urlretrieve(url,'dataset.csv')
```

6.Converting API data into CSV Format:

```
df= pd.read_csv('dataset.csv',skiprows=14)

df.head()
```

	YEAR	MO	DY	T2M_MAX	T2M_MIN	RH2M	PRECTOTCOR
0	1983	1	1	31.29	13.09	49.25	0
1	1983	1	2	30.44	13.89	50.56	0
2	1983	1	3	28.63	10.81	41.00	0
3	1983	1	4	28.73	10.23	41.38	0
4	1983	1	5	28.59	12.59	52.75	0

Figure 6.2 Data in csv format.

6.4.3 Future engineering

Future engineering is a crucial step in creating precise and trustworthy machine learning models for predicting the weather. We extract features from our dataset in this stage, convert them into appropriate formats, and create new features that can enhance model performance.

6.4.4 Data splitting

Data splitting is a key stage in creating machine learning models for forecasting the weather. In this stage, we separate our dataset into two or more subsets, often a training set and a testing set. These sets are used, respectively, to train and test our machine learning models.

It is crucial to randomly divide our dataset and make sure that the distribution of the input features and the target variable is the same in both the training and testing sets so that the machine learning models can generalize to new data effectively. Typically, 70%–80% of the total dataset is utilized for training, and the remaining 20%–30% is used for testing.

Code:

```
from sklearn.preprocessing import MinMaxScaler

scaler = MinMaxScaler()

df_scaled = scaler.fit_transform(df)

print('Scaled df:\n' , df_scaled,'\n', df_scaled.shape)
```

Output:

```
Scaled df:
[[0.42557483 0.2071251  0.48927466 0.        0.80586081 0.1270936 ]
 [0.39128681 0.24026512 0.5037594  0.        0.77289377 0.24137931]
 [0.3182735  0.11267606 0.39805396 0.        0.81318681 0.06502463]

 ...

 [0.78055668 0.64250207 0.31789031 0.        0.7032967  0.3408867 ]
 [0.790238   0.68765534 0.29301194 0.        0.67032967 0.30049261]
 [0.81040742 0.77920464 0.29234852 0.        0.65934066 0.44630542]]
 (14711, 6)
```

Code:

```
x_train, y_train = np.array(x_train), np.array(y_train)

x_test, y_test = np.array(x_test), np.array(y_test)

x_train.shape, y_train.shape, x_test.shape, y_test.shape
```

Output:

```
((11767, 6), (11767, 6), (2942, 6), (2942, 6))
```

Code:

```
from sklearn.model_selection import KFold
cv = KFold (n_splits=10, shuffle=False)
results = []
for tr, tt in cv.split(x_train,y_train):
    model.fit(x_train[tr], y_train[tr])
    prediction = model.predict(x_train[tt])
    results. append ((prediction, tt))
```

Output:

```
329/329 [==============================] - 0s 1ms/step - loss: 0.0046 - accuracy: 0.8777
37/37 [==============================] - 0s 927us/step
329/329 [==============================] - 0s 1ms/step - loss: 0.0045 - accuracy: 0.8784
37/37 [==============================] - 0s 1ms/step
329/329 [==============================] - 0s 1ms/step - loss: 0.0045 - accuracy: 0.8770
37/37 [==============================] - 0s 1ms/step
329/329 [==============================] - 0s 1ms/step - loss: 0.0046 - accuracy: 0.8797
37/37 [==============================] - 0s 1ms/step
329/329 [==============================] - 1s 2ms/step - loss: 0.0046 - accuracy: 0.8835
37/37 [==============================] - 0s 2ms/step
329/329 [==============================] - 1s 2ms/step - loss: 0.0045 - accuracy: 0.8773
37/37 [==============================] - 0s 1ms/step
329/329 [==============================] - 1s 2ms/step - loss: 0.0046 - accuracy: 0.8801
37/37 [==============================] - 0s 2ms/step
329/329 [==============================] - 1s 2ms/step - loss: 0.0045 - accuracy: 0.8786
37/37 [==============================] - 0s 2ms/step
329/329 [==============================] - 1s 2ms/step - loss: 0.0045 - accuracy: 0.8801
37/37 [==============================] - 0s 2ms/step
329/329 [==============================] - 1s 2ms/step - loss: 0.0046 - accuracy: 0.8830
37/37 [==============================] - 0s 2ms/step
```

6.5 MODEL BUILDING

The use of neural networks in weather prediction models has grown because of their capacity to understand intricate correlations between inputs and outputs. An input layer, many hidden layers, and an output layer make up the typical multilayer perceptron (MLP) design. We used a four-layer neural network as the foundation of our weather prediction model. Six neurons make up the input layer to indicate meteorological characteristics such as temperature, humidity, precipitation, surface pressure, and wind speed. Eight neurons are included in each of the intermediate hidden layers, which enables the network to learn complex connections. Six neurons make up the output layer, which forecasts the climatic characteristics. The neural network was built and trained using the Keras API on top of TensorFlow, making it effective and convenient for applications like weather prediction.

```
model = Sequential()
model.add(Dense(6, input_dim= 6 ,activation= 'relu'))
model.add (Dense (8, activation= 'relu'))
model.add (Dense (8, activation= 'relu'))
model.add (Dense (6))
model. summary ()
model. compile (optimizer = 'adam', loss = 'mse', metrics = ['accuracy'])
```

Output:

```
Model: "sequential"

 Layer (type)                Output Shape              Param #
=================================================================
 dense (Dense)               (None, 6)                 42

 dense_1 (Dense)             (None, 10)                70

 dense_2 (Dense)             (None, 10)                110

 dense_3 (Dense)             (None, 6)                 66

=================================================================
Total params: 288
Trainable params: 288
Non-trainable params: 0
```

6.5.1 Model training

The provided code snippet illustrates how a machine learning model is trained using Keras. The following is the result of the training and validation process:

```
history=model.fit(x_train,y_train,validation_data=(x_tes    t,    y_test), epochs=200, batch_size=15, shuffle=True)
```

With a batch size of 15, the model goes through 200 epochs of training, and the data is shuffled to improve generalization. The "history" variable, which offers insightful details on the model's performance during the training process, can be used to get the validation findings.

6.6 CROSS-VALIDATION

To ensure reliable model performance on new data, K-fold cross-validation, a key machine learning technique, divides data into subsets for several rounds of training and evaluation. This method offers substantial benefits, such as improved accuracy and reliability, leading to more efficient machine learning model construction, despite its high computing cost and implementation difficulties.

Code:

```python
from sklearn.model_selection import KFold

cv = KFold (n_splits=10, shuffle=False)

results = []

for tr, tt in cv.split(x_train,y_train):

    model.fit(x_train[tr], y_train[tr])

    prediction = model.predict(x_train[tt])

    results. append ((prediction, tt))
```

Output:

```
329/329 [==============================] - 0s 1ms/step - loss: 0.0046 - accuracy: 0.8777
37/37 [==============================] - 0s 927us/step
329/329 [==============================] - 0s 1ms/step - loss: 0.0045 - accuracy: 0.8784
37/37 [==============================] - 0s 1ms/step
329/329 [==============================] - 0s 1ms/step - loss: 0.0045 - accuracy: 0.8770
37/37 [==============================] - 0s 1ms/step
329/329 [==============================] - 0s 1ms/step - loss: 0.0046 - accuracy: 0.8797
37/37 [==============================] - 0s 1ms/step
329/329 [==============================] - 1s 2ms/step - loss: 0.0046 - accuracy: 0.8835
37/37 [==============================] - 0s 2ms/step
329/329 [==============================] - 1s 2ms/step - loss: 0.0045 - accuracy: 0.8773
37/37 [==============================] - 0s 1ms/step
329/329 [==============================] - 1s 2ms/step - loss: 0.0046 - accuracy: 0.8801
37/37 [==============================] - 0s 2ms/step
329/329 [==============================] - 1s 2ms/step - loss: 0.0045 - accuracy: 0.8786
37/37 [==============================] - 0s 2ms/step
329/329 [==============================] - 1s 2ms/step - loss: 0.0045 - accuracy: 0.8801
37/37 [==============================] - 0s 2ms/step
329/329 [==============================] - 1s 2ms/step - loss: 0.0046 - accuracy: 0.8830
37/37 [==============================] - 0s 2ms/step
```

6.7 RESULTS

The performance of the model in forecasting maximum temperature, minimum temperature, relative humidity, wind speed, precipitation, and surface pressure is shown in the training versus validation visual graphs, which offer a thorough analysis of how well the model captures these weather parameters during the training and validation stages (Figures 6.3 and 6.4).

Telangana's geography and districts are shown in Figure 6.5, which uses Power BI to showcase the data. A forecast value tooltip that displays the expected range of minimum and maximum temperatures, humidity levels, precipitation amounts, surface pressure, and wind speed for each district is shown while the cursor is over it. Users can explore and get insights into Telangana's weather patterns with this interactive visualization, which provides a thorough overview of the region's expected weather conditions.

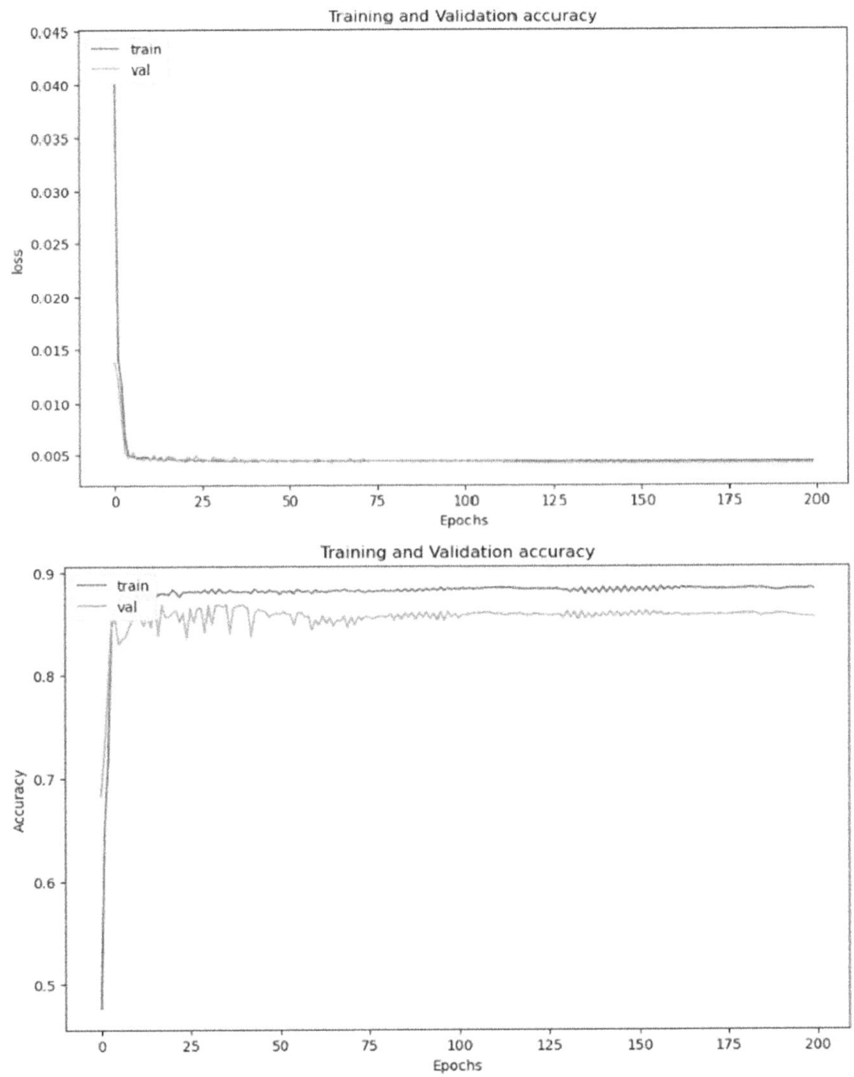

Figure 6.3 Visuals of testing and validation of machine learning algorithm.

6.8 CONCLUSIONS

The accuracy of weather forecasts has been significantly improved thanks to machine learning. The project's model worked well and caught complex interactions between many meteorological characteristics. This study offers a solid validation of the effectiveness of machine learning in weather fore-casting, even though there is still room for improvement. Machine learning

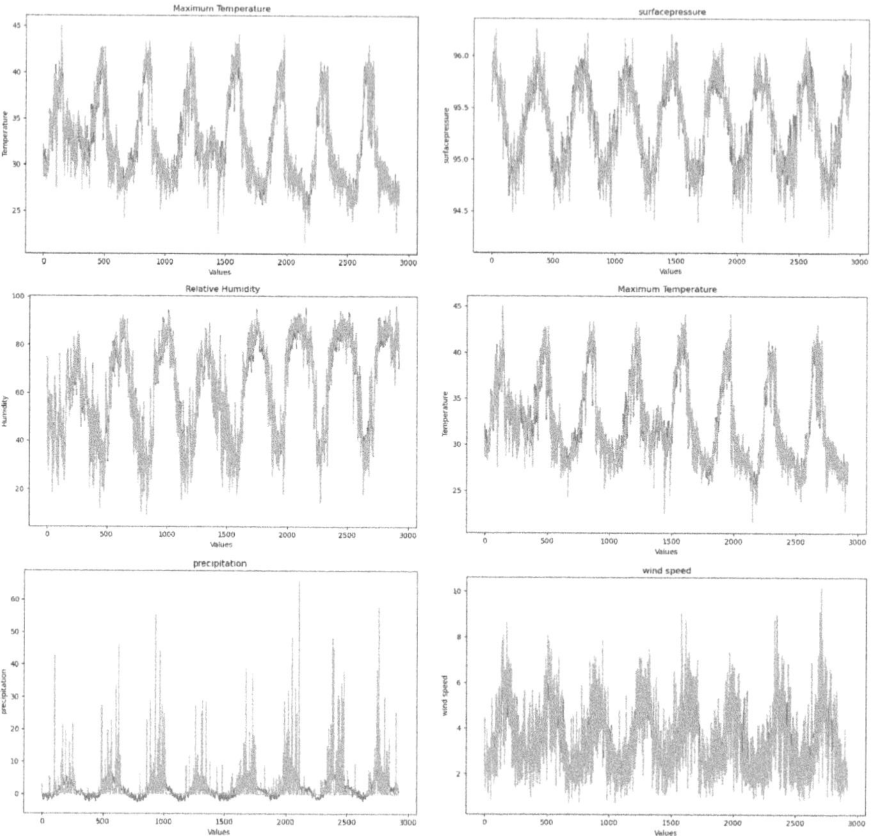

Figure 6.4 Training and validation visuals of maximum temperature, minimum temperature, relative humidity, wind speed, precipitation, and surface pressure.

models are anticipated to evolve into more reliable, accurate, and crucial tools for efforts to predict the weather and manage disasters as the field of machine learning advances and more research and development is carried out.

REFERENCE

https://power.larc.nasa.gov/data-access-viewer/.

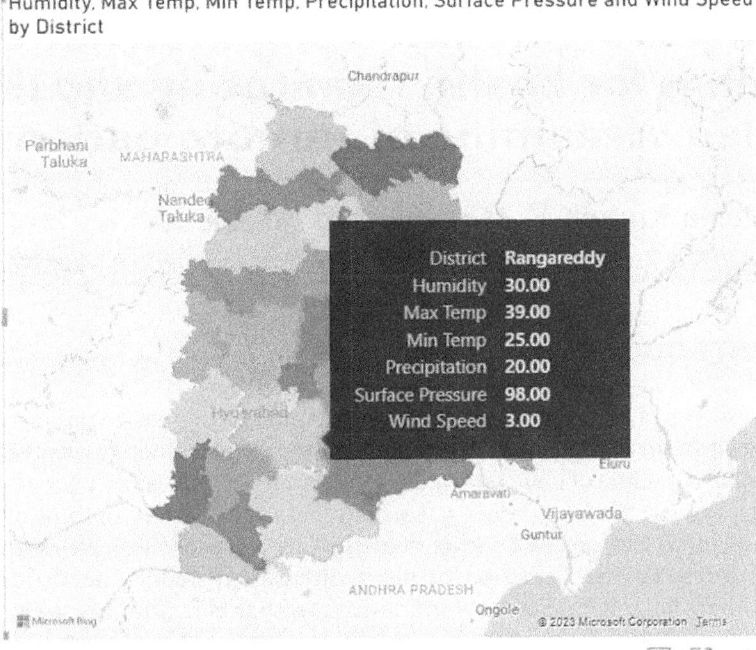

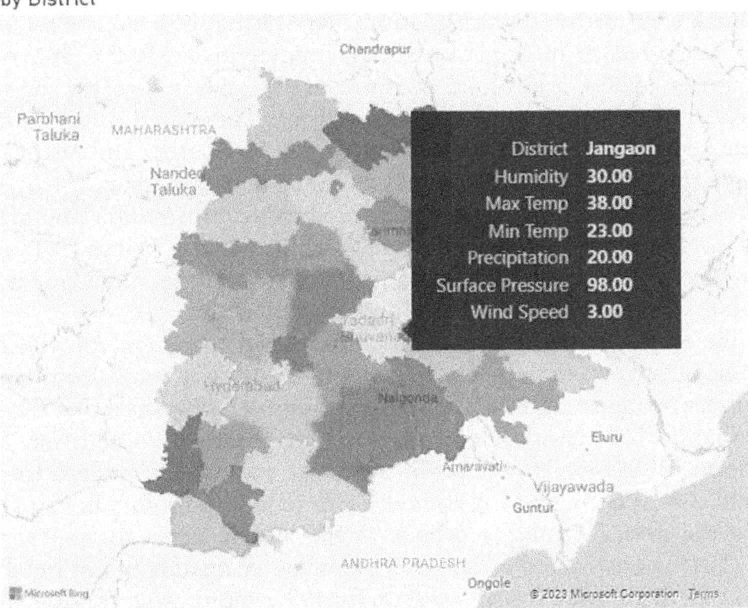

Figure 6.5 Predicted weather parameters on Telangana map using Power BI.

Chapter 7

RTOS and task–role-based access control algorithm using Internet of Things for health monitoring and live video streaming of remote patients

M. Lenin Kumar, V. Malathy, and M. Anand

7.1 INTRODUCTION

The research is based on designing a smart and secure remote patient monitoring system (RPMS) for knowing healthcare parameters. Recently, it has gained popularity and moved to the forefront of concern for research groups and the medical industry. Unawareness about the symptoms of heart disease and the absence of earlier detection are a few of the reasons for high death rates. Earlier detection of heart disease can reduce death rates by more than half. RPMS allows monitoring of patients who are away from clinical settings. The growth of semiconductor technology allows the collecting and sending of patients' medical records to doctors in remote places. It enables continuous monitoring and helps doctors to make easy analyses and quick treatment recommendations. RPMS requires sensors for sensing the patient's health information such as temperature, heart rate, electrocardiogram (ECG), and blood pressure. RPMS needs a server for storing the collected information for future analysis. The stored information can be accessed from the medical server database by doctors and medical staff through the internet for further assessment and treatment. Also, RPMS requires software automation to analyze health information and alert the doctor and others to any crucial state. The Internet of Things (IoT) plays a major role in designing RPMS and it provides anytime, anywhere connectivity between the patient, RPMS, and medical staff.

Uddin et al. (2018) explained and simulated a system for continuous monitoring of remote patients. It is designed using a central unit called a patient-centric agent and the simulation results demonstrated that the design can enhance the security and privacy in remote patient monitoring. Verma and Sood (2018) explained about a remote monitoring system for home patients. The system used fog computing to reduce the communication delay among the device, cloud, and back to the mobile application. Jegadeesan et al. (2019) constructed a remote monitoring device for infant incubators. Chen et al. (2019) developed a system for heart monitoring for early prediction of heart problems. El Kafhali and Salah (2018) modeled and analyzed the performance of an IoT-enabled healthcare system that uses fog or edge

DOI: 10.1201/9781003510420-7

nodes to reduce the communication delay between the remote cloud server and the device. Satija et al. (2017) presented a real-time signal quality-aware ECG telemetry system for IoT-based healthcare monitoring. Solosko et al. (2017) designed a continuous outpatient ECG monitoring system. Clarke et al. (2018) implemented a remote patient monitoring platform based on the IEEE standards corresponding to personal health devices. A Zigbee wireless communication protocol is used in this design and found that it provides reliable wireless communication and good performance. Choudhary and Kumar (2016) presented the analysis of heart rate and patient monitoring by using a wireless sensor network. Kioumars and Tang (2011) presented a wireless network for monitoring heart rate and body temperature. Majumder et al. (2018) constructed a remote monitoring system to evaluate and diagnose the cognitive health of elderly people at home by using a wearable microelectromechanical systems (MEMS) inertial sensor. Park et al. (2016) presented an ECG monitoring system using an Android smartphone. Mahbub et al. (2019) demonstrated a completely integrated charge amplifier architecture with minimal power consumption for monitoring respiration rate.

Al Disi et al. (2018) investigated the performance of compressive sensing and reconstruction of ECG signals on an IoT platform. Pathinarupothi et al. (2018) designed a system in which the wearable sensors passed the health information to two software engines running in IoT smart edge. Sood and Mahajan (2018) suggested a system to continuously monitor blood pressure in real time for the analysis of hypertension attacks. The system uses IoT and fog for remote monitoring and an artificial neural network for risk prediction. Kiani et al. (2015) from Masimo Corp. designed and developed a patient safety system for use with an automatically adjusting bed. Wang et al. (2019) designed a non-interactive privacy-preserving priority classification algorithm for the privacy of users in a wireless body area network based on on-the-go healthcare services. Chen et al. (2018) used hardware-based trustworthy computing technology in the design of a secure and efficient remote patient monitoring architecture. Dietrich et al. (2017) explained two new semaphore-based methodologies for task synchronization in RTOS. Synchronization turns out to be more critical in RTOS. Indersain and Singh (2013) implemented the kernel of a microcontroller powered by an ARM processor for the implementation of scheduling and multitasking. Jae Hwan Koh and Byoung Wook Choi (2013) analyzed the execution of RTOS-based constant frameworks utilizing a real-time application interface and Xenomai.

7.2 METHODOLOGY

For patient healthcare monitoring, this study introduces a clever and energy-efficient medical sensor design. The health parameters measured in this design are heart rate, body temperature, and body movements. Temperature is measured to confirm whether the patient has a fever. An ECG is used to measure the heart rate and its rhythm to know whether the heart is functioning normally. Body movements are monitored to know about the physical activity of the patient. The following sensors are employed:

a) **ECG sensor:** The electrical activity of the heart can be charted and it is known as electrocardiogram (ECG). The sensor consists of an integrated signal conditioning block AD8232. It is economic and operates at low power. Generally, ECG signals picked up from electrodes are noisy. From this noisy signal, the AD8232 extracts, amplifies, and filters the correct ECG signals.

b) **Body temperature sensor:** The body temperature due to cell metabolism is not evenly distributed across the body. The average body temperature measured using a thermometer at the ear of an adult is somewhere between 36.4°C and 37.3°C. If the temperature is higher than this, it is considered a fever. The body temperature is gathered using a temperature sensor (LM35).

c) **Body position sensor:** Sensing body movement can be used to identify the activity of the patient. A three-axis MEMS technology-based accelerometer is used in this work to identify the patient's body motion. In an MEMS accelerometer, the mechanical sensing structures are created in micro sizes on semiconductor material. It can be coupled to microelectronic circuits to measure the acceleration. It works based on the variable capacitive principle. The low-power three-axis digital linear accelerometer LIS302DL is used in this work.

d) **Sensor signal processing:** The body position sensor is a digital output sensor, whereas the ECG and temperature sensor are analog outputs. Analog-to-digital converter (ADC) peripherals are used to process analog outputs, while inter-integrated circuit (IIC or I2C) peripherals are used to measure digital output. In this architecture, the STM32F429 microcontroller is preferred for processing sensor output. This microcontroller offers the popular and extensively used successive approximation type ADC.

7.3 SYSTEM ARCHITECTURE

A local health monitoring system (LHMS) is constructed to measure the patient's health information using the sensors chosen for this design. The sensors are connected to a microcontroller and the values from the sensors are processed using the ADC peripheral of that microcontroller. The processed information is displayed on a personal computer for testing and analysis. Various serial port reading software are available in the market. For this design, the RealTerm software is used to monitor the text form of

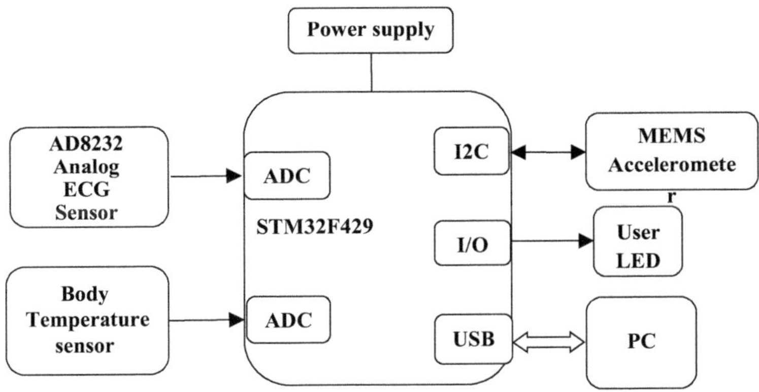

Figure 7.1 Block diagram of LHMS.

information. For plotting the ECG, data serial plotting software is used. The system does not allow remote patient monitoring, because the sensors and computer are connected to the microcontroller through wires. Figure 7.1 shows the block diagram representation of the constructed system and Figure 7.2 shows a graphical view of the same.

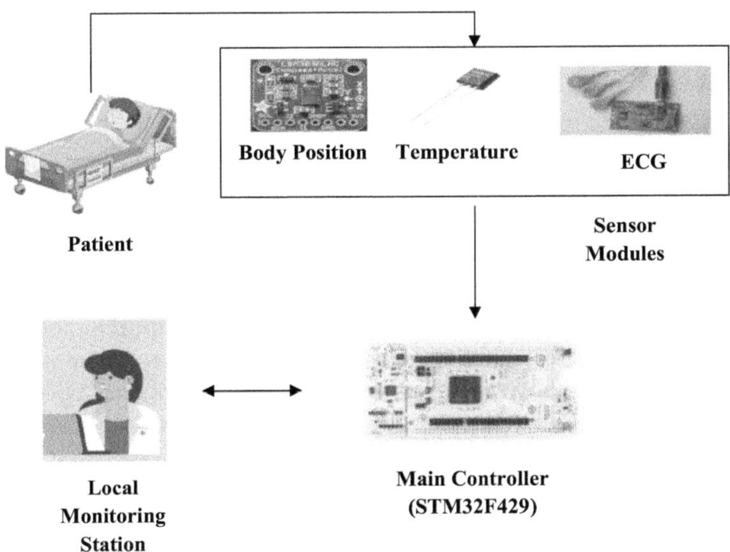

Figure 7.2 Graphic view of LHMS.

7.3.1 STM32F429ZI microcontroller

The STM32 is one of the microcontroller series from STMicroelectronics based on ARM Cortex M4, 32-bit RISC processor architecture. The core has a floating-point unit and runs at a maximum frequency of 180 MHz. The NUCLEO board designed using the STM32F429ZI microcontroller does not require a separate programmer or debugger, because it is incorporated with the ST-LINK debugger/programmer.

7.3.2 Software design

The user needs an integrated development environment (IDE) and a compiler to translate the higher-level language program into machine-level language when writing the code. And for downloading the machine language code into the microcontroller, it needs a programmer, or a debugger to debug the working of the code. Atollic TrueSTUDIO IDE is used and RealTerm software is used to monitor the sensor values and also for basic testing and debugging.

7.3.3 Atollic TrueSTUDIO IDE

Atollic TrueSTUDIO IDE is an all-in-one development tool for the STM32 series microcontroller and it is used by small-, medium-, and large-sized companies, students, researchers, and hobbyists all over the world. It supports code debugging for single and multi-core. The ST-Link debugger is employed for code debugging.

7.3.4 RealTerm

The RealTerm software is used in personal computers for capturing serial data coming from external devices. It can be used for testing, debugging, and data logging purposes. The software will receive data in various formats such as ASCII, Hex, Binary, or Integer.

The ECG, heart rate, body temperature, and body movements are sensed and communicated to the RealTerm software running on a personal computer for further analysis. Since the device and computer communicate through wires, the restriction is that the patient has to be monitored by doctors nearby, probably within a building. Due to reduced code size, the size of ROM and RAM is reduced. Hence, the processing power is also reduced resulting in less power consumption.

7.3.5 E-health monitoring

In this work, an internet-based e-healthcare monitoring system with access control using a body sensor network is developed.

Devices incorporated with sensors and other actuators can be connected to the internet and can be monitored from anywhere in the world. Since these devices have limited computing resources and memory, a small TCP/IP is developed for implementation at the internet layer, which is called Lightweight Internet Protocol (LwIP). Internet connectivity using LwIP, making LHMS into an RPMS, and implementation of an access control algorithm for secure accessing of patient health information is necessary.

To overcome the limitation, the system is connected to the internet to make it available online anytime. The internet connectivity to the system is provided using an internet modem. The system acts as an embedded web server, and it communicates to the internet modem using LwIP. Clients can access the embedded web server from anywhere in the world. The patient's medical information should be access restricted and protected.

LwIP is comprised of a Packet Buffer layer layer, buffers, and a memory management section.

To offer end-to-end communications, data must be structured, transmitted, routed, and received according to the TCP/IP model architecture. Four abstraction levels are included in this paradigm, and they are used to group comparable protocols according to the degree of networking that is involved. The layers, from lowest to highest, are described next.

The link layer manages communication in a single network segment (link) of a local area network between a host and a router. The internet layer links distinct networks and is responsible for logical data transmission between them over the internet. The transport layer is responsible for host-to-host communications. The application layer consists of protocols for data exchange between processes across an IP network. Figure 7.3 presents the system architecture to support the network.

7.3.6 Server and client communication

The ARM Cortex-M4-based STM32F429ZI microcontroller acts as the web server. The code for the webpage is created using HTML and loaded into the microcontroller. The LwIP stack runs inside the controller and connects the microcontroller to the internet modem. The patient's medical parameters are read from the sensor and recorded inside the controller memory. The web server is assigned a separate IP address to access the system. The client is only allowed access to the monitoring server, per the access control mechanism, if they enter the proper username and password when logging in.

7.3.7 T-RBAC algorithm

The user may visit a system to access information from it. The access control algorithm must decide whether the user is authorized and whether the

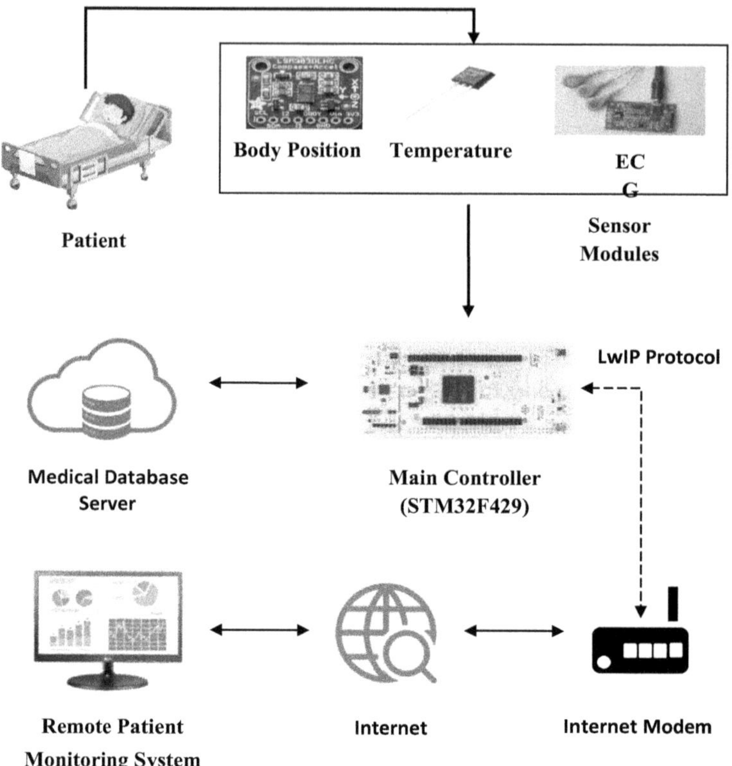

Figure 7.3 System architecture.

access request is valid. The access control algorithm makes decisions based on various factors like the user's role, designation, access rights, permissions, and rules. Based on this, the algorithm is classified into various types. T-RBAC is one of the various types. T-RBAC is formed by combining role-based access control (RBAC) and task-based access control (TBAC).

In the RBAC algorithm, each user is allocated a role and each role is allocated with a set of permissions. If the user's role consists of permissions to access information, then the user can access it. Each user must be assigned to one or multiple roles. In RBAC, there is no separation between roles and tasks. But roles and tasks are separately considered in T-RBAC. Tasks are classified into two things: tasks that are part of workflow and tasks that are not a part of workflow. The structure of T-RBAC is shown in Figure 7.4.

The role group defines the tasks that are allowed to be executed by the user and the functional role defines the access permissions that are needed for accessing the secured health information. In this cloud-based healthcare monitoring system, tasks support active access control, while roles support passive access control. The number of clients is contained in an occupant,

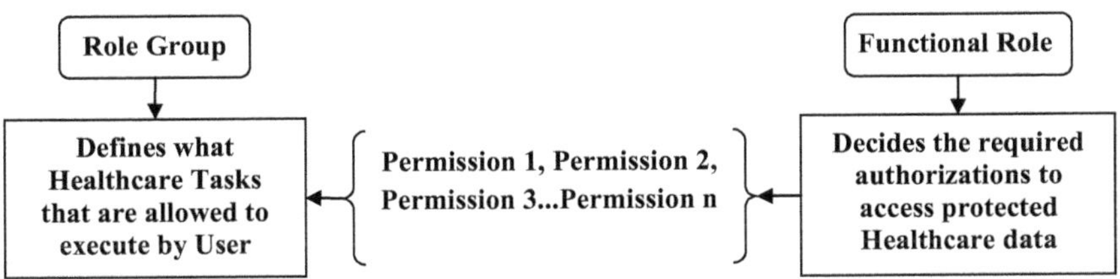

Figure 7.4 Role group and functional role process in T-RBAC.

which is situated in the cloud computing system. Each client is given a role, roles are assigned to tasks with or without workflow, and tasks are distributed based on authorization.

Clients who have established roles can perform several jobs while performing tasks with workflow or tasks without processes that are assigned to them. The roles are allowed in accordance with their tasks, and the rights allocated to them might change dynamically depending on the work at hand. What duties can be performed in what capacities and under what circumstances is decided by authorization for the clients.

7.3.8 Least privilege method

Permission will exist only during the period of tasks. If it gets completed, the permission will be canceled. For example, a doctor can be assigned permission to see a patient record only during a task like a general checkup.

7.3.9 Separation of duty

For tasks that are part of a workflow instance in the T-RBAC system, it may be applicable. It does not apply to task classes that focus on supervision and approval.

7.3.10 Assignment

Permission will get canceled if an assignment is completed. For example, consider assigning a special doctor to get opinions about further treatment for a patient. The administrative user will assign this permission. After the assignment is completed, automatically the permission will be denied.

7.3.11 Location-based restriction

It is possible to consider the time and location of customers when granting access to a job. Example: {Name, doctor, information of patient being monitored, read, day and location of hospital}. This example set tells that

Name as a doctor can monitor patient information only from inside the hospital during day hours.

In RPMS, the patient's health information is measured and transferred to a remote computer via the internet upon a request received from the client. LwIP stack is used in ARM cortex M4 architecture microcontroller for enabling internet connection to it. For providing a security feature to the patient monitoring system, the T-RBAC algorithm is included in the system software. According to the objective of providing secure access, the system verifies the user's role and task during their log in and gives access based on the rules set in the T-RBAC algorithm. This ensures the security and privacy of patients by restricting access to unknown users and modification of medical data. The system is programmed for doing multiple tasks such as patient health monitoring, data recording, handling internet communication, and giving access to patient's health information according to the T-RBAC algorithm. To effectively run these tasks and to make the system into real real-time system, RTOS is planned to be used in software.

7.4 SJF PRIORITY SCHEDULED RTOS

An IoT-based e-healthcare system is developed with live video monitoring using shortest job first (SJF) priority scheduled RTOS. If RPMS gives an additional feature of live video monitoring, it will be very helpful to extract a lot of information related to patient activity. By visually seeing a patient, the doctor or other medical personnel can decide whether the patient is normal or abnormal. Here RTOS in software architecture, its various scheduling methods and benefits are explained, and an IoT-based e-health care system is developed including live video streaming of the patient's condition.

7.4.1 Architecture of RTOS

Kernel space and user space are the two components of an RTOS system that is operational. The smallest and most important part of an operating system (RTOS) is its kernel. Devices and memory management are included in its control, and it also offers an interface via which applications can be programmed to make use of the resources. Additionally, based on OS architecture, services like program protection management and multitasking are included.

The RTOS kernel offers the abstraction layer between the programming application and hardware. The FreeRTOS is attached to the microcontroller's system bus in order to monitor signals like Tick, Start, and Interrupt as well as the addresses of RAM that were accessed during the execution of the application code. There are five functional blocks in the RTOS.

Depending on the address at which the microcontroller was accessible during the execution of the program, the Controller of Task (CT) detects the task that is currently being performed. For each clock cycle, CT compares the bus address to the related task addresses. The Task signal receives the corresponding task's number if the address reached is task-related. The Identifier for Function (IF) has conducted an analysis of the functions implemented through the process of task scheduling in order to verify that the scheduling process is carried out in the order of executions. Last but not least, IF will identify the event based on the order that caused the scheduling.

The Monitoring of List and Error Generator (MLEG) block receives the signal for the event scheduler and the task that was in progress. Using this information, the MLEG divides each task into two categories: tasks that are ready to start and those that are blocked to start. Each job is then sorted according to its priority and state. The MLEG block will be used to spot scheduling misbehavior. The lists produced by the MLEG module are stored in the final two blocks, Content-Addressable Memory 1 (CAM1) and Content-Addressable Memory 2 (CAM2). The tasks that are in the ready condition are saved in CAM1, while the tasks that are in the blocked state are saved in CAM2.

7.4.2 Task management

To ensure continuity in the real-time application program, the application is broken up into basic, sequential, and schedulable components of a program called "Task." In a real-time perspective, the task is the fundamental unit of implementation, and it is controlled by three key time-related properties: the deadline, the release date, and the execution time. The timeline is described as the period during which the task must be finished.

Time of release is the point at which a function can begin to be used, while time of execution is the point at which a task is completed. Each task could be in any one of the four possible states: ready, running, dormant, and blocked. During an application program's execution, certain tasks continuously transition from one state to another. However, only one function can be in the running mode at any given time when an execution is taking place.

7.4.3 Preemptive scheduling

The algorithm with the support priority contains a list of all tasks in the ready state for use in preemption scheduling. The tasks are arranged according to priority. As a result, whenever a scheduling event occurs, the job that has the highest priority and is in the ready state is executed. The scheduler can be used to track the status of each task, and it selects the task from those that are ready to be executed. The CPU is then allotted to the task that the scheduler has selected. It reduces waiting times while assisting the

CPU in making the most of its utilization across several tasks in a multi-tasking program. The processor control that is assigned to the task with the highest priority over all times is required for preemptive scheduling based on priority. The current task is promptly suspended if a job with a higher priority is prepared to run, and the task with the higher priority is given processor control.

The preemptive-based scheduling algorithm used in the suggested method is SJF-based priority scheduling. Every process with the same priority is run according to the burst time. This indicates that the procedures with the lowest burst time will be carried out first; this type of algorithm lowers the average waiting time as well as the turnaround time.

The following formulas are used to determine turnaround and waiting times:

Turnaround time = Completion time of process – Time of arrival

Time of waiting = Turnaround time – Burst time

In this algorithm, the following steps will be performed:

1. The task is placed in the ready queue.
2. The order in which the processes access the CPU will depend on their priority; the processes with higher priorities will access the CPU before those with lower priorities.
3. SJF is used to break the tie if two processes have comparable kinds of priorities.

These three operations will be carried out again until the ready queue is empty.

Calculation of waiting time and turnaround time will be processed separately.

7.4.4 Overall view of e-health monitoring system

The hardware part of this system consists of sensors for physiological parameter monitoring, a camera for live video feed, and a microcontroller for processing the sensor data and camera output and relaying the information to remote clients upon request.

The sensor unit consists of three sensors: ECG, body position, and temperature sensor. The ECG and temperature sensor are analog output sensors, and the body position sensor is a digital output sensor. AD8232 (ECG sensor) is used to observe the patient's ECG. By using this ECG wave, the heart rate of the patient is calculated. The temperature sensor LM35 is used to measure the patient body temperature. The analog output of these two sensors is converted to digital form using ADC peripherals available in the microcontroller and processed further. This microcontroller uses a successive approximation type ADC, which is popular and widely used.

No matter how large the analog input voltage is, conversion takes considerably less time. Additionally, the conversion time is kept constant. Three electrodes are used to collect patient ECG signals at different locations, such as the right arm, left arm, and right leg. These signals are processed to a single analog voltage before being fed into a microcontroller's ADC channel to be converted to digital data. Using the IIC peripheral included in the STM32F429 microcontroller, the digital output of the MEMS accelerometer used to determine body position is measured. It monitors acceleration characteristics, and when readings are compared to thresholds, patient positions like supine, right, left, or sitting are determined.

The device includes an upgraded camera module in addition to biomedical sensors to monitor the patient in real time. STMicroelectronics' STM32F429 was chosen for this design because it has a large amount of RAM and is one of the most potent microcontrollers available. Every time a client requests a video feed, the microcontroller obtains the JPEG images from the camera via a built-in peripheral (DCMI), starts to broadcast them in MJPEG format over the internet, and plays them at a fast enough pace. The image has a resolution of 470 × 272. A significant amount of RAM (approximately 256 KB) is provided by the STM32 controller, which is a requirement for this kind of application.

RTOS is used to create the system software, and priority scheduling is based on SJF. SJF is one of the preemptive-based scheduling methods that overcomes the shortcomings of first come, first serve (FCFS)-based scheduling by cutting down on the average wait time for task processing as well as the average turnaround time of jobs to the CPU. For efficient control with the FreeRTOS, this system provides data with low interrupt latency and low thread switching delay.

7.4.5 RPMS web server performance testing

In continuous patient monitoring systems, large amounts of data need to be transferred between the embedded web server and client systems at any time throughout the day.

Latency is a parameter that represents, how fast a server replies to the request from the client.

Throughput represents how many requests per second a server can handle. Load testing of an embedded web server helps to identify these parameters by sending simulated HTTP traffic to the server from testing software installed on one or multiple computers.

The stress test is conducted using a simulation platform known as the "web server stress tool." It is an effective HTTP client/server testing tool. The performance can be analyzed under normal and excessive loads. Detailed test logs and easily understandable graphs help to identify the performance and speed of response. Results can point out strange performance issues on the web server. This tool can be used to benchmark any HTTP web servers. The software can be installed on any client computer for the analysis. In the simulation software, simulation time, number of users, and test type are configured before the test starts. Once the test starts, the software analyzes the performance of the web server for the assigned loads and gives detailed test logs and graphs that show the performance against various simulated inputs.

The different test types the software supports are CLICKS, TIME, and RAMP. The server performance test is conducted with a constant number of users in CLICKS, until each user is generating a specified number of clicks. In TIME, the test is conducted with a constant load for a specific time. In RAMP mode, the test is conducted with an increasing load during run time for a specific duration. The number of users, delay between each click, time outs, simulation data rate, and the URL to be tested can be configured in the software. The simulation can be done for up to 10,000 users with more than 1 Gigabit/sec network throughput. The software gives results in text format and graphical format, which can be easily understandable.

The configuration parameters used in the simulation for testing the designed remote patient monitoring web server are listed in Table 7.1.

Wearable sensors such as a temperature sensor, body position sensor, and ECG sensor collect the patient's temperature, body movements, ECG, and heart rate. They send the data to the dedicated website that is designed in the system software. The system also effectively streams live video on a separate webpage. The sensor readings are with acceptable precision. Since the proposed system is a medical device, the system should respond immediately without any delay, therefore FreeRTOS with SJF scheduling is used. Compared to FCFS scheduling, the SJF scheduling algorithm reduces the average waiting time and turnaround time with acceptable precision. The security feature made with the T-RBAC algorithm efficiently protects the system from security threats. The LwIP stack consumes little memory and RTOS with SJF-based priority schedule effectively manages the task

Table 7.1 Parameters configured for testing the server

Parameter	Setting
Test type	Ramp (load increases during the test time)
Number of users	5
Click delay	5 seconds
Test duration	1 minute
Tested embedded web server	ARM Cortex M4 based 32 bit RISC processor architecture microcontroller STM32F429ZI, Core operates at the frequency of up to 180 MHz, 2 MB flash, 256 KB SRAM
Test client	Dual Core Pentium Processor @ 2.9 GHz, 4 GB RAM, 32 bit Windows OS
Time out	120 seconds (request taking longer time than this are discarded)

and reduces the execution time. Finally, the designed RPMS web server is included for load testing using the web server stress test tool to determine its performance such as response time and throughput against various load capacities.

7.5 RESULTS AND DISCUSSION

The system designed in the first level consists of the feature of local monitoring of health parameters using a personal computer. It means the device cannot be accessed from a remote place. It is capable of measuring temperature, ECG, heart rate, and body position of a patient.

The sensors are tested initially to identify their working and to read and test their output. While testing the LM35 temperature sensor, it produced a 10 mv increase in output voltage for every 1 degree rise in temperature. Using this relation, the output voltage is converted to degrees Celsius. The output voltage ECG sensor is tested using cathode ray oscilloscope (CRO) and the peak voltage is identified. By counting this peak point for a minute, the heart rate is calculated in the microcontroller. The acceleration parameters corresponding to the X, Y, and Z axes are read from the MEMS accelerometer in digital format via IIC protocol. All this measured information is passed via the UART serial communication protocol. Through the serial port of any personal computer, this information can be read and seen using serial data monitoring software. It is shown in Figure 7.5.

RealTerm serial monitoring software is a very useful tool for feeding or capturing data to or from a system for testing and analysis purposes. The first line in the black color area (see Figure 7.5) shows the measured heart rate from the ECG sensor data. The second line shows the temperature in degrees Celsius, and the values X, Y, and Z are the values corresponding to the patient's body position that are read from the MEMS accelerometer.

Figure 7.6 shows the ECG waveform plotted using the data obtained from the microcontroller. The waveform is plotted using serial data plotting software. The observed values can be compared to the standard values for identifying whether the heart is functioning normally. For the sake of illustration, two ECG wave cycles are chosen, and the RR interval (time in seconds) between the two neighboring R waves is 0.69 s. The absence of any irregularity between the two adjacent R waves is shown. The heart's proper operation is confirmed by this finding.

The sensors are designed to operate in low power and the microcontroller is also configured to operate in low-power mode to reduce energy consumption. Figure 7.7 shows the comparison between the energy consumption in low-power mode with normal mode. It can be understood that the power consumption will be less in low-power mode compared with normal mode and it is in the ratio of 1:2. While implementing RTOS in software, it can

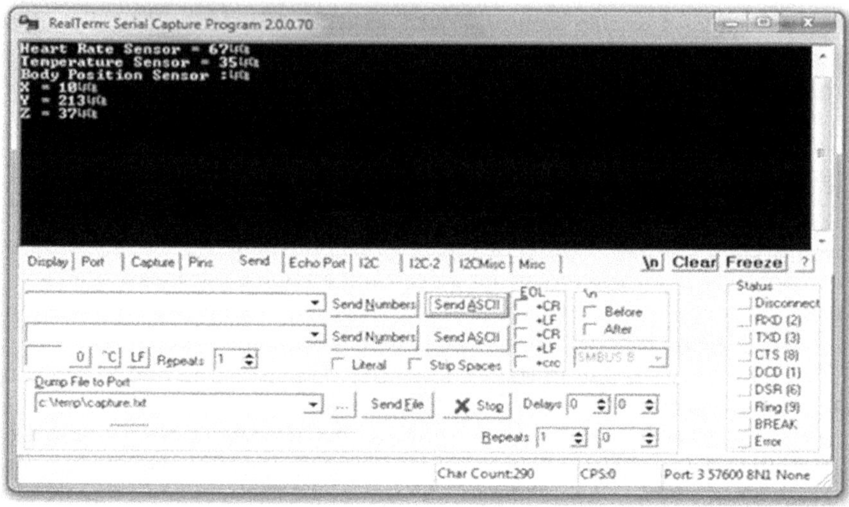

Figure 7.5 Data reading using real-term serial data monitoring software.

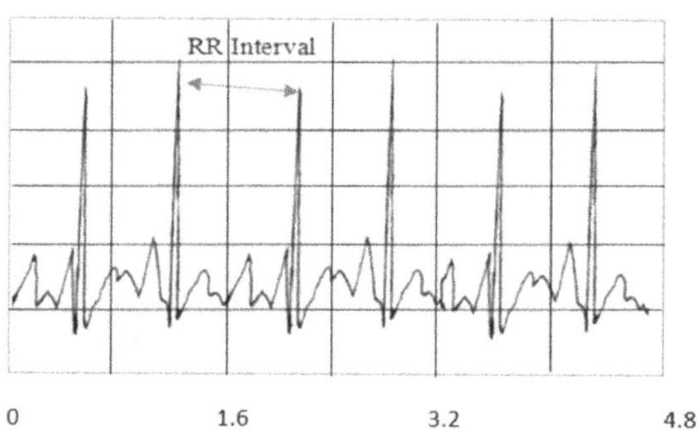

Figure 7.6 Sample ECG data plotted using serial plotter.

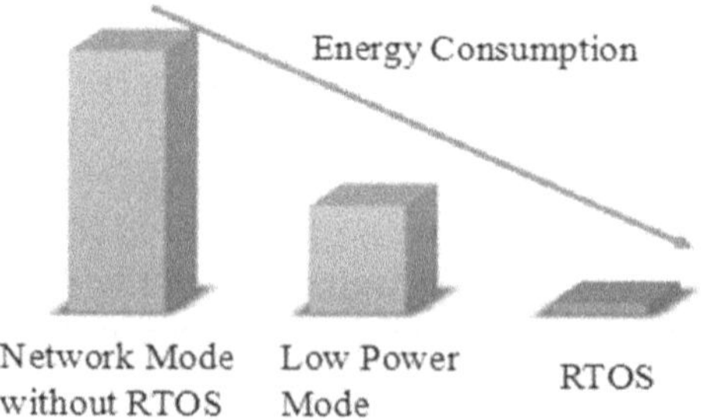

Figure 7.7 Comparison chart for energy consumption.

Table 7.2 Body temperature and heart rate for 20 persons

Sample	Temperature (°C)	Heart rate (bpm)
Person 1	35	70
Person 2	36	84
Person 3	36	82
Person 4	38	78
Person 5	37	90
Person 6	38	88
Person 7	36	74
Person 8	35	76
Person 9	35	82
Person 10	37	80
Person 11	38	94
Person 12	36	76
Person 13	37	78
Person 14	36	82
Person 15	37	92
Person 16	35	76
Person 17	35	78
Person 18	38	80
Person 19	37	84
Person 20	36	74

be noted that the power consumption is reducing further which is almost ten times less than normal mode.

For testing the system's accuracy, the temperature and heart rate of 20 middle-aged persons are collected using this device. Table 7.2 shows the results. For adults, the normal temperature will lie between 36.4°C and 37.3°C and the heart rate will be somewhere between 60 to 100 bpm. The results can be inferred that the readings taken from the device matched with the standard readings of normal people.

Using a personal computer or mobile phone and the internet, the second level of the designed system accumulates a feature of remote monitoring of the health. This device is capable of measuring a few parameters such as the temperature, heart rate, and body position of a patient. Illustrating the webpage as a reference, we can get an ECG waveform. The system is designed as an embedded web server along with the T-RBAC algorithm. Figure 7.8 depicts healthcare monitoring.

The RBAC algorithm contains a greater advantage, hence it is preferred. Figure 7.9 illustrates the contrast between the RBAC and T-RBAC models. From the chart, it can be stated that one role/permission can be created at a

Figure 7.8 Hardware model of internet-based e-healthcare monitoring system.

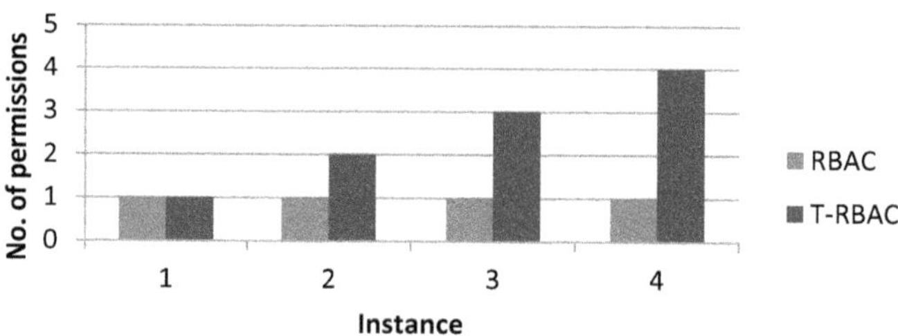

Figure 7.9 Comparisons of RBAC and T-RBAC.

time in the RBAC system, whereas with the T-RBAC system, several tasks and permissions can be created.

The system administrator is responsible for adding or removing customers, establishing tasks, setting rights and limits, etc. Here, the service provider grants the system admin role to an occupant and allots the access privileges so that they can manage the verification and authorization procedure. The STM32F429 microcontroller used to build the web server stores the T-RBAC parameters, such as clients, tasks, roles, restrictions, resources, permissions, and approval policies, in memory. Table 7.3 displays a sample user list and passwords.

When a client logs in, the system verifies their identity and assigns them to certain roles. Figure 7.10 and Figure 7.11 describe the T-RABC algorithm's procedure and flowchart.

The designed embedded web server can be accessed remotely using an internet-connected computer or mobile phone. The unique IP address of the web server is used to access the hardware. Upon calling up the system using its IP address, it will show a login screen as in Figure 7.12.

Table 7.3 Sample use list and passwords

User	Password	Role	Task
Admin	Admin	Maintains users list	Adding, deleting and modifying user settings
Doctor	Doctor	Doctor	Viewing patients' health information, prescribing medicines, writing report
Nurse	Nurse	Nurse	Viewing prescriptions and follow treating the patient according to it
Relation	Relation	Caretaker	Can view the reports to know the status of patient

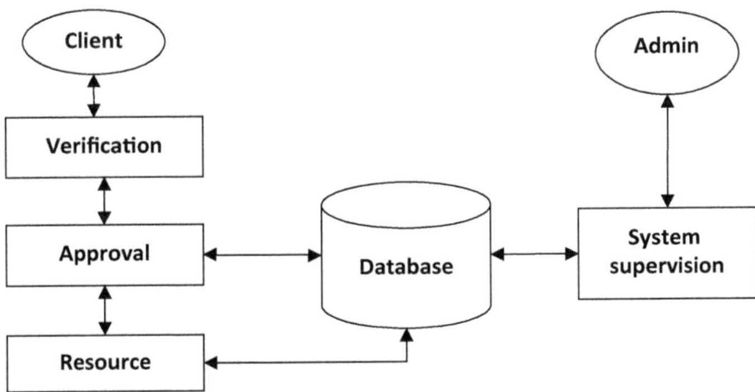

Figure 7.10 Process of T-RBAC.

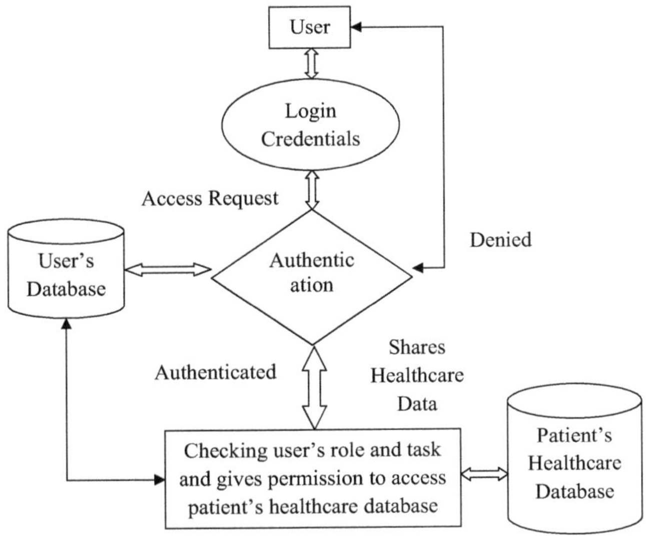

Figure 7.11 Flowchart of T-RBAC.

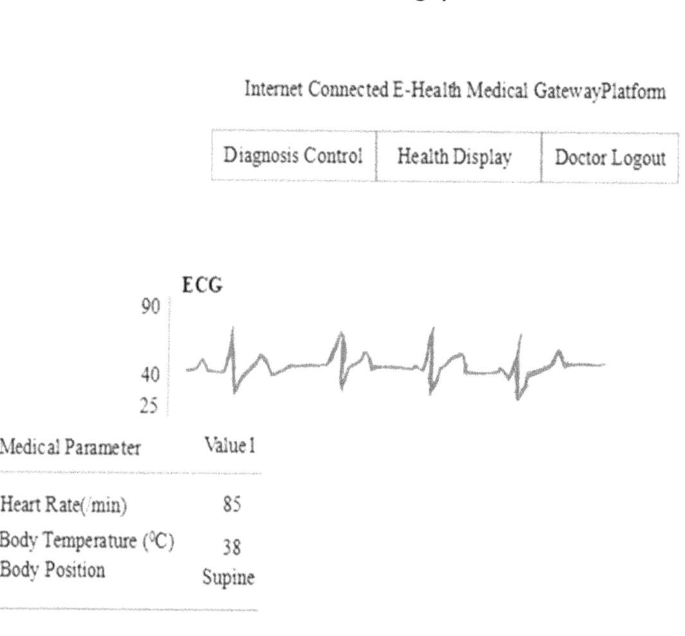

Figure 7.12 Login page of patient's health monitoring system.

Figure 7.13 Task- and role-based patient's health monitoring page for doctor.

Users with different user IDs and passwords are assigned to doctors and caregivers. The patient's health monitoring page for the doctor and caregiver is displayed in Figures 7.13 and 7.14. It is task- and role-based. The doctor can view patient medical records, prescribe medications, and add inferences as needed, all in accordance with the roles, tasks, and permissions specified in the T-RBAC database. However, the caretaker is only permitted to examine the medical records; they cannot add or change any information on that page.

The individual pixels on the page can be effectively controlled with the Canvas API. We use JavaScript to draw and vitalize objects on what appears to be a blank slate.

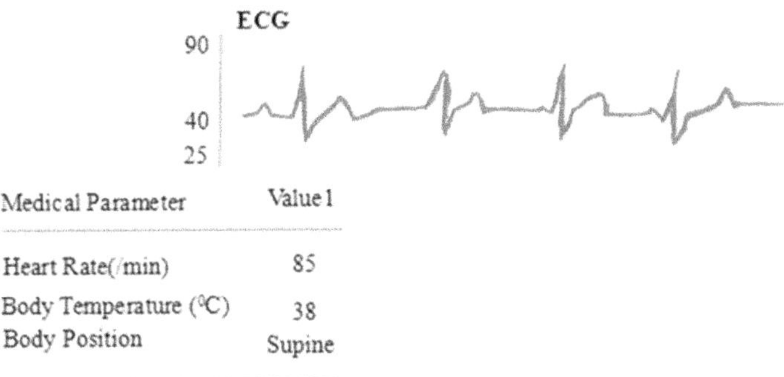

Figure 7.14 Health monitoring page for caregiver with task- and role-based access.

7.5.1 Results of IoT-based e-healthcare system with live video monitoring using SJF priority scheduled RTOS

As shown in Figure 7.15, the microcontroller's input/output ports are coupled to a three-lead ECG sensor, a body temperature sensor, a three-axis MEMS accelerometer, a 2 MP camera, and a buzzer. An Ethernet connection is used to link the system to a nearby internet modem. The system

Figure 7.15 Implementation of overall hardware.

Table 7.4 FCFS-based priority scheduling for ten tasks

Tasks	Burst time	Priority
T1	160	2
T2	95	1
T3	175	3
T4	59	1
T5	180	4
T6	190	5
T7	200	6
T8	184	5
T9	195	6
T10	210	7

Table 7.5 SJF-based priority scheduling for ten tasks

Task	Burst time	Waiting time (ms)	Turnaround time (ms)
T1	160	0	160
T2	95	160	255
T3	175	255	430
T4	59	430	489
T5	180	489	669
T6	190	669	859
T7	200	859	1059
T8	184	1059	1243
T9	195	1243	1438
T10	210	1438	1648
Average		660.2	825

receives internet connectivity, allowing access from anywhere. Through a 9V DC converter, the system is powered from an AC source.

For choosing a suitable scheduling, a simple analysis is made, and results are tabulated and shown in Table 7.4, Table 7.5, and Table 7.6. From the comparison of results, SJF scheduling is chosen for the design. SJF-based priority scheduling, which is used in this instance in the RTOS component of microcontroller programming, is thus determined to be appropriate for this application. This priority scheduling states that if many tasks with the same priority are received, the task with the shortest burst time is carried out first. Figure 7.16 presents the Gantt chart for tasks according to SJF.

The LwIP stack handles limiting the maximum length of the data that can be delivered. Figure 7.17 shows FCFS and SJF-based priority scheduling waiting times. Figure 7.18 shows FCFS and SJF-based priority scheduling

Table 7.6 SJF-based turnaround time for ten tasks

Task	Burst time	Priority	Waiting time (ms)	Turnaround time (ms)
T4	59	1	0	59
T2	95	1	59	154
T1	160	2	154	314
T3	175	3	314	489
T5	180	4	489	669
T8	184	5	669	853
T6	190	5	853	1043
T9	195	6	1043	1238
T7	200	6	1238	1438
T10	210	7	1438	1648
Average			625.7	790.5

T4	T2	T1	T3	T5	T8	T6	T9	T7	T10

0   59   154   314   489   669   853   1043   1238   1438   1648

Figure 7.16 Gantt chart for tasks according to SJF.

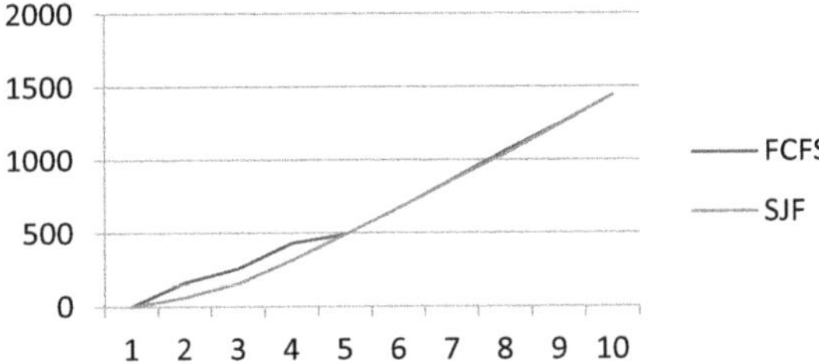

Figure 7.17 FCFS and SJF-based priority scheduling waiting times.

turnaround times. As seen in Figure 7.19, the web client receives the same number of packets as the server sends.

After login, the user can navigate to any page according to the role and task rights. The user list stored in the database helped to achieve this. In the event of an aberrant heart rate or body temperature, the system also sounds a bell to inform nearby caregivers. The webpage displaying sensor information is shown in Figure 7.20. Figure 7.21 shows the webpage that streams the live video. The doctors have full access to these pages.

The embedded web server runs on a microcontroller which has limited memory and CPU performance. It is comparatively lower than the normal

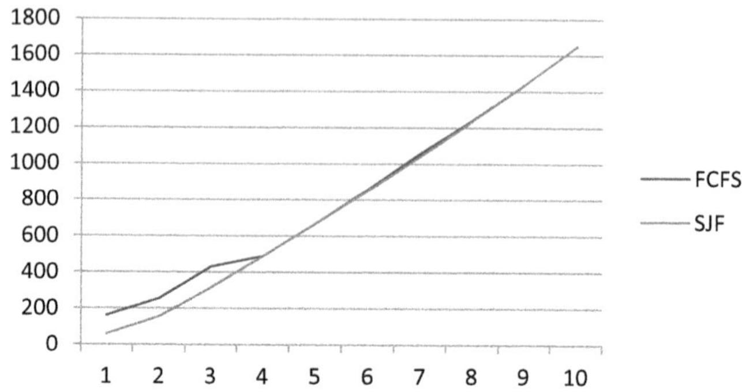

Figure 7.18 FCFS- and SJF-based priority scheduling turnaround times.

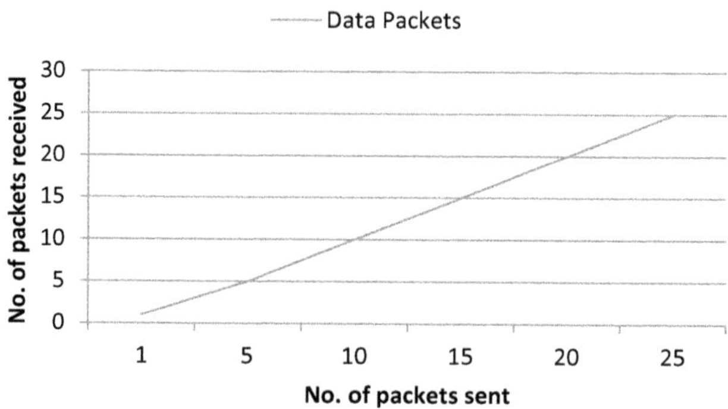

Figure 7.19 Data packets sent and received.

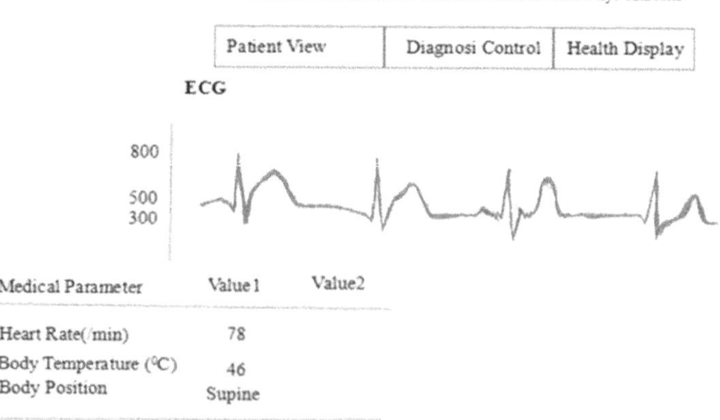

Figure 7.20 Monitoring streaming page.

Figure 7.21 Video streaming page.

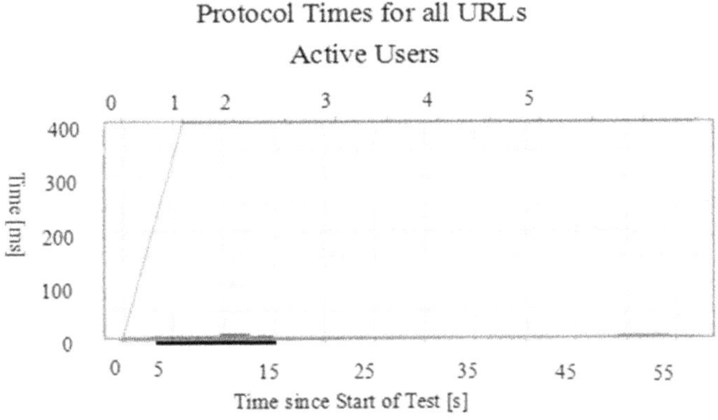

Figure 7.22 The protocol times for all URLs.

web server running on a limitless computer. Therefore, conducting stress tests for embedded web servers is a good practice.

The test on the RPMS web server was conducted for 57 seconds. The protocol times for all URLs are shown in Figure 7.22. The server and user bandwidth graph are shown in Figure 7.23. In Figure 7.24, a graph of the data transferred, system memory usage, and CPU load is displayed. The graph of open requests and transferred data are shown in Figure 7.25. The spectrum of click times, and the hierarchy and times of all hits are shown in Figure 7.26 and Figure 7.27, respectively. Click time, hits, and users for the given URL are shown in Figure 7.28. Click times and errors for the URL are shown in Figure 7.29.

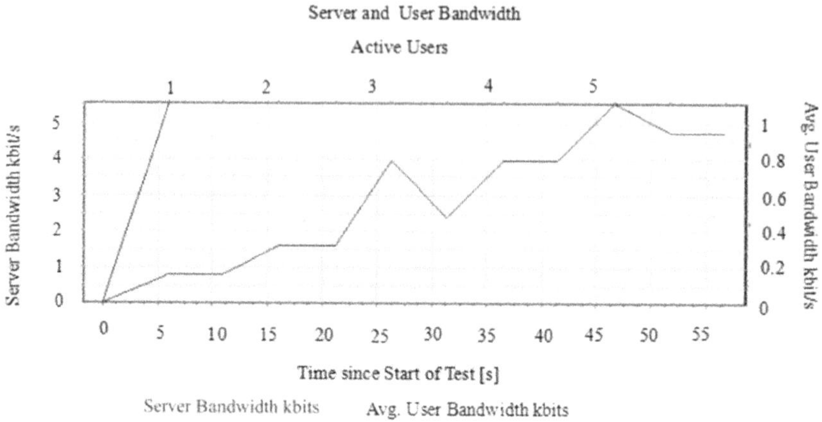

Figure 7.23 Server and user bandwidth.

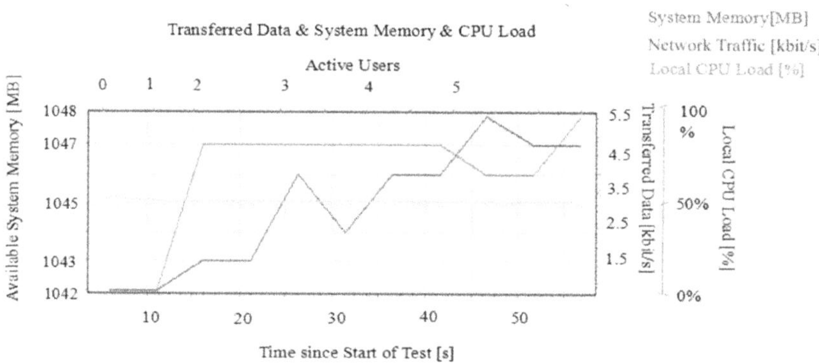

Figure 7.24 Transferred data, system memory, and CPU load.

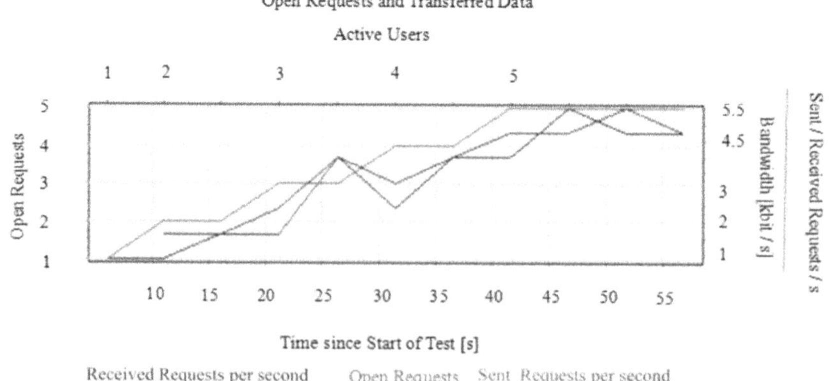

Figure 7.25 Open requests and transferred data.

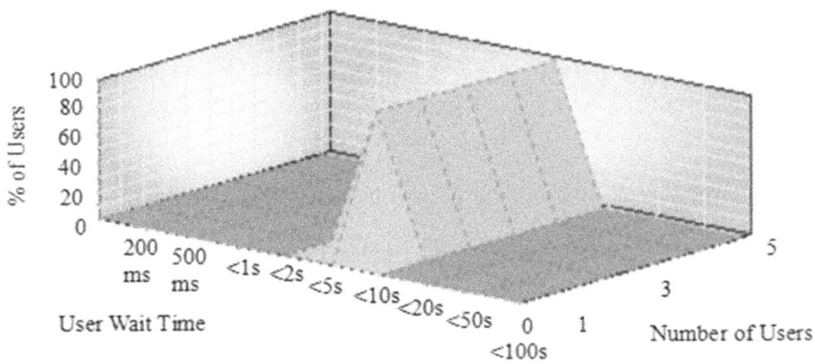

Figure 7.26 Spectrum of click times.

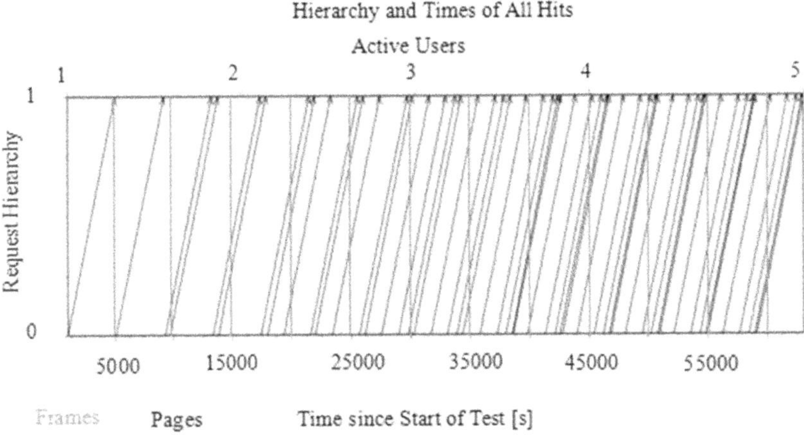

Figure 7.27 Hierarchy and times of all hits.

Table 7.7 lists the summary results per URL. The basic standalone patient monitoring device is enhanced with an internet facility and security features using the T-RBAC algorithm. In addition to various sensors that measure the health parameters of a patient, a camera is added to follow the patient's physical activity.

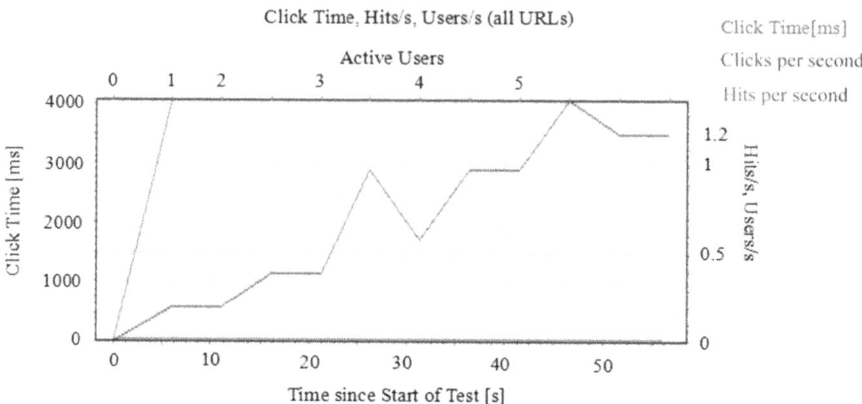

Figure 7.28 Click time, hits, users (all URLs).

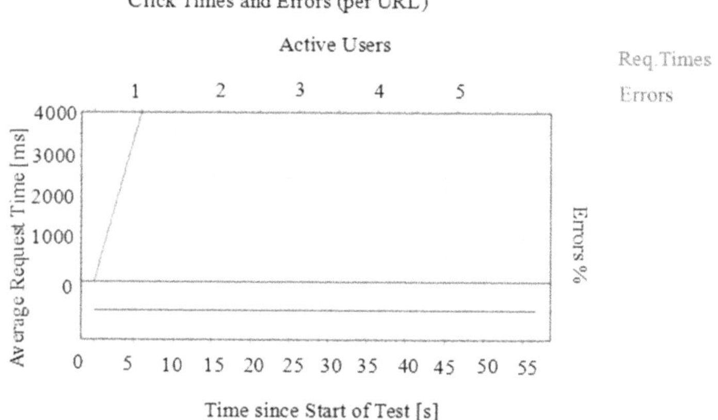

Figure 7.29 Click times and errors (per URL).

Table 7.7 Summary results per URL

Name	Clicks	Errors	Errors (%)	Time spent (ms)	Avg. click time (ms)
	43	0	0.00	173687	4039

7.6 CONCLUSIONS

The research work's main objective is to design an internet-based e-health-care monitoring system using an RTOS and T-RBAC-based access control algorithm.

The device is designed for monitoring a single person in a remote area. For multi-patient monitoring, such as in a hospital environment, the same device can be designed in multiple numbers and the data can be routed to a central cloud server. An access control algorithm can be implemented to provide access restrictions. Instead of accessing from a web browser, mobile applications can be designed for monitoring.

LIST OF ABBREVIATIONS

ADC:	analog-to-digital converter
CAM:	Content-Addressable Memory
CT:	Controller of Task
ECG:	electrocardiogram
FCFS:	first come, first serve
IF:	Identifier for Function
IoT:	Internet of Things
LHMS:	local health monitoring system
LwIP:	Lightweight Internet Protocol
MEMS:	microelectromechanical systems
MLEG:	Monitoring of List and Error Generator
RPMS:	remote patient monitoring system
RTOS:	real-time operating system
SJF:	shortest job first
T-RBAC:	task–role-based access control
TCP/IP:	Transmission Control Protocol/Internet Protocol

REFERENCES

Al Disi, M, Djelouat, H, Kotroni, C, Politis, E, Amira, A, Bensaali, F, et al. 2018, 'ECG Signal Reconstruction on the IoT-Gateway and Efficacy of Compressive Sensing Under Real-Time Constraints', *IEEE Access*, vol. 6, pp. 69130–69140

Chen, J, Valehi, A & Razi, A 2019, 'Smart Heart Monitoring: Early Prediction of Heart Problems through Predictive Analysis of ECG Signals', *IEEE Access*, vol. 7, pp. 120831–120839

Chen, Y, Sun, W, Zhang, N, Zheng, Q, Lou, W & Hou, YT 2018, 'Towards Efficient Fine-Grained Access Control and Trustworthy Data Processing for Remote Monitoring Services in IoT', *IEEE Transactions on Information Forensics and Security*, vol. 14, no. 7, pp. 1830–1842

Choudhary, D & Kumar, R 2016, February 12–13, '2016: Heart Rate Analysis and Monitoring of Patients from Offsite through Wireless Sensor Network', IEEE 2nd International Conference on Computational Intelligence & Communication Technology (CICT), Ghaziabad, India, pp. 248–253

Clarke, M, de Folter, J, Verma, V & Gokalp, H 2018, 'Interoperable End-to-End Remote Patient Monitoring Platform based on IEEE 11073 PHD and ZigBee Health Care Profile', *IEEE Transactions on Biomedical Engineering*, vol. 65, no. 5, pp. 1014–1025

Dietrich, C, Hoffmann, M & Lohmann, D 2017, 'Global Optimization of Fixed-Priority Real-Time Systems by RTOS-Aware Control-Flow Analysis', *ACM Transactions on Embedded Computing Systems (TECS)*, vol. 16, no. 2, pp. 1–25

El Kafhali, S & Salah, K 2018, 'Performance Modeling and Analysis of Internet of Things Enabled Healthcare Monitoring Systems', *IET Networks*, vol. 8, no. 1, pp. 48–58

Indersain, NS & Singh, D 2013, 'Design and Implementation of μc/Os II Based Embedded System Using ARM Controller', *International Journal of Engineering and Technical Research*, vol. 1, no. 2, pp. 1–4

Koh, JH & Choi, BW 2013, 'Real-time Performance of Real-time Mechanisms for RTAI and Xenomai in Various Running Conditions', *International Journal of Control and Automation*, vol. 6, no. 1, pp. 235–246

Jegadeesan, S, Dhamodaran, M, Azees, M & Shanmugapriya, SS 2019, 'Computationally Efficient Mutual Authentication Protocol for Remote Infant Incubator Monitoring System', *Healthcare Technology Letters*, vol. 6, no. 4, pp. 92–97

Kiani, MJE, Al-Ali, A, Muhsin, B, Sampath, A Olsen, GA & Masimo C 2015, 'Medical Monitoring System', U.S. Patent 9, 218, 454

Kioumars, AH & Tang, L 2011, 'Wireless Network for Health Monitoring: Heart Rate and Temperature Sensor', IEEE 5th International Conference on Sensing Technology (ICST), November 28–December 1, Palmerston North, New Zealand, pp. 362–369

Mahbub, I, Shamsir, S, Pullano, SA, Fiorillo, AS & Islam, SK 2019, 'Design of a Charge Amplifier for a Low-Power Respiration Monitoring System', *IET Circuits, Devices & Systems*, vol. 13, no. 4, pp. 499–503

Majumder, S, Mondal, T & Deen, MJ 2018, 'A Simple, Low-Cost and Efficient Gait Analyzer for Wearable Healthcare Applications', *IEEE Sensors Journal*, vol. 19, no. 6, pp. 2320–2329

Park, J, Seong, K, Noh, HK, Lee, WC, Kim, B, Sim, JY et al. 2016, 'An ECG Monitoring System using Android Smart Phone', IEEE International Conference in Consumer Electronics-Asia (ICCE-Asia), October 26–28, Seoul, South Korea, pp. 1–4

Pathinarupothi, RK, Durga, P & Rangan, ES 2018, 'IoT based Smart Edge for Global Health: Remote Monitoring with Severity Detection and Alerts Transmission', *IEEE Internet of Things Journal*, vol. 6, no. 2, pp. 2449–2462

Satija, U, Ramkumar, B & Manikandan, MS 2017, 'Real-Time Signal Quality-Aware ECG Telemetry System for IoT-Based Health Care Monitoring', *IEEE Internet of Things Journal*, vol. 4, no. 3, pp. 815–823

Solosko, T, Gehman, S, Herleikson, E, Lyster, T, Fong, S, Hansen, K et al. 2017, 'Continuous Outpatient ECG Monitoring System', U.S. Patent 9, 615, 793

Sood, SK & Mahajan, I 2018, 'IoT-Fog-based Healthcare Framework to Identify and Control Hypertension Attack', *IEEE Internet of Things Journal*, vol. 6, no. 2, pp. 1920–1927

Uddin, MA, Stranieri, A, Gondal, I & Balasubramanian, V 2018, 'Continuous Patient Monitoring with a Patient Centric Agent: A Block Architecture', *IEEE Access*, vol. 6, pp. 32700–32726

Verma, P & Sood, SK 2018, 'Fog Assisted-IoT Enabled Patient Health Monitoring in Smart Homes', *IEEE Internet of Things Journal*, vol. 5, no. 3, pp. 1789–1796

Wang, G, Lu, R & Guan, YL 2019, 'Achieve Privacy-Preserving Priority Classification on Patient Health Data in Remote eHealthcare System', *IEEE Access*, vol. 7, pp. 33565–33576

Chapter 8

Role of IoT in communications

*Shweta Gupta, Akula Praveen Kumar,
and Bukka Sanjay*

8.1 INTRODUCTION

The Internet of Things (IoT) (Kumar et al.2019) refers to the connection of physical devices attached to sensors, tools, appliances, and other devices with software and networking that enable them to store and share information. The main goal of the Internet of Things is to improve efficiency, intelligence, and decision-making capabilities by connecting the physical and digital worlds. This chapter explores the changing role of the Internet of Things in communication and its many impacts on the way we interact, gather information, and navigate the world around us. The Internet of Things has many applications in various fields. These include device connectivity and machine-to-machine (M2M) communication (Verma et al. 2016) with each other through wired or wireless means, often without human intervention. M2M allows devices to send and receive information, share information, and network by collaborating on projects. The benefits of M2M communication are:

- Increased productivity: M2M communication allows instant data exchange, allowing devices to perform better operations, reduce time, and make informed decisions independently. This increases efficiency and reduces dependence on manual intervention.
- Increased accuracy and precision: Direct communication between devices eliminates human error and data delays, ensuring data is accurate and true. This is important for projects such as electronics manufacturing and remote monitoring.
- Scalability: M2M communication is scalable and allows for connecting many devices. It can be easily scaled up or down to meet changing needs; this is important in IoT scenarios where multiple devices need to work together seamlessly. This makes them ideal for dynamic environments and large deployments.
- Cost reduction: M2M communications can deliver significant cost savings in every aspect of operations by streamlining operations and processes.

DOI: 10.1201/9781003510420-8

8.2 M2M COMMUNICATION TYPES

8.2.1 Cellular network

LTE-M and NB-IoT Sigfox, known for their versatility, low power consumption, and ability to connect M2M communications, are increasingly moving to other devices and becoming more popular. Cellular connectivity is common in applications such as asset tracking, smart devices, and remote monitoring (Attaran 2023). Bluetooth, Wi-Fi, and Zigbee are mainly used for short-range M2M communication and provide high data and reliable connections in a short range. Satellite radios provide remote connections for equipment in remote or isolated areas, enabling M2M communication even in very challenging environments. They can provide global service but will require higher costs and energy consumption. Figure 8.1 shows how mobile devices can connect to Wi-Fi networks.

Challenges of M2M networks include:

- Safety and security: Protecting sensitive information transferred between devices and ensuring secure communication is the challenge of M2M (Kim et al.2013).
- Interoperability: Standardizing communication protocols and ensuring compatibility between various devices from different manufacturers is crucial for the widespread use of M2M.
- Scalability and reliability: Creating robust and scalable M2M systems that can handle large amounts of data and provide reliable communication across complex communications is an ongoing challenge.

M2M communication is developing rapidly and has great potential in the future. Artificial intelligence (AI) and machine learning will play an important role in making M2M systems smarter and more autonomous. Advances in communication technology will make data exchange between

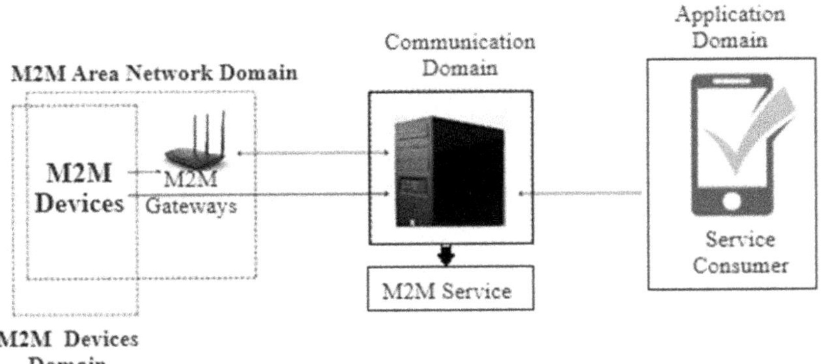

Figure 8.1 Wi-Fi connects to mobile.

devices faster, more reliable, and more secure. As these technologies evolve, M2M communications will become the foundation of a truly connected world, transforming businesses, and changing the way we live and work.

The development of 5G networks is also expected to play an important role in supporting high-speed data and low-latency M2M communications.

8.2.2 Data networks

Sensor networks provide a decentralized method of data collection, allowing transmission to many sensors in an area. Sensor networks often include more than one type of sensor and allow different data to be collected. For example, a smart agricultural machine might have soil moisture sensors, temperature sensors, light sensors, and cameras.

8.2.3 Sensor networks

Sensors are devices that can measure physical properties such as temperature, humidity, light, and movement. Sensors constantly monitor the environment, collect data, and then send it to a central location for analysis and integration. This network is called a sensor network. The types of sensor networks are:

- Wired sensor networks: These networks rely on physical cables and wires to connect sensors to a central hub, providing reliability and bandwidth but limited scalability.
- Wireless sensor networks (WSNs): These networks use wireless technologies such as Bluetooth, Wi-Fi, or LPWAN to transmit data, providing greater flexibility and power, but at the expense of power, electricity, and greater competition.

Applications of sensor networks include the following:

Smart city: Monitoring of air purity, traffic, waste management, noise, soil, crop health measurement, moisture measurement, etc.

Energy-efficient communication: Difficulties and choices to be made in WSNs include energy boosting. The amount of energy battery power that is available for an extended period is essential and frequently necessitates low-level messaging and data processing.

Data security and privacy: Protecting sensitive data collected by sensors is very important and requires security measures and encryption protocols.

Data management and analysis: Making the most of the large amounts of data generated by sensor networks requires advanced data

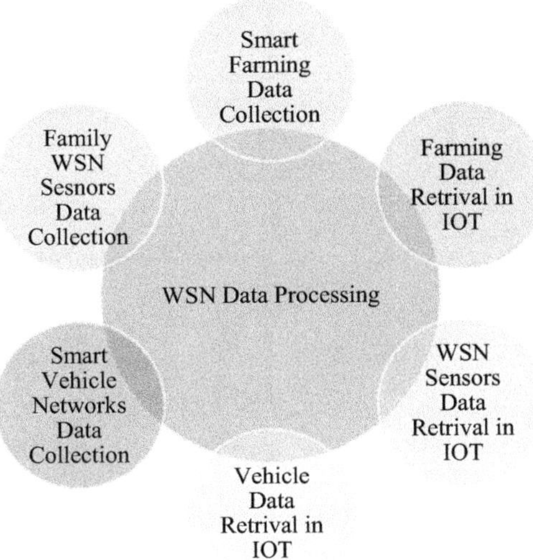

Figure 8.2 IoT and WSN communication.

management and analysis tools to turn information into insight. Figure 8.2 shows how sensors collect data and process the data for the user in an IoT environment.

Advances in miniaturization, artificial intelligence, and energy efficiency will make sensor networks smaller, smarter, and more efficient. Next-generation networks will provide more services, increase data efficiency and security, further unlock the potential of the Internet of Things in many areas, and make the human world more accessible.

8.3 SMART CITIES AND INFRASTRUCTURE

8.3.1 Smart cities

Since many things in cities are connected to the internet, and old methods are being transformed into new solutions. For a smart city, let's examine how core areas (traffic management, public services, and infrastructure) can be reinvented through the magic of the internet:

Smart traffic management is a process. Smart traffic is a tool that uses technology to manage traffic on the road. It uses sensors on roads, vehicles, and infrastructure to collect information about speed, vehicles, lines, and accidents. This information is sent to a central location that displays traffic times, suggests alternative routes, and even reroutes traffic based on current

Figure 8.3 Traffic using IoT sensors.

conditions. This will help reduce traffic congestion and pollution and make the city peaceful.

Figure 8.3 shows how IoT (traffic sensors) controls the vehicles to manage the traffic.

The Internet of Things is a technology that facilitates the operation of public services. It helps keep streets safe and clean. Smart water meters monitor water usage and detect leaks so they can be quickly fixed (Velsberg et al. 2020). Lights adjust their brightness according to human activity and charge themselves with solar energy, saving energy and making them safer. The waste management system reduces the environmental impact of disposal by instantly optimizing collection by level. IoT is also making the infrastructure of cities smarter. Bridges and buildings are equipped with sensors that provide information about their condition, so we can learn about their health and repair them before they get damaged. The voluntary plan is adjusted according to need, so it is less expensive. This network uses data to work together to make cities work better. Figure 8.4 presents the different IoT applications enabled in cities.

8.3.1.1 Challenges and decisions

IoT can make cities better, but there are still some problems. Data security and privacy are important, so we need strict rules and ethics. We also need to ensure that different systems work well together. This will help cities operate effectively. A smart city with IoT can be a place where technology and people work together to create a better future. They can be more efficient, effective, and sustainable. The use of IoT in smart cities can increase

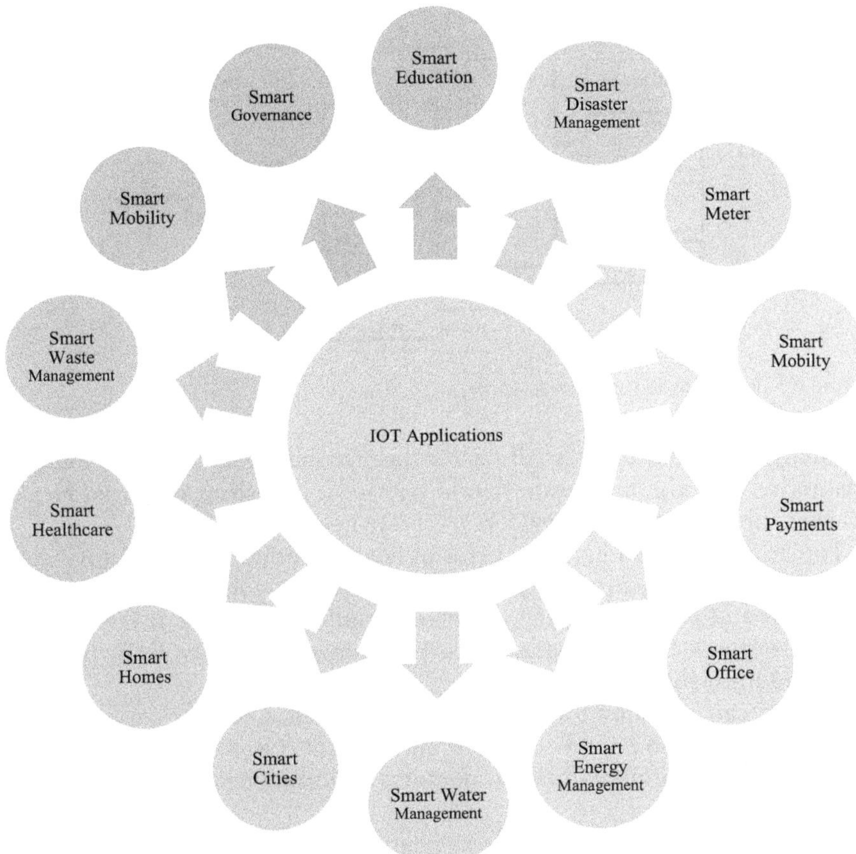

Figure 8.4 IoT applications enabled in cities.

energy efficiency, reduce traffic, and manage waste. It can control traffic lights, monitor traffic, find closed routes, and detect traffic problems. By using the Internet of Things in the city, we can create a system where everything works, making life and the environment better.

8.3.2 Healthcare

There is more information in the human body. The Internet of Things tracks this data and aids healthcare and can monitor patients remotely. This helps improve health. It can also help rural people stay healthy. The IoT can collect patient information and send it to doctors. This may help people with long-term health problems (Akhtar et al.2021). These systems have gained importance in the field of medicine and can help provide better healthcare for everyone.

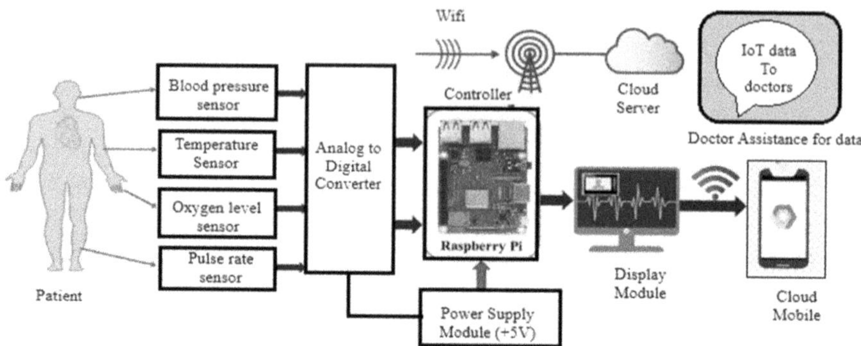

Figure 8.5 IoT helps in health monitoring.

Smart wearable devices allow for health management by helping to instantly track the heart rate, blood pressure, and sleep patterns. Health trackers help healthcare providers to create health plans.

Figure 8.5 presents how IoT helps in health monitoring to doctors using IoT communications in which the patient data is converted from analog to digital and processed using a controller unit, then to the IoT cloud, and finally to the doctor for checking the parameters of oxygen level, temperature, bold pressure, and pulse rate.

Sensors can also monitor diseases remotely such as diabetes or high blood pressure in order to intervene in time and prevent complications. Measuring blood sugar can send readings to doctors, providing better insulin therapy.

- Smart sensors can also monitor patients at home and alert caregivers to possible falls, nonadherence to medications, or sudden changes in vital signs. They also promote quality care in a familiar environment by identifying beds that recognize sleep patterns and alert nurses to risks.

8.3.2.1 Benefits of using IoT in healthcare

Early detection and prevention: Real data helps doctors detect health problems at an early stage with devices such as smartwatches. For example, if a smartwatch detects an abnormal heart rate, it can alert people to take steps to prevent a heart attack. It's like having an alarm system that alerts you to your health (Azzawi et al.2016).

- Improve patient outcomes: Remote care helps patients and doctors communicate better. This helps develop a personalized treatment plan and helps patients stick to their medication schedule. Imagine getting instant advice on healthy habits that will encourage you to stay healthy. It's like having a personal cheerleader for your health.

- Reduced healthcare costs: Remote care can help you avoid unnecessary hospitalizations and emergency room visits. This saves patients' lives and healthcare money. Consider effectively managing chronic diseases from the comfort of your home by eliminating hospitals and their resources. It's like having a superhero healthcare professional take you out of the hospital.
- Challenge of information security and privacy: Protecting health information is important and requires cybersecurity measures and clear details.
- Accessibility and affordability: Ensuring equitable access to these technologies is crucial to preventing existing healthcare disparities.
- Integration and interoperability: Different devices and systems need to communicate with one another to provide effective remote monitoring, which requires technical procedures, rights, and open platforms.

8.3.2.2 The future of IoT in healthcare

The future of IoT in healthcare looks good. This means smart sensors that can track vital signs in our clothes, devices that use artificial intelligence to detect diseases at an early stage, and even robots that can perform remote surgery according to doctors' instructions. As technology and ethical issues are addressed, these advances will help us better monitor our health, again leading to better healthcare. It's like a new medical song, and the Internet of Things is the key to a healthy future for everyone.

8.4 INDUSTRIAL INTERNET OF THINGS (IIOT)

The Internet of Things is changing the way businesses and industries operate. It is like a dance where machines connect and can share information. This helps companies become more efficient and effective, which is a game changer for companies around the world. It's like a high-tech stereo system that makes everything work (Schneider 2017).

Imagine a factory equipped with sensors that can monitor energy use, improve production processes, and predict when machines will experience problems. Industrial IoT (IIoT) is like magic. Robots can work together effectively, machines can detect where their problems are, and production lines can be changed on the fly. Businesses can get more done, spend less time, and make the best use of resources. It sounds like a powerful workout.

8.4.1 Benefits of using IoT in business

There is no need to wait for the machine to stop working. With IIoT sensors, we can pay attention to equipment vibration, temperature changes, and power consumption. This helps us predict when problems will arise before they occur and cause more problems. Like a good song, it seems to avoid downtime, save on maintenance costs, and keep everything running smoothly.

8.4.2 Challenges

As we embrace this technological revolution, we also need to take care of some important things. Keeping data safe, ensuring different systems work well together, and finding the right balance between people and machines are challenges we need to solve. It's like writing music with IIoT. We need to think about ethics to ensure employees can update and manage data responsibly. It's all about finding the right path to success and future responsibilities.

8.4.3 The future of IIoT

The dance of IoT has begun, and the emergence of artificial intelligence and technology will be spectacular. This means factories can improve themselves and production lines can work autonomously (Chhaya et al. 2017). Imagine robots learning from what has gone before, adapting to different needs, and working together like a perfect orchestra. It's like technical music.

8.5 IOT IN CONSUMER ELECTRONICS

The Internet of Things has integrated into our daily lives like magic and changed the way we enjoy our devices. Two areas where this is evident are smart homes and smart devices. They make us so easily, personally, and inextricably connected to the world around us (Ozadowicz et al.2018). Think of lighting that suits your mood, a thermometer that measures your preferences, or a device that sounds an alarm. It's like everything is at your fingertips to make life easier and happier again.

8.5.1 Wearable technology and smart homes

Fitness trackers silently monitor your heart rate, smartwatches display notifications, and headphones provide instant translation. It's like the perfect compromise between technology and the blurring of our bodies and devices. This smart communication allows us to see the digital world comfortably while wearing it (Wu and Luo 2019).

8.5.2 Advantages and challenges

Smarter living: Smart homes can be more efficient, save energy, and increase security. Wearable devices, on the other hand, can increase awareness about consumer health, exercise, and constant connectivity.

- Personal awareness: Both smart homes and wearable devices can learn travel preferences, adapt to specific needs, and create truly personalized experiences.
- Improve safety and security: Smart homes provide alarms, automatic emergency response, and remote monitoring, while wearable devices can detect falls, notify individuals of an emergency, and track locations.
- Privacy issues: Data security and ethics are important. Devices require strong encryption and clear user control over the recorded data ledger.
- Interoperability: Ensuring different devices and systems speak the same language and avoid technology silos for seamless collaboration.
- Accessibility and affordability: Making technology accessible and affordable for different groups of people is important for innovation.

8.6 FUTURE

Advances in artificial intelligence, advanced computing, and microcomputers can help support more smart homes and smart devices. Think smart refrigerators optimized for brand names, glasses that display personal health information, and headsets that instantly translate entire conversations. The future of consumer electronics is the music of innovation, blurring the lines between technology and everyday life.

- Environmental monitoring: The Earth provides a constant stream of environmental information, from the whispers of air currents to ocean waves. IoT is based on listening, and deploying sensor networks, satellites, and other technologies to monitor air and pollution and sound an important message for the future (Ullo and Sinha 2020).
- Global climate monitoring: A network of sensors and satellites that monitor temperature, humidity, precipitation, and greenhouse gases provides critical data for understanding and predicting climate change. Imagine a global orchestra of weather stations and satellites working together to paint a detailed picture of global climate change.

- Local microclimate management: Sensor networks monitor the microclimate of cities, forests, and landscapes to warn against interventions such as tree planting and urban heat island mitigation. Imagine small perimeter sentinels whispering accurate information about local events, thereby increasing local security (Schena et al. 2016).
- Air quality monitoring: Sensor networks in cities and industrial areas track pollution levels over time, providing alerts for pollution control and recommendations for public health.
- Water quality monitoring: Meters measure pollutants, bacteria, and chemicals in water bodies, alert authorities to hazards, and recommend water purification strategies.
- Decision-making: Environmental information enables policymakers to implement strategies to reduce household electricity emissions and protect fragile ecosystems.
- Improving public health: Timely monitoring of air pollution to remind people to take preventive measures to reduce air pollution and water pollution.
- Sustainable resource management: Environmental information demonstrates best practices in agriculture, forestry, and urban planning that enable better use of resources and reduce environmental impact.
- Data integration and analysis: Combining data from different sources and turning them into insights requires data management and tools (Gupta and Kumar 2023).
- Access and cost: Ensuring fair environmental monitoring and sharing information across borders is very difficult.
- Ethical decision-making: Protecting sensitive environmental information and ensuring the information is used responsibly requires maintaining safety and ethics.

Advances in technology, artificial intelligence, and data analysis will bring greater sophistication and accuracy to environmental monitoring. Consider small smart sensors placed on trees, animals, and even buildings that provide a rough picture of the health of the planet. By listening carefully to the environmental symphony and acting on its insights, we can create an efficient and technologically compatible future.

Communication is the leader in the Internet of Things orchestra, directing the flow of information and the harmony of communication between devices. The emergence of 5G and next-generation wireless technologies promises to transform communications by delivering powerful advances in speed, reliability, and innovation.

Researchers are exploring 6G and beyond, focusing on faster speeds, lower latency, and wider bandwidth. These advances are leading to holographic communication, universal intelligence, and online experiences that blur the lines between the physical and digital worlds.

8.7 MAJOR DEVELOPMENTS IN IOT

- Improved sensor function: Low power means you can send data more frequently. Effectively enable instant monitoring and control across different IoT applications.
- Edge computing: Processing data on edge devices closer to the data center, rather than relying on central cloud servers, reduces latency and increases performance, which is important for real-time IoT applications.
- Satellite integration: Besides 5G, people are also exploring the integration of satellite communications into IoT networks. This could expand IoT connectivity to remote and underserved areas, leading to a more comprehensive global IoT ecosystem.
- Network slicing: 5G networks can be segmented into virtual devices, each with specific resources for specific applications, improving performance and security for different IoT applications.

Electrification: Widespread use of 5G and other infrastructure will require significant investment and collaboration.

Cross-device compatibility: Devices must be standardized and functioning to maximize 5G and other network capabilities.

Security: Protecting IoT devices' enormous datasets and fixing 5G and other network vulnerabilities is crucial. Internet of Things music will get more complicated and fascinating as communication technology develops. 5G and other technologies will unleash billions of linked devices, creating a world of seamless data flow, edge intelligence, and faultless innovation. The future of IoT communications is a burst of power that will alter enterprises, redefine experiences, and improve the future by integrating gadgets and people.

AI as IoT symphony conductor, improving experiences and driving improvements: IoT is a bustling orchestra whispering data and completing duties. However, a specialist is needed for true harmony. AI algorithms and magic knowledge boost the IoT experience and promote constant improvement.

Research and understanding: AI's ability to analyze large amounts of data on connected devices can uncover hidden patterns. Consider sensors that silently report changes in temperature in the workplace. Smart intelligence prevents inefficiencies before they occur by organizing predictive maintenance.

Instant decision-making: Artificial intelligence enables devices to make instant decisions by processing information at the speed of light. Imagine a smart home that adjusts the lighting according to your mood, thanks to intelligence that determines your mood through information such as your tone of voice and the patterns of your interactions.

Adaptive and personalized experience: Artificial intelligence learns from user behavior and preferences to tailor the IoT experience to the user's needs. Consider wearable devices that make recommendations based on health goals and use intelligence to analyze past performance and health data for self-improvement.

Anomaly detection and security: Artificial intelligence ensures the security of the IoT symphony by detecting unusual patterns and potential threats in big data (Goel et al. 2023). Think of AI as a vigilant listening device that protects the network from cyberattacks by detecting malicious communication devices. Artificial intelligence algorithms optimize processes and resource allocation for optimal results and energy savings. Check out linked devices that whisper energy utilizing intelligent patterns to save energy. Artificial intelligence predicts equipment breakdown using sensor data, enabling better maintenance, and saving downtime (Rawat et al. 2018). Aircraft image sensors work with AI to schedule preventive maintenance before vital components wear out.

Automate tasks: AI can automate repetitive and complex processes, freeing up human resources for higher-level tasks. Consider smart buildings that handle ventilation and temperature to let personnel focus on guest comfort and service.

Data privacy and security issues: IoT devices must have strong data governance and security to use sensitive data responsibly through intelligence.

Access and transparency: To minimize IoT data discrepancies, AI systems must be trained on diverse and biased data. Innovation and user trust require equitable access and openness to AI-powered IoT devices and the IoT promises an exciting future. Imagine AI-powered sensors in plants that instantaneously signal food demands or cities that modify traffic based on pedestrian and vehicle data from AI algorithms. AI and the Internet of Things will tango together and make our world smarter, more efficient, and more personal. Development and ethics will determine the success of the Internet of Smart Things. Together, we can use AI to improve everyone's IoT experience and create a future where machines respect people, benefits, and expectations. We are considering embedding clever AI chips into IoT devices to better integrate them. This chip speeds up the device. IoT device bids are required. This ID will be recorded on the AI chip and can be used to log into multiple IoT products to share data. The ID helps gadgets stay integrated.

8.8 SUMMARY

In summary, the role of IoT in communication is not only to send information but also to create dynamic conversations, foster innovation, shape our relationships with the world for good, and ultimately define the future of the human–computer.

As we navigate this transition, it is important to embrace the opportunities while addressing the challenges to ensure that IoT discussions contribute to building a better future for everyone. IoT, AI, and WSN can develop strong, intelligent systems with numerous applications. IoT is a network of internet-connected physical devices that share data. Sensors, actuators, cameras, and other smart devices collect and transmit data as IoT devices. A central node or gateway receives wireless data from spatially scattered sensors that monitor physical or environmental variables. WSN sensors measure temperature, humidity, pressure, light, and more. AI creates algorithms and models to help machines execute human tasks.

The AI subsets of machine learning (ML) and deep learning (DL) train models using data to predict or decide. In real time, WSN sensors collect data from the actual environment. IoT devices can collect data from numerous sources for this network. Data is analyzed by AI systems for patterns, trends, and anomalies. ML and DL are used to forecast, classify, and optimize. The system can make intelligent decisions independently based on analysis. AI algorithms learn from fresh data and improve decision-making. AI can command IoT devices and actuators, for instance, smart home temperature control, industrial process control, and energy optimization. AI optimizes system parameters and adapts to changes. This adaptability improves smart city, agriculture, healthcare, and industrial automation efficiency and responsiveness. AI can schedule sensor activity or employ low-power modes to improve energy consumption in WSN nodes that use batteries. AI can improve IoT security by spotting irregularities and potential breaches. IoT, AI, and WSN enable intelligent systems to make autonomous decisions and optimize across domains.

REFERENCES

Akhtar N, Rahman S, Sadia H, Perwej Y. A holistic analysis of Medical Internet of Things (MIoT). *Journal of Information and Computational Science.* 2021 Apr 16;11(4), 209–222.

Attaran M. The impact of 5G on the evolution of intelligent automation and industry digitization. *J Ambient Intell Human Comput.* 2023;14(5), 5977–5993. https://doi.org/10.1007/s12652-020-02521-x.

Azzawi MA, Hassan R, Bakar KA. A review on Internet of Things (IoT) in healthcare. *International Journal of Applied Engineering Research.* 2016 Nov;11(20), 10216–10221.

Chhaya L, Sharma P, Bhagwatikar G, Kumar A. Wireless sensor network based smart grid communications: Cyber-attacks, intrusion detection system, and topology control. *Electronics.* 2017;6(1), 5.

Goel A, Goel AK, Kumar A. Performance analysis of multiple input single layer neural network hardware chip. *Multimedia Tools and Applications*. 2023a, 1–22.

Goel A, Goel AK, Kumar A. The role of artificial neural network and machine learning in utilizing spatial information. *Spatial Information Research*. 2023b;31(3), 275–285.

Gupta S, Kumar A. Study on early accurate diagnosis and treatment of COVID-19 with smartphone tracking using bionics. *Security and Privacy*. 2023, e303.

Kim J, Lee J, Kim J. Yun J. M2M service platforms: Survey, issues, and enabling technologies. *IEEE Communications Surveys and Tutorials*. 2013 Oct 23;16(1), 61–76.

Kumar S, Tiwari P, Zymbler M. Internet of Things is a revolutionary approach for future technology enhancement: A review. *Journal of Big Data*. 2019;6(1), 111. https://doi.org/10.1186/s40537-019-0268-2.

Ożadowicz A, Grela J, Wisniewski L, Smok K. Application of the Internet of Things (IoT) technology in consumer electronics-case study. In 2018. IEEE 23rd International Conference on Emerging Technologies and Factory Automation (ETFA). 2018 Sep 4 (Vol. 1, pp. 1037–1042). IEEE.

Rawat AS, Rana A, Kumar A, Bagwari A. Application of multi layer artificial neural network in the diagnosis system: A systematic review. *IAES International Journal of Artificial Intelligence*. 2018;7(3), 138.

Sairam KV, Gunasekaran N, Redd SR. Bluetooth in wireless communication. *IEEE Communications Magazine*. 2002 Jun;40(6), 90–96.

Schena E, Tosi D, Saccomandi P, Lewis E, Kim T. Fiber optic sensors for temperature monitoring during thermal treatments: An overview. *Sensors*. 2016 July 22;16(7), 1144.

Schneider S *The Industrial Internet of Things (IIoT) Applications and Taxonomy. Internet of Things and Data Analytics Handbook*. 2017 Feb 17. 41–81.

Ullo SL, Sinha GR. Advances in smart environment monitoring systems using IoT and sensors. *Sensors*. 2020 May 31;20(11), 3113.

Velsberg O, Westergren UH, Jonsson K. Exploring smartness in public sector innovation-creating smart public services with the Internet of Things. *European Journal of Information Systems*. 2020 Jul 3;29(4), 350–368.

Verma PK, Verma R, Prakash A, Agrawal A, Naik K, Tripathi R, Alsabaan M, Khalifa T, Abdelkader T, Abogharaf A. Machine-to-Machine (M2M) communications: A survey. *Journal of Network and Computer Applications*. 2016;66, 83–105, ISSN 1084-8045. https://doi.org/10.1016/j.jnca.2016.02.016.

Wu M, Luo J. Wearable technology applications in healthcare: A literature review. *Online Journal of Nursing Informatics*. 2019 Nov;23(3).

Chapter 9

Smart healthcare system using machine learning and IoT

Vaccine delivery and monitoring

Raj Gusain, Sushant Shekhar, Anurag Vidyarthi, Rishi Prakash, and R. Gowri

9.1 INTRODUCTION

The COVID-19 pandemic has triggered an unprecedented global healthcare crisis, demanding innovative and agile solutions for vaccine distribution. The urgency to vaccinate billions of individuals worldwide has underscored the need for an efficient and equitable allocation process (Basset et al. 2018, 2021; Adarsh et al. 2021; Adida et al. 2013). Traditional methods have struggled to keep pace with the scale and complexity of this endeavor, particularly in verifying recipient eligibility in real time. This chapter introduces a groundbreaking approach that leverages the synergies of embedded systems, machine learning, and the Internet of Things (IoT) to revolutionize vaccine delivery (Angeles 2005; Badhotiya et al. 2021; Bag et al. 2019; Banerjee et al. 2018; Bangura et al. 2020). By integrating real-time data streams from Twitter, a powerful platform for user-generated content, we embark on a transformative journey to swiftly and accurately identify eligible vaccine recipients. This interdisciplinary amalgamation promises to redefine how we approach healthcare and emergency response, heralding a new era in vaccine distribution.

At the heart of our approach lies the convergence of three pivotal technologies: embedded systems, machine learning, and IoT (Figure 9.1). Embedded systems, characterized by their compact size and low power consumption, have long been the backbone of numerous critical applications (Baryannis et al. 2019; Bhardwaj 2020; Carbonneau et al. 2008). In the context of vaccine delivery, they serve as the linchpin for real-time decision-making. By embedding our machine learning model within a microcontroller unit, we empower it to process incoming tweet data and promptly verify user eligibility. This seamless integration bridges the gap between virtual analysis and real-world action, ensuring that vaccines are allocated with the utmost efficiency. Machine learning, a discipline within artificial intelligence, plays a central role in our innovative solution. Specifically, a bidirectional long short-term memory (LSTM) network forms the nucleus of our predictive model. LSTMs excel in handling sequential data, making them exceptionally well-suited for processing natural language text, such as tweets. The

DOI: 10.1201/9781003510420-9

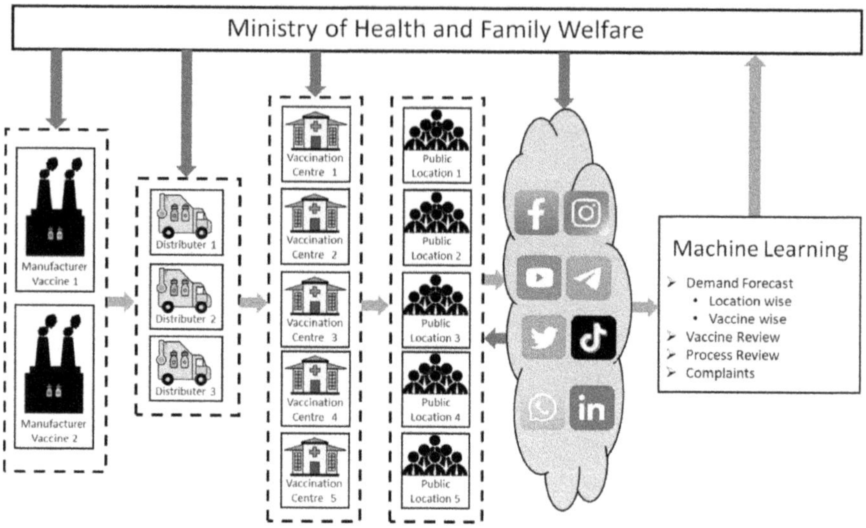

Figure 9.1 Block diagram of proposed system.

bidirectional architecture endows the model with a contextual understanding that extends both forward and backward along the sequence (Casado-Vara et al. 2018; Catalini and Gans Jul 2020; Chick et al. 2017; Chiu et al. 2008; Chowdhury et al. 2021; Ćirović et al. 2014). This comprehensive comprehension enables it to make nuanced and accurate predictions about user eligibility based on the content of their tweets. As a result, our model acts as a powerful filter, sifting through the vast ocean of information to identify relevant data points that determine vaccine allocation (Cole et al. 2019; Cui et al. 2021; Decouttere et al. 2021).

Twitter, as a social media platform, emerges as a crucial data source for our system. During the pandemic, it has become a hub for real-time discussions, updates, and information sharing related to COVID-19 and vaccination efforts. By tapping into this resource, we gain access to a rich tapestry of user-generated content (Dizbay and Ozturkoglu et al. 2020; Duijzer et al. 2018; Dwivedi et al. 2020, 2021; Eccleston-Turner and Upton Jun 2021; Edwards et al. 2021; El Baz and Ruel 2021). We employ natural language processing (NLP) techniques to parse and extract meaningful insights from these tweets. This approach not only enables us to identify potential vaccine recipients but also allows us to analyze sentiment, gauge public perception, and detect emerging trends in vaccine discourse. The IoT serves as the connective tissue that binds our system together. IoT devices, equipped with sensors and communication capabilities, facilitate seamless data flow between the microcontroller unit and the external environment. This dynamic interplay ensures that the model receives a constant stream of real-time data, enabling it to adapt and respond to evolving situations (Ellis et al. 2015; Fahimnia et al. 2018; Gautam et al. 2021; Guan et al. 2020; Gumus et al. 2010; Haidari et al. 2016; Hanson et al. 2017). Moreover, IoT technology empowers our system to scale effortlessly, accommodating diverse deployment scenarios and operational environments. The integration of embedded systems, machine learning, and IoT for vaccine delivery represents a monumental leap forward in healthcare technology. By harnessing the collective power of these technologies, we embark on a mission to optimize vaccine allocation, ensuring that doses reach those who need them most. This interdisciplinary approach not only holds the potential to reshape how we respond to crises but also sets a precedent for leveraging advanced technology in healthcare initiatives. As we navigate the complexities of the COVID-19 pandemic and look toward a future of global health challenges, this innovative solution serves as a beacon of hope and a testament to human ingenuity (Hasan et al. 2019; Hobbs 2020; Karmaker et al. 2021; Kartoglu and Milstien 2014; Kinra et al. 2020).

9.2 DATA SOURCE

Social media has evolved into a massive data repository, offering a goldmine for diverse fields ranging from business and marketing to public health and social sciences. Given the global embrace of platforms like Facebook, Twitter, and Instagram, they present a unique blend of both qualitative and quantitative insights. There are primarily two categories of data culled from social media. First, there's structured data – numerical or categorical details that lend themselves to straightforward analysis, such as likes, retweets, follower counts, and post timestamps. On the other hand, unstructured data, predominantly textual but also inclusive of images and videos, requires more intricate methods of examination. This data comprises textual content from posts and comments, as well as multimedia elements (Korpela et al. 2017; Kouzinopoulos et al. 2018; Krau et al. 2021; Lee and Trimi 2021).

Social media applications offer a plethora of benefits across various domains. For businesses, they present a robust platform for market research, enabling the analysis of consumer preferences, feedback, and trends, which aids in product development, targeted advertising, and strategic planning. Additionally, sentiment analysis of textual content helps companies gauge public feelings toward specific topics, brands, or products. Beyond business, these platforms serve as valuable tools for sociological and political research, facilitating the study of societal patterns and political views. Furthermore, in times of crisis, whether they be natural disasters or public relations blunders, monitoring social media allows organizations to detect and swiftly react. Even in the realm of public health, keeping tabs on social media posts concerning specific diseases can yield real-time data about outbreaks or even counter the propagation of misinformation (Li et al. 2017; Shekhar et al. 2021).

The dataset used in this study is sourced from Kaggle, a renowned platform for data science competitions and datasets. Specifically, the dataset is obtained from the Kaggle dataset. This dataset contains a diverse collection of tweets, each annotated with essential metadata, including keywords, locations, and tweet content. The dataset is a rich resource for training and evaluating the proposed model, as it encompasses a wide range of tweet scenarios, reflecting real-world situations.

Figure 9.2 illustrates the method of retrieving real-time tweets through the Twitter streaming API. These tweets are collected and stored in the Hadoop Distributed File System (HDFS). Flume serves as an intermediary, connecting to the Twitter streaming API and fetching tweets based on predetermined keywords set in its configuration. The acquired tweets are recorded in the JSON format and subsequently shifted to HDFS. Considering the intricate nature of tweet content, which spans various languages and detailed structures, it's essential to refine this data for in-depth

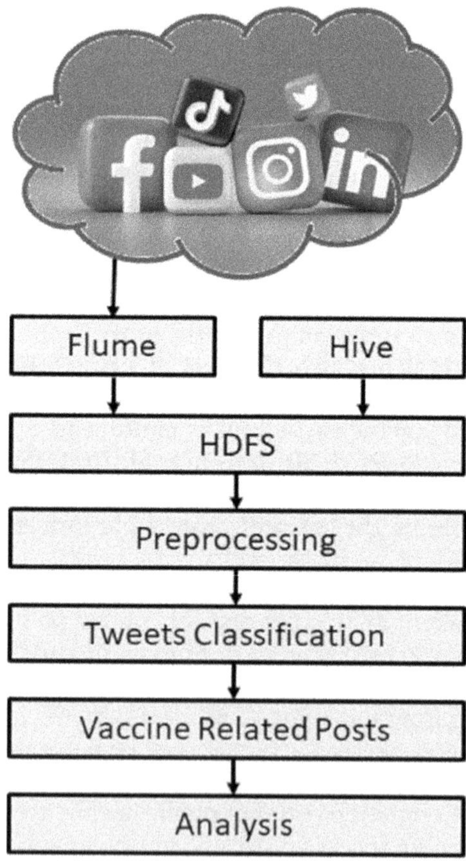

Figure 9.2 Extracting and analyzing the tweets.

analysis. Apache Hive, an integral part of Hadoop, aids this refinement process by utilizing Serializer–Deserializer (Shekhar et al. 2023; Verma et al. 2019; Virdi 2016). Through HiveQL, the system isolates tweets in English and categorizes them efficiently for the subsequent analysis engine.

9.3 DATA PREPROCESSING

Data preprocessing is a pivotal phase in the data analysis and machine learning workflow, focusing on refining raw data into a more digestible and efficient form. This involves cleaning the data to remove outliers or inaccuracies, handling missing values, normalizing and standardizing data scales, and sometimes even feature engineering to derive new attributes. These modifications aim to make the data more suitable for analysis and ensure that any subsequent modeling or analysis leverages the most relevant and clean data, leading to more accurate and trustworthy outcomes.

9.3.1 Importance of data preprocessing

(a) **Improved accuracy:** One of the main reasons data preprocessing is of paramount importance is the significant enhancement it provides to the accuracy of algorithms. Raw, unprocessed data often contains noise, outliers, and missing values. These anomalies can be misleading, causing algorithms to derive inaccurate or biased conclusions. They might prevent algorithms from capturing the true underlying patterns and relationships within the data. By preprocessing data – cleaning out the noise, treating the outliers, and imputing missing values – we present algorithms with a more accurate representation of reality. Consequently, these algorithms can then make more precise predictions and offer sharper insights.

(b) **Efficiency:** Another vital aspect of data preprocessing is the efficiency it introduces in the training process. When we talk about machine learning or data analytics, we often deal with large datasets. Dirty or unclean data can heavily burden the computational processes, causing algorithms to take longer to converge to a solution or, in some cases, not converge at all. By preprocessing and cleaning the data, irrelevant or redundant information is eliminated. This streamlined version of the dataset can significantly speed up the training process, ensuring that algorithms run smoothly and in a more timely fashion.

(c) **Interpretability:** Beyond just accuracy and efficiency, data preprocessing enhances the interpretability of results. Raw data, being messy and unstructured, can yield results that are hard to understand or insights that are obscured by the data's inherent noise. However, when data is well-prepared and preprocessed, the results derived from it tend to be clearer. This clarity makes it easier for data scientists, stakeholders, and decision-makers to interpret the patterns, insights, and relationships that emerge from the data. In the end, better interpretability means that the insights derived from the data can be effectively communicated and can lead to informed decisions.

9.3.2 Common data preprocessing techniques

(a) **Data cleaning:** Data cleaning, also known as data cleansing or preprocessing, is essential for enhancing data quality by identifying and rectifying errors and inconsistencies. This process involves tasks like removing duplicates, filling in missing values, correcting inaccuracies, and standardizing formats, ensuring the data is primed for analysis and provides valid insights. Properly cleaned data ensures accurate outcomes, optimized decision-making, and resource efficiency in data-driven operations. Common challenges include missing values, which can arise for various reasons and are addressed using techniques such as imputation, leveraging algorithms like XGBoost, or predictive filling. Another concern is outliers, which can skew analyses. Techniques like the interquartile range (IQR) method, Z-score calculations, and visualization tools help detect and manage such deviations, ensuring data reflects genuine patterns and trends.

(b) **Data transformation:** Data transformation encompasses various techniques to modify data for improved processing and analysis. Among these techniques, normalization rescales features to fit within a specific range, typically [0, 1], benefiting algorithms that are sensitive to feature magnitudes. Standardization, on the other hand, adjusts features to possess a mean of 0 and a variance of 1, catering to algorithms that presuppose features are centered on zero with comparable variances. Additionally, log transformation is employed to stabilize variance and render the data more aligned with a normal distribution.

(c) **Feature engineering:** Feature engineering and data reduction are critical aspects of the data preprocessing pipeline in machine learning. The process of transforming continuous variables into discrete bins, known as binning, allows for better visualization and sometimes improved model performance by categorizing continuous data. Another method, one-hot encoding, converts categorical variables into a numerical format, ensuring machine learning algorithms can utilize them more effectively. In scenarios where the dimensionality of the dataset is large and perhaps cumbersome, feature extraction techniques like PCA (principal component analysis) come into play. PCA can distill the essence of the original features into a smaller set, preserving as much variance and information as possible. Furthermore, sometimes it's not just about tweaking the existing features but about creating new ones that provide additional insight or fill gaps in the original dataset, which is where feature creation is beneficial. On the other hand, dimensionality reduction techniques like t-SNE (t-distributed stochastic neighbor embedding) and LDA (linear discriminant analysis) are also powerful tools for preserving information while reducing the feature space. For visual representation and understanding of the distribution of continuous data, histograms provide a snapshot of the frequency distributions, giving an overview of the data's spread and central tendencies. In sum, a comprehensive approach to data preprocessing integrates various methods to prepare data, ensuring that the subsequent machine learning models are both efficient and effective.

(d) **Handling imbalanced data:** In the domain of data analysis and machine learning, one of the prominent challenges that experts often confront is imbalanced data distribution, where the classes of the target variable are not equally represented. Imbalanced data can skew the predictions of a model, frequently resulting in a bias toward the majority class and diminishing the predictive power for the minority class. To combat this, several strategies are employed. First, upsampling can be utilized wherein the number of instances in the minority class is artificially increased to match the majority class, often by duplicating the existing instances. Conversely, downsampling involves reducing the number of instances in the majority class to balance it with the minority class, albeit at the risk of the potential loss of information. Another innovative approach to tackle this problem is through the generation of synthetic data. For instance, the Synthetic Minority Over-Sampling Technique (SMOTE) creates synthetic instances of the minority class by interpolating between existing ones. By doing so, it aids in enhancing the diversity of the minority class data without resorting to mere replication. However, it's essential to consider that each of these techniques has its advantages and disadvantages, and the choice of method should be predicated on the specific nature and needs of the dataset at hand. Moreover, sometimes combining multiple methods or incorporating domain-specific knowledge can provide better results than relying on a single technique.

(e) **Data splitting:** Data splitting is a crucial step in the process of building and validating machine learning models, ensuring that models have not only been trained adequately but also have the capacity to generalize well to new, unseen data. The primary objective behind data splitting is to partition a dataset into distinct subsets, typically comprising a training set, a validation set, and a test set. The training set is used to train the model and adjust its parameters. It forms the foundation upon which a machine learning model learns the patterns and relationships inherent within the data. The validation set, on the other hand, is instrumental in tuning hyperparameters, selecting among different model architectures, and preventing overfitting. It provides a sandbox for model evaluation during the training phase without risking information leakage from the test set. The test set, which remains untouched during the training phase, serves the ultimate purpose of gauging a model's performance in the real world. This separation ensures an unbiased evaluation, allowing practitioners to assess how well the model is likely to perform on entirely new and unobserved data. It's important to note that the proportions in which data is split can vary based on the size of the dataset and the specific problem context, but a common approach is the 70–15–15 or 80–10–10 split for training, validation, and testing, respectively. Additionally, techniques like k-fold cross-validation further refine this process by repeating the splitting multiple times, providing a more robust evaluation of model performance. Proper data splitting not only bolsters the confidence in a model's predictions but also lays down a structured pathway for iterative model improvement.

Data preprocessing is a crucial, yet often overlooked, aspect of the data analysis pipeline (Figure 9.3). Investing time in this step ensures that the

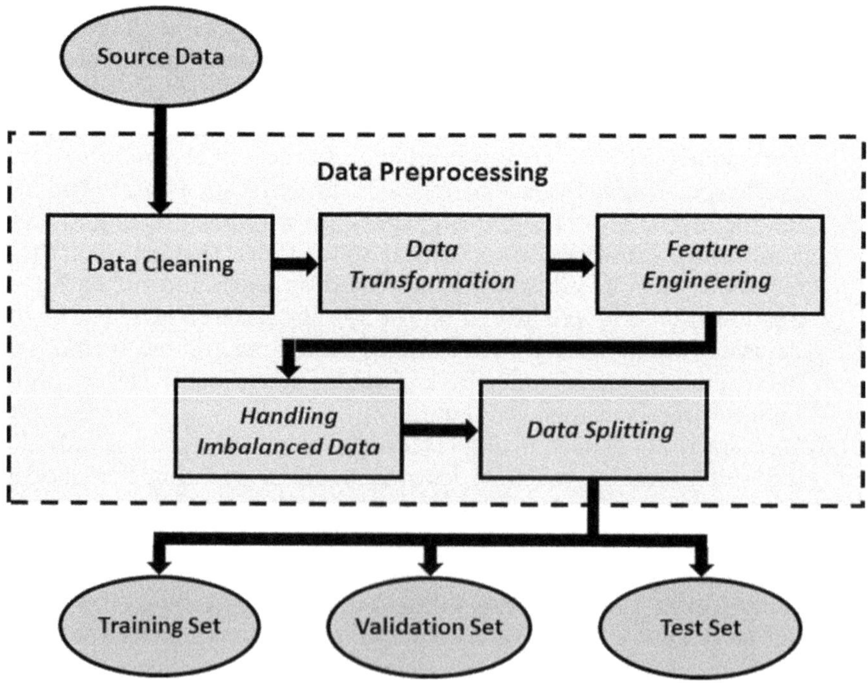

Figure 9.3 Block diagram of data preprocessing.

data fed to analytical models is of high quality, leading to more accurate and meaningful outcomes. Properly preprocessed data can save time in the modeling stage, reduce the chances of arriving at misleading conclusions, and increase the overall effectiveness of the analysis or machine learning task at hand.

9.4 METHODOLOGY

The development and evaluation of the model for vaccine delivery and monitoring involves a systematic approach, encompassing data preparation, model construction, and evaluation metrics. This section delineates the steps undertaken to create an accurate and reliable predictive system.

9.4.1 Data preparation

Before feeding data, especially textual data, into a machine learning model or any analytical algorithm, it's essential to prepare and preprocess it to ensure maximum efficacy of the model. Data preparation can involve several processes to clean and format the data, making it easier for models to discern patterns and make predictions (Figure 9.4).

(a) **Lowercasing:** Textual data, especially from sources like social media, can have varying cases due to user input methods, personal choices, or platform-specific rules. Having different cases (e.g., "HELLO," "Hello," "hello") can lead the model to mistakenly consider them as different words. By converting all characters in the text to lowercase, you eliminate the potential redundancy and discrepancies arising from character cases. This ensures that the words "Hello" and "hello" are treated the same.

(b) **Removing special characters:** In many textual analyses, especially sentiment analysis, the presence of special characters and punctuation

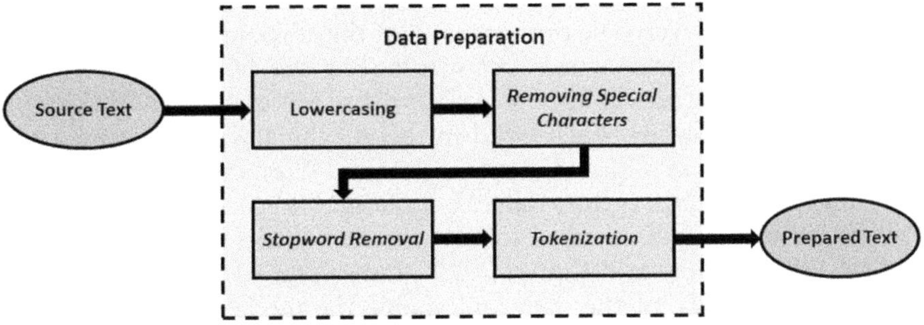

Figure 9.4 Block diagram of data preparation.

like @, #, and ! can be considered noise. They may not necessarily convey sentiment by themselves and can clutter the input data. By removing these characters, you are narrowing down the input to only those components that have meaningful semantic content, making it easier for models to discern patterns.

(c) **Stopword removal**: Stopwords are commonly occurring words in a language that don't provide significant information on their own. In English, words like "and," "the," and "is" can be considered stopwords. While they're essential for sentence construction, they often don't carry meaningful sentiment by themselves. Removing stopwords can reduce the dimensionality of the data, making computational tasks faster and more efficient. Also, it helps in focusing on words that genuinely carry sentiment or meaning relevant to the analysis.

(d) **Tokenization**: Tokenization involves breaking down text into smaller pieces, typically words or subwords. In the context of tweets or any text, tokenization usually means splitting the text into individual words. Once the text is tokenized, each token (or word) can be analyzed individually, making it feasible for models to understand and learn patterns. Tokenization also prepares the text for further steps like stemming, lemmatization, or embedding, which might be essential for advanced analyses.

These data preparation steps are pivotal to refine the raw data into a structured format, making it more digestible and interpretable by machine learning models. These processes also help in eliminating potential noise, ensuring that the algorithms work on clean and meaningful data, leading to better and more accurate outcomes.

9.4.2 Model construction

The section talks about constructing a deep learning model for a certain task, likely text classification based on the context (Figure 9.5). Here's a more detailed explanation of each component:

(a) **Embedding layer**: The primary goal of the embedding layer is to convert sparse representations of words (like one-hot encoded vectors) into dense vectors of fixed size. This dense representation can capture semantic meaning. Each word in the vocabulary is represented as a dense vector in a high-dimensional space. Words with similar meanings tend to have vectors that are closer in this space. During training, these vectors get adjusted to capture the contextual relationships in the data (Gupta and Kumar 2023). Dense representations reduce the dimensionality of the data and enable the model to learn semantic relationships between words. For example, "king" and "queen" might have vectors that are closer together compared to "king" and "apple."

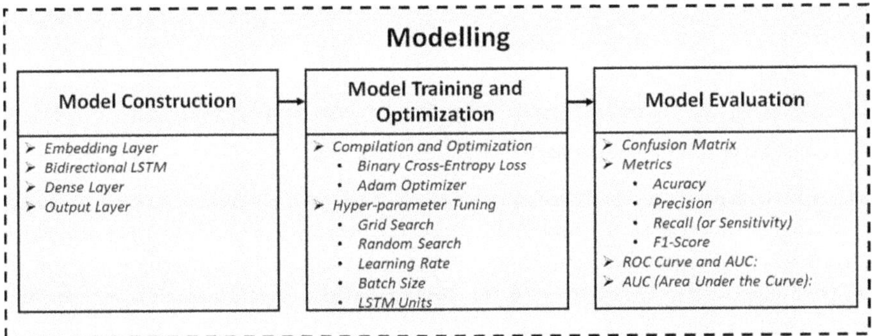

Figure 9.5 Block diagram of model construction.

(b) **Bidirectional LSTM:** This captures information from both past (left-to-right) and future (right-to-left) states in a sequence. Traditional LSTMs process data from the start to the end of a sequence. Bidirectional LSTMs, on the other hand, process data in both directions with two separate internal memories, and then merge their outputs. This is particularly useful for textual data, as the meaning of a word can be dependent on its subsequent words. For example, in the sentence "He went to Paris and visited the Eiffel _____," knowing that "Tower" is likely to fill the blank becomes clearer after reading the entire sentence.

(c) **Dense layer:** To further process the features extracted by the LSTM layers and make them suitable for the final prediction, a dense layer, also known as a fully connected layer, connects each input to each output within its layers. It takes the outputs of the LSTM layer (which are sequences) and transforms them into a fixed-size vector by applying a linear transformation followed by a nonlinear activation function. The dense layer can learn patterns across the different features extracted by the LSTMs, making the model more expressive.

(d) **Output layer:** This layer makes the final prediction. Given the context, it seems the task is binary classification, e.g., predicting whether a tweet is related to a vaccine or not. This layer contains a single neuron with a sigmoid activation function. The sigmoid function outputs a value between 0 and 1, which can be interpreted as the probability of the input belonging to the positive class. The simplicity of this layer makes it suitable for binary classification tasks. By setting a threshold (e.g., 0.5), the model's output can be converted to either of the two classes (e.g., vaccine or not).

This model aims to classify textual data (like tweets) by first converting words into dense vectors (embedding layer), capturing sequential relationships in both directions (bidirectional LSTM), refining the features (dense layer), and finally making a binary classification decision (output layer).

9.4.3 Model training and optimization

Model training and optimization refers to the process of teaching a machine learning or deep learning model to make predictions on data and adjusting it to make those predictions as accurate as possible. This involves two main phases: the actual training and the optimization of the model's parameters and hyperparameters.

9.4.3.1 Compilation and optimization

Compilation is the process of configuring the model for training. This includes defining the loss function, the optimizer, and evaluation metrics.

(a) **Binary cross-entropy loss:** Since the context mentions binary classification, the binary cross-entropy loss is appropriate. In binary classification, there are only two possible outcomes. The binary cross-entropy loss measures the difference between the actual and predicted probabilities for the positive class. Lower values of this loss signify a better model.

(b) **Adam optimizer:** The choice of an optimizer plays a crucial role in model training. Optimizers are algorithms used to adjust internal parameters (like weights) to minimize the loss. The Adam optimizer, short for Adaptive Moment Estimation, is a popular choice as it combines the advantages of two other extensions of stochastic gradient descent: AdaGrad and RMSProp. It's known for achieving good results fast.

9.4.3.2 Hyperparameter tuning

Hyperparameters are parameters that are not learned from the data. Instead, they are set prior to the commencement of the training process and govern the overall behavior of the model. To achieve the best possible model performance, we need to choose the right set of hyperparameters.

(a) **Grid search:** This is a brute-force approach to hyperparameter tuning. It involves trying every possible combination of hyperparameters from a predefined list. For instance, if learning rates of 0.001, 0.01, and 0.1 and batch sizes of 32, 64, and 128, the grid search will try every possible combination, leading to a total of $3 \times 3 = 9$ different combinations.

(b) **Random search:** Rather than trying every combination like grid search, a random search samples hyperparameters from a distribution over a fixed number of iterations. It might not be as thorough as a grid search, but it can be quicker and, in many cases, find a better or equally good combination in less time.

(c) **Learning rate:** This hyperparameter determines the step size at each iteration while moving toward a minimum of the loss function. Too high, and you might overshoot the optimal value; too low, and training might be very slow or get stuck.

(d) **Batch size:** Refers to the number of training examples utilized in one iteration. A smaller batch size often provides a regularizing effect and lower generalization error, while a larger batch size speeds up the learning process.

(e) **LSTM units:** If the model uses LSTM layers, the number of LSTM units determines the capacity of the model or the width of the memory cell. More units allow the model to remember more information but also increase computational complexity and risk of overfitting.

Model training and optimization is a two-fold process. First, compile the model with a suitable loss function and optimizer. Then, fine-tune its hyperparameters using techniques like grid search and random search to ensure the best performance.

9.4.4 Model evaluation

Evaluating a model helps us understand its effectiveness and areas of improvement. There are several metrics and tools available to evaluate the performance of classification models, some of which are discussed next.

(a) **Confusion matrix:** A confusion matrix is a table used to describe the performance of a classification model. It contains the following four components:

    i.   True positive (TP): Actual positive correctly predicted as positive.

    ii.   True negative (TN): Actual negative correctly predicted as negative.

    iii.   False positive (FP): Actual negative incorrectly predicted as positive.

    iv.   False negative (FN): Actual positive incorrectly predicted as negative.

From the confusion matrix, we can derive metrics like precision, recall, and accuracy.

(b) **Metrics**

    i.   *Accuracy* is the ratio of the number of correct predictions to the total number of input samples.

$$\text{Accuracy} = \frac{\text{Number of correct predictions}}{\text{Total number of predictions}}$$

While accuracy is a simple and intuitive metric, it might not be a good indicator for imbalanced datasets where one class significantly outnumbers the other.

ii. *Precision* evaluates the correctness of positive predictions.

$$\text{Precision} = \frac{\text{True Positives}}{\text{True Positives} + \text{False Positives}}$$

High precision means that there is a low rate of false-positive prediction. It essentially answers: "Of all the positive predictions, how many were correct?"

iii. *Recall* (or *sensitivity*) measures the model's ability to correctly identify actual positives.

$$\text{Recall} = \frac{\text{True Positives}}{\text{True Positives} + \text{False Negatives}}$$

High recall indicates that the model correctly identifies a high percentage of the positive instances. It answers: "Of all the actual positives, how many did the model correctly predict?"

iv. *F1-score* is the harmonic mean of precision and recall and provides a single score that balances both the concerns of precision and recall.

$$\text{F1} = 2 \times \frac{\text{Precision} \times \text{Recall}}{\text{Precision} + \text{Recall}}$$

In scenarios where there's an uneven class distribution, the F1-score might be a better metric than accuracy.

(c) **ROC curve (receiver operating characteristic curve):** The ROC curve is a graphical plot that illustrates the diagnostic ability of a binary classifier system as its discrimination threshold varies. The curve is plotted with the true positive rate (recall) against the false-positive rate.

(d) **AUC (area under the curve):** This is the area underneath the ROC curve. An AUC of 1 indicates perfect classification, while an AUC of 0.5 indicates that the classifier is no better than random guessing. The AUC provides an aggregate measure of the model's performance across all possible classification thresholds. A higher AUC indicates better model performance.

Overall, the model evaluation helps us assess the reliability and robustness of our model, ensuring that it functions well not just on training data but also on unseen data.

Table 9.1 Performance analysis

Parameter	Value
Accuracy	95.2%
Precision	94.8%
Recall	95.7%
F1-score	95.2%
AUC	94%

9.5 RESULTS OF PROPOSED MODEL

The developed model exhibits remarkable performance in classifying vaccine-related tweets. The model is evaluated based on various metrics to comprehensively assess its effectiveness, as shown in Table 9.1 and Table 9.2.

The system achieves an impressive accuracy of 95.2%, indicating its proficiency in correctly classifying vaccine-related tweets. The model exhibits high precision at 94.8%, signifying the low rate of false positives. With a recall of 95.7%, the system effectively identifies a significant portion of actual vaccine-related tweets. The F1-score, a harmonic mean of precision and recall, stands at 95.2%. This metric confirms the system's balance between precision and recall. The ROC curve exhibits an AUC value of 0.94, highlighting the system's discriminatory power. This metric emphasizes the model's ability to minimize both false positives and false negatives.

The results underscore the efficacy of the proposed model in accurately identifying vaccine-related tweets. The high precision and recall values validate the system's capacity to minimize erroneous classifications, ensuring that emergency response teams receive reliable information. The substantial AUC value further affirms the model's capability to distinguish between true positives and false positives. These findings have profound implications for vaccine-related responses and management. By automating tweet categorization, emergency response teams can promptly assess the severity of a vaccine crisis, enabling a more coordinated and efficient response. The public also benefits, as accurate information dissemination becomes a priority during critical times.

Moreover, the high-performance metrics lay the groundwork for further advancements. As the model undergoes continuous refinement and

Table 9.2 Confusion matrix

	Predicted negative	Predicted positive
True negative	1410	90
True positive	130	1370

augmentation, its capabilities are poised to improve, further enhancing vaccine-related response efforts. In conclusion, the proposed model's exceptional performance, as evidenced by its high accuracy, precision, recall, and AUC values, marks a significant advancement in this technology. This system is poised to become an indispensable tool for emergency response teams globally, augmenting their ability to manage a vaccine crisis effectively and efficiently.

9.6 EMBEDDED SYSTEM INTEGRATION

In order to operationalize the proposed model for real-time verification and decision-making on vaccine allocation, a crucial step involves its integration into a microcontroller unit. This integration is pivotal in enabling swift and autonomous decision-making based on the predictions generated by the model. The microcontroller unit serves as the central processing hub, where the model's computations are executed efficiently and rapidly.

Furthermore, to ensure a seamless and uninterrupted flow of information, the integrated system is interfaced with IoT devices. These IoT devices act as the conduit for data exchange between the embedded system and the broader network infrastructure. By leveraging IoT technology, the system gains the capability to collect and transmit relevant information in real time, thereby facilitating timely decision-making processes. One key advantage of this integration is the immediate response capability it affords. When a tweet is received, the embedded system processes it through the integrated model within milliseconds, swiftly determining its vaccine-related classification. This rapid assessment ensures that crucial decisions regarding vaccine allocation are made promptly, optimizing the allocation process. Moreover, the integrated system maintains a continuous connection with the central database, enabling it to retrieve updated information and adapt to evolving circumstances. This dynamic interaction with the database ensures that the model is equipped with the latest data, enhancing its accuracy and reliability in classifying vaccine-related tweets.

The integration of the proposed model into an embedded system, coupled with its seamless connection to IoT devices, marks a significant advancement in this technology. It not only streamlines the decision-making process but also enables a more agile and responsive approach to vaccine allocation, particularly during critical situations. The manufacturer and vaccination centers are already equipped with communication devices. The vaccine-related data is always logged and shared with the authorities. The requirement is just to integrate the IoT devices and include the embedded system. On the other hand, the vaccine containers and the vaccine van have to be upgraded with smart facilities, so that authorities can manage and monitor the complete system. This integration is poised to revolutionize vaccine crisis response efforts, allowing for more efficient and effective allocation of resources to those in need. By combining the power of machine learning, embedded systems, and IoT technology, this integrated system represents a formidable tool in vaccine crisis management.

9.6.1 Smart container for vaccines

A smart container designed for transporting vaccines must ensure the safe and efficient delivery of the vaccines without compromising their potency (Figure 9.6). Here are some facilities and features that a smart container for vaccines should have:

(a) **Robust physical design:** The smart container for vaccines boasts a robust physical design. Its durable construction ensures maximum protection against any physical damage, making it resilient in various conditions. Additionally, the container is designed to be both water-proof and dustproof, ensuring the vaccines remain uncompromised and safe in any environment.

(b) **Temperature control and monitoring:** Smart containers for vaccines incorporate advanced temperature control and monitoring features. These containers are equipped with active cooling systems that ensure specific temperature ranges are maintained, which are crucial for pre-serving the vaccine's viability. To ensure constant oversight, they come integrated with sensors that offer real-time temperature monitoring.

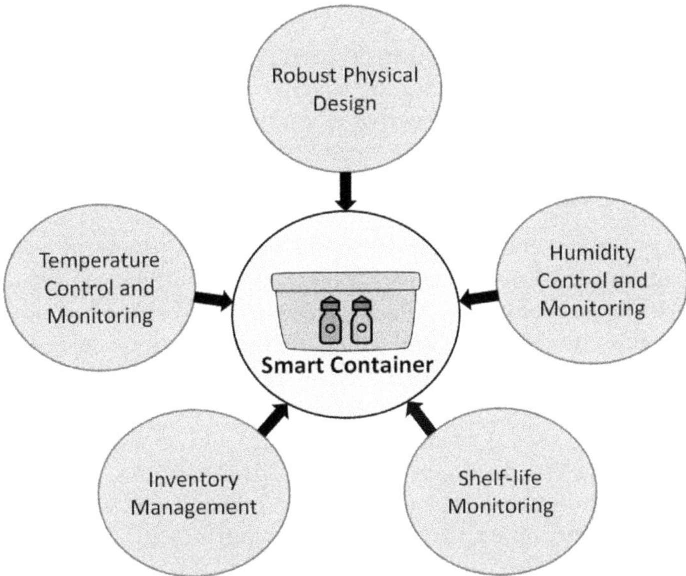

Figure 9.6 Smart container diagram of data preparation.

Moreover, for added accountability and quality assurance, these containers have a temperature logging function that records temperature data over time. Furthermore, should there be any deviation from the designated temperature range, an alert system is in place to promptly notify the concerned personnel.

(c) **Humidity control and monitoring:** The smart container for vaccines comes equipped with advanced humidity control and monitoring features, which are crucial for ensuring the stability of certain vaccines. Just as with temperature controls, this system provides real-time data monitoring and has logging capabilities to ensure the vaccines are maintained under optimal conditions.

(d) **Inventory management:** Smart containers for vaccines incorporate advanced inventory management features. These containers are equipped with sensors or systems that deliver real-time data on the quantities of vaccines, ensuring timely replenishments and effective distribution. Additionally, each vaccine batch within the container is labeled with RFID tags or barcodes. This not only ensures traceability but also aids in the accurate tracking of individual batches, ensuring safety and authenticity throughout the distribution chain.

(e) **Shelf-life monitoring:** The smart container for vaccines incorporates advanced shelf-life monitoring systems. These systems meticulously track the expiration dates of the vaccines stored within. As an added safeguard, the container is designed to send out alerts when any vaccine is approaching its expiration date, ensuring that the integrity and efficacy of the vaccines are maintained.

Incorporating these features into a smart container would significantly enhance the safety and efficacy of vaccine transportation, ensuring that vaccines remain potent and effective when they reach their destination.

9.6.2 Smart van for vaccines

A smart van for vaccines, designed for the transportation and potential on-the-spot administration of vaccines, plays a pivotal role in public health. By merging the latest technological advances with cold chain best practices, it can ensure the safe, efficient, and timely delivery of vaccines. Following are the indispensable features of such a sophisticated vehicle.

(a) **Temperature management:** Ensuring the right temperature is essential, especially when it comes to transporting vaccines where consistent cooling can be the difference between efficacy and waste. Addressing this need, the smart vaccine van incorporates IoT sensors that continuously monitor and document the van's internal temperature. This not only offers real-time temperature insights but also records historical data, helping businesses identify and study any inconsistencies. For added safety, if temperatures sway outside the optimal range, immediate alerts are sent to the responsible team. Such rapid alerts empower prompt actions, reducing risks of vaccine spoilage. Additionally, backup cooling systems stand by to jump into action, guaranteeing an uninterrupted cold chain even if the main cooling mechanism malfunctions. With these integrated features, the smart vaccine van ensures vaccines are transported in a secure, timely, and highly dependable environment.

(b) **Navigation and route management:** For the smart van transporting vaccines, navigation and route management encompasses a suite of advanced tools and systems. Central to this is real-time GPS tracking, offering immediate insights into the van's current position. This allows for continuous monitoring of the van's journey, ensuring it remains on its planned path and reduces the chances of detours or potential risks. Enhancing this capability is route optimization. Instead of strictly adhering to a predetermined route, this feature leverages sophisticated data analytics to identify the most efficient path, considering elements like traffic, ongoing construction, and other possible obstructions. This leads to a journey that's not only safer but also faster and more resource-efficient (Figure 9.7).

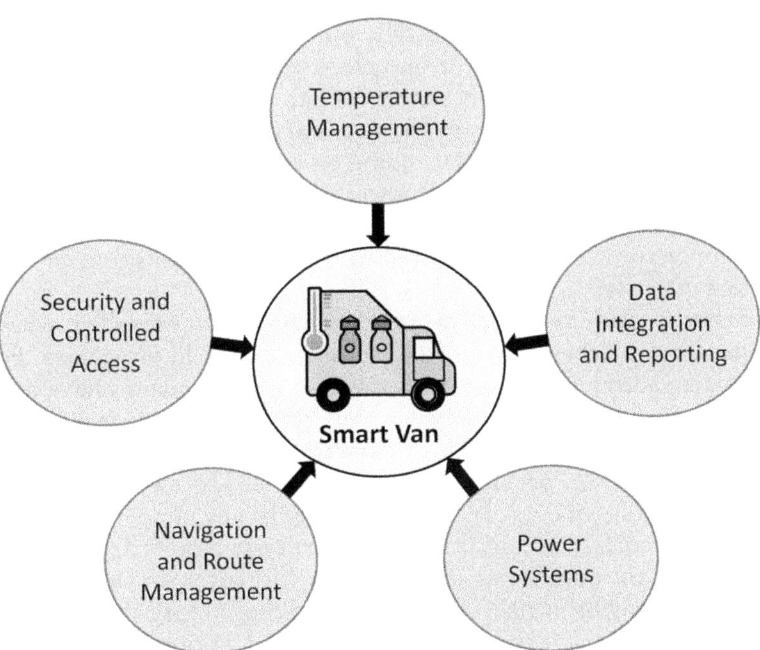

Figure 9.7 Smart navigation diagram of data preparation.

(c) **Data integration and reporting:** Smart vans equipped for vaccine transportation incorporate data integration and reporting, essential in today's digital era, to monitor vital parameters throughout transit. These vans utilize digital records to document crucial data, such as temperature, humidity, and the exact travel route. This diligent recording provides thorough insight into the conditions endured during transportation. Further enhancing this system is cloud connectivity, allowing for immediate data synchronization. As data is recorded, it's swiftly uploaded to a cloud database. This not only safeguards the data through redundancy but also guarantees anytime, anywhere access for timely review. In sum, these features present a comprehensive solution for the efficient collection, storage, and analysis of transportation data.

(d) **Power systems:** In the evolving landscape of healthcare logistics, ensuring the safe and consistent transportation of vaccines is of paramount importance. Smart vaccine vans have emerged as a critical tool for this purpose. Central to their operation is a robust power system that guarantees the integrity of the vaccines during transit. One of the most crucial features of these vans is their redundant power systems. These systems serve as fail-safes, ensuring that even in the event of a primary power outage or failure, the van's operations are not compromised. Backup systems can range from additional battery reserves to harnessing renewable energy sources like solar panels. Solar panels, for instance, can be installed on the roof of the van, absorbing sunlight and converting it into electricity. This not only provides an environmentally friendly power backup but also ensures that the van can operate and maintain the necessary refrigeration levels for prolonged periods without depending on conventional power sources. By integrating such redundant power options, these smart vaccine vans can ensure that the vaccines they transport remain at their required temperatures and are delivered to their destinations uncompromised and effective.

(e) **Security and controlled access:** Security and controlled access are paramount when transporting sensitive goods like vaccines. For the smart vaccine van, several advanced security measures have been put in place to ensure the safety and integrity of the vaccines during transportation. First, surveillance cameras are installed both inside and outside the van, providing a continuous feed to monitor the vehicle's surroundings and its cargo. Should there be any tampering or attempts at unauthorized access, immediate alert systems are triggered to notify the relevant authorities or personnel. Moreover, in situations where the van may be at risk, there's the capability to lock the van remotely, ensuring its contents remain untouched. Additionally, to gain entry into the van's vaccine storage areas, an individual must pass through biometric or RFID security systems, ensuring that only authorized personnel can access the vaccines. This multilayered approach guarantees that the vaccines are transported in the most secure environment possible.

9.7 CONCLUSION

The development and integration of the smart healthcare system, coupled with its embedding into a microcontroller unit and interface with IoT devices, represents a significant advancement in this technology. Through this comprehensive system, we have demonstrated the potential to revolutionize the way vaccine-related information is processed, classified, and responded to in real time. The model's performance, with an accuracy of 95.2% and a robust ROC curve exhibiting an AUC value of 0.94, underscores its efficacy in accurately identifying vaccine-related tweets. This level of precision is paramount in ensuring that critical resources, such as vaccines, are allocated to the most urgent situations, optimizing response efforts. The integration of the smart healthcare system into an embedded system brings the power of machine learning directly into the heart of vaccine-related response operations. This enables the system to autonomously and swiftly process incoming tweets, making instantaneous decisions on their vaccine-related classification. The immediate response capability of the embedded system significantly reduces the time lag between information receipt and action, thus enhancing the efficiency of response efforts. The interface with IoT devices further enhances the system's capabilities by establishing a seamless flow of data. This connectivity ensures that the embedded system is continuously updated with the latest information, allowing it to adapt to evolving circumstances and make informed decisions based on the most current data available. In practical terms, this integrated system holds the potential to save lives and resources during critical moments. By accurately identifying vaccine-related tweets and swiftly allocating resources, we can improve the overall effectiveness of the smart healthcare system. Additionally, the system's adaptability and real-time decision-making capabilities make it a valuable tool for handling dynamic and rapidly evolving situations.

9.8 FUTURE ENHANCEMENTS

While the smart healthcare system has demonstrated significant potential in revolutionizing vaccine crisis management, there are several avenues for further enhancement and refinement like expanding the system's capabilities to process and classify tweets in multiple languages would increase its applicability on a global scale. This would be particularly beneficial in regions where vaccine-related information may be communicated in languages other than English. Incorporating a temporal analysis component would enable the system to assess the urgency of vaccine-related tweets based on their timestamp. This would allow for prioritization of response efforts in rapidly evolving situations. Enhancing the model's ability to understand context and sarcasm in tweets would further improve its accuracy. This could involve incorporating natural language processing techniques and sentiment analysis to gain deeper insights into the meaning behind tweets. Establishing direct integration with emergency services and response teams would enable the system to trigger immediate action based on its classifications. This could involve automated alerts, resource allocation, or even coordinating with first responders. Augmenting the system with geospatial analysis capabilities would allow for location-specific response. By pinpointing the geographical context of tweets, the system could direct resources to areas most in need. Continual refinement of the machine learning model through techniques like hyperparameter tuning, ensembling, and exploring advanced model architectures can further improve accuracy and response time. Implementing a feedback loop mechanism that allows human operators to verify the system's classifications and provide corrective input would serve as a valuable tool for ongoing model training and validation. Ensuring the system is capable of handling a high volume of incoming tweets during peak vaccine crisis is crucial. This may involve cloud-based solutions, load balancing, and redundancy measures to guarantee uninterrupted service.

REFERENCES

Abdel-Basset M., Chang V., Nabeeh N.A. An intelligent framework using disruptive technologies for COVID-19 analysis. *Technological Forecasting and Social Change*. 2021.

Abdel-Basset M., Manogaran G., Mohamed M. Internet of Things (IoT) and its impact on supply chain: A framework for building smart, secure and efficient systems. *Future Generation Computer Systems-The International Journal of Escience*. 2018, September.

Adarsh S., Joseph S.G., John F., Lekshmi M., Asharaf S. A transparent and traceable coverage analysis model for vaccine supply-chain using blockchain technology. *IT Professional*. 2021.

Adida E., Dey D., Mamani H. Operational issues and network effects in vaccine markets. *European Journal of Operational Research*. 2013.

Angeles R. RFID technologies: Supply-chain applications and implementation issues. *Information Systems Management*. 2005.

Badhotiya G.K., Sharma V.P., Prakash S., Kalluri V., Singh R. Investigation and assessment of blockchain technology adoption in the pharmaceutical supply chain. *Materials Today: Proceedings*. 2021.

Bag S., Tiwari M.K., Chan F.T.S. Predicting the consumer's purchase intention of durable goods: An attribute-level analysis. *Journal of Business Research*. 2019, January.

Banerjee M., Lee J., Choo K.-K.-R. A blockchain future for internet of things security: A position paper. *Digital Communications and Networks*. 2018.

Bangura J.B., Xiao S., Qiu D., Ouyang F., Chen L. Barriers to childhood immunization in sub-Saharan Africa: A systematic review. *BMC Public Health*. 2020.

Baryannis G., Dani S., Antoniou G. Predicting supply chain risks using machine learning: The trade-off between performance and interpretability. *Future Generation Computer Systems*. 2019.

Bhardwaj S.C. Satellite navigation and sources of errors in positioning: A review. International Conference on Advances in Computing, Communication & Materials (ICACCM), Dehradun, India. 2020, 43–50.

Carbonneau R., Laframboise K., Vahidov R. Application of machine learning techniques for supply chain demand forecasting. *European Journal of Operational Research*. 2008.

Casado-Vara R., Prieto J., De la Prieta F., Corchado J.M. How blockchain improves the supply chain: Case study alimentary supply chain. *Procedia Computer Science*. 2018.

Catalini C., Gans J.S. J. Some simple economics of the blockchain. *Communications of the ACM*. 2020.

Chick S.E., Hasija S., Nasiry J. Information elicitation and influenza vaccine production. *Operations Research*. 2017.

Chiu R.K., Chang C.M., Chang Y.C. A forecasting model for deciding annual vaccine demand. 4th International Conference on Natural Computation (ICNC 2008), Jian, China. 2008, October 18–20.

Chowdhury P., Paul S.K., Kaisar S., Moktadir M.A. COVID-19 pandemic related supply chain studies: A systematic review. *Transportation Research: Part E: Logistics and Transportation Review*. 2021.

Ćirović G., Pamučar D., Božanić D. Green logistic vehicle routing problem: Routing light delivery vehicles in urban areas using a neuro-fuzzy model. *Expert Systems with Applications*. 2014.

Cole R., Stevenson M., Aitken J. Blockchain technology: Implications for operations and supply chain management. *Supply Chain Management-An International Journal*. 2019, June.

Cui L., Xiao Z., Wang J., Chen F., Pan Y., Dai H., Qin J. Improving vaccine safety using blockchain. *ACM Transactions on Internet Technology (TOIT)*. 2021.

Decouttere C., De Boeck K., Vandaele N. Advancing sustainable development goals through immunization: A literature review. *Globalization and Health*. 2021.

Dizbay I.E., Öztürkoğlu Ö. Determining significant factors affecting vaccine demand and factor relationships using fuzzy DEMATEL method. International Conference on Intelligent and Fuzzy Systems. 2020.

Duijzer L.E., Van Jaarsveld W., Dekker R. Literature review: The vaccine supply chain. *European Journal of Operational Research*. 2018.

Dwivedi Y.K., Hughes D.L., Coombs C., Constantiou I., Duan Y.Q., Edwards J.S., Upadhyay N. Impact of COVID-19 pandemic on information management research and practice: Transforming education, work and life. *International Journal of Information Management*. 2020, December.

Dwivedi Y.K., Hughes L., Ismagilova E., Aarts G., Coombs C., Crick T., Williams M.D. Artificial Intelligence (AI): Multidisciplinary perspectives on emerging challenges, opportunities, and agenda for research, practice and policy. *International Journal of Information Management*. 2021, April.

Eccleston-Turner M., Upton H. International collaboration to ensure equitable access to vaccines for COVID-19: The ACT-accelerator and the COVAX facility. *Milbank Quarterly*. 2021, June.

Edwards B., Gloor C.A., Toussaint F., Guan C., Furniss D. Human factors: The pharmaceutical supply chain as a complex sociotechnical system. *International Journal of Quality in Health Care*. 2021.

El Baz J., Ruel S. Can supply chain risk management practices mitigate the disruption impacts on supply chains' resilience and robustness? Evidence from an empirical survey in a COVID-19 outbreak era. *International Journal of Production Economics*. 2021.

Ellis S., Morris H.D., Santagate J. *IoT-Enabled Analytic Applications Revolutionize Supply Chain Planning and Execution*. International Data Corporation (IDC) White Paper. 2015.

Fahimnia B., Jabbarzadeh A., Sarkis J. Greening versus resilience: A supply chain design perspective. *Transportation Research: Part E: Logistics and Transportation Review*. 2018.

Gautam A., Verma G., Qamar S. et al. Vehicle pollution monitoring, control and challan system using MQ2 sensor based on Internet of things. *Wireless Pers Commun*, 116(2). 2021, 1071–1085.

Guan D. et al. Global supply-chain effects of COVID-19 control measures. *Nature Human Behaviour*. 2020.

Gumus A.T., Guneri A.F., Ulengin F. A new methodology for multi-echelon inventory management in stochastic and neuro-fuzzy environments. *International Journal of Production Economics*. 2010.

Gupta S., Kumar A. Study on early accurate diagnosis and treatment of COVID-19 with smart phone tracking using bionics. *Security and Privacy*. 2023, e303.

Haidari L.A. et al. The economic and operational value of using drones to transport vaccines. *Vaccine*. 2016.

Hanson C.M., George A.M., Sawadogo A., Schreiber B. Is freezing in the vaccine cold chain an ongoing issue? *A literature review. Vaccine*. 2017.

Hasan H., AlHadhrami E., AlDhaheri A., Salah K., Jayaraman R. Smart contract-based approach for efficient shipment management. *Computers and Industrial Engineering*. 2019.

Hobbs J.E. Food supply chains during the COVID-19 pandemic. *Canadian Journal of Agricultural Economics/Revue Canadienne d'Agroeconomie*. 2020.

Karmaker C.L., Ahmed T., Ahmed S., Ali S.M., Moktadir M.A., Kabir G. Improving supply chain sustainability in the context of COVID-19 pandemic in an emerging economy: Exploring drivers using an integrated model. *Sustainable Production and Consumption*. 2021.

Kartoglu U., Milstien J. Tools and approaches to ensure quality of vaccines throughout the cold chain. *Expert Review of Vaccines*. 2014.

Kinra A., Ivanov D., Das A., Dolgui A. Ripple effect quantification by supplier risk exposure assessment. *International Journal of Production Research*. 2020.

Korpela K., Hallikas J., Dahlberg T. Digital supply chain transformation toward blockchain integration. *Proceedings of the 50th Hawaii International Conference on System Sciences.* 2017.

Kouzinopoulos C.S. et al. Using blockchains to strengthen the security of internet of things. International ISCIS Security Workshop. 2018.

Kraus S., Schiavone F., Pluzhnikova A., Invernizzi A.C. Digital transformation in healthcare: Analyzing the current state-of-research. *Journal of Business Research.* 2021.

Lee S.M., Trimi S. Convergence innovation in the digital age and in the COVID-19 pandemic crisis. *Journal of Business Research.* 2021.

Li Z., Liu G., Liu L., Lai X., Xu G. IoT-based tracking and tracing platform for prepackaged food supply chain. *Industrial Management and Data Systems.* 2017.

Shekhar S et al. Comparative analysis of NavIC multipath observables for soil moisture over different field conditions. *Progress in Electromagnetics Research M*, 117. 2023, 25–35.

Shekhar S et al. Sensitivity of multipath peak frequency of navigation with Indian constellation (NavIC) towards surface soil moisture over bare land. IEEE International Geoscience and Remote Sensing Symposium IGARSS, Brussels, Belgium. 2021, 7000–7003.

Verma G., Singhal T., Kumar R. et al. Heuristic and statistical power estimation model for FPGA based wireless systems. *Wireless Pers Commun*, 106(4). 2019, 2087–2098.

Virdi S.K. Implementation of crossbar switch for NOC on FPGA. 3rd International Conference on Computing for Sustainable Global Development (INDIACom), New Delhi, India. 2016, 2087–2091.

Chapter 10

IoT in agile management and banking sector

Worakamol Wisetsri, Vijai, and Deval Verma

10.1 INTRODUCTION

In the contemporary speedy-paced and rapidly evolving business landscape, companies face the challenge of turning in first-rate products and services that meet purchaser expectations (Wells, 2014; Nowotarski and Pasławski, 2016; Muhammad et al., 2021). Traditional, plan-pushed tactics regularly warfare to evolve to changing necessities and market conditions, mainly to not on-time deliveries and dissatisfied clients. This has caused the upward push of the agile technique as an extra flexible and consumer-centric approach to undertaking management and software program development (Ahmed and Mohammed, 2019). The agile methodology emphasizes iterative and incremental improvement, client collaboration, adaptability, and continuous improvement. It prioritizes handing over prices to customers in shorter time frames and embracing change as an herbal part of the development system (McDonald and Welland, 2005).

Agile groups work in cross-functional units, fostering collaboration, communication, and shared ownership. By regarding clients all through the improvement manner, agile methodologies ensure that the very last product meets their wishes and expectations (Bardolet and Dhanwani, 2020). Furthermore, we can explore a variety of gear that aid agile methodologies, facilitating undertaking management, collaboration, communication, model control, and monitoring progress (Fauska et al., 2014). Agile methodologies offer advantages along with flexibility, quicker time to marketplace, stepped-forward customer pride, more desirable group collaboration, transparency, continuous development, and powerful risk mitigation. The specific, agile approach offers a flexible and client-centric approach to venture control and software program software development. It promotes collaboration, adaptability, and non-stop improvement, permitting agencies to reply effectively to converting necessities and supply first-rate products that meet purchaser goals and expectations (Gromova, 2018).

DOI: 10.1201/9781003510420-10

10.2 CONCEPT OF AGILE MANAGEMENT

Agile control is an idea and approach to handling initiatives and organizations that prioritize adaptability, collaboration, and iterative progress. It originated inside the software program development enterprise but has been adopted in diverse sectors, including banking. The central idea of agile control is to include change in place of face up to it (De Borba et al., 2019). Traditional management practices regularly observe a linear and predictive approach, where special plans are created prematurely and strictly adhered to. In contrast, agile management acknowledges that necessities and instances can alternate hastily, and therefore emphasizes flexibility and responsiveness (Philbin, 2015).

Agile control encourages cross-functional teams to work together in a collaborative and self-organizing way (Tokel et al., 2019). These teams are empowered to make choices, plan, and execute work in quick iterations or sprints, and continuously research and adapt based on feedback. The iterative method allows for the delivery of incremental fees and offers opportunities for stakeholders to enter and influence the direction of the assignment. Another vital factor of agile control is its patron-centric attention. Agile methodologies prioritize information and meeting customer needs and expectancies. Regular feedback loops, a person trying out, and non-stop engagement with clients are crucial to the agile technique, ensuring that the products or services are aligned with consumer requirements (Upadhyay et al., 2020).

The concept of agile management also promotes a culture of non-stop development. Through retrospectives and normal mirrored images, teams and businesses can identify regions for enhancement and implement adjustments consequently (Cojocaru et al., 2022). The emphasis on gaining knowledge from experience and experimentation fosters innovation and drives continuous increase. While the agile concept is rooted in software improvement, its ideas and practices were efficiently applied within the banking sector. By embracing agile management, banks can improve mission effects, enhance purchaser enjoyment, accelerate time to marketplace, and better adapt to the rapidly changing scenario of the banking industry (Olowe et al., 2020).

10.3 AGILE METHODS IN GENERAL

Agile strategies are hard and fast principles and practices used in undertaking control and software program improvement to decorate flexibility, collaboration, and responsiveness. These strategies prioritize iterative and incremental improvement, adaptive plans, and purchaser involvement.

10.3.1 Iterative and incremental development

Agile techniques sell an iterative and incremental approach to assignment execution. Instead of attempting to define and supply the whole undertaking prematurely, projects are split into smaller increments or iterations. Each generation outcomes in a doubtlessly shippable product increment, taking into account early shipping of value and frequent feedback (Nomeer, 2021).

10.3.2 Adaptive planning

Agile strategies understand that requirements and priorities can change at some stage in a mission. Rather than creating special, rigid plans at the start, agile teams keep a flexible and adaptive planning approach. Plans are constantly refined and changed primarily based on emerging insights and converting desires (Ahmed and Elali, 2021).

10.3.3 Cross-functional teams

Agile methods emphasize the formation of cross-functional teams that possess all of the necessary competencies and expertise required to supply the task. These teams are self-organizing and collaborate closely throughout the assignment, selling powerful communication, shared ownership, and collective decision-making.

10.3.4 Customer involvement

Agile methods prioritize purchaser involvement and remarks for the duration of the undertaking lifecycle. Regular engagement with customers enables making certain that the very last product meets their desires and expectancies. Customers are regularly worried about defining necessities, reviewing iterations, and providing comments, bearing in mind non-stop development and consumer pleasure.

10.3.5 Continuous integration and delivery

Agile strategies inspire continuous integration and delivery practices. Code modifications are regularly integrated into a shared repository, making sure that any conflicts or issues are addressed directly (Alipour et al., 2022).

Continuous integration allows for early identification and determination of integration problems, even as continuous transport guarantees that running software can be launched to users quickly and reliably (Labarrère et al., 2017).

10.3.6 Embracing change

Agile strategies embody change as a natural and predicted part of the undertaking lifecycle. They are aware that requirements may evolve, new insights might also arise, and market conditions may additionally shift. Rather than resisting trade, agile teams adapt and reply to it, making sure that the mission remains aligned with cutting-edge wishes and goals (Pozhidaeva, 2022).

10.3.7 Empirical process control

Agile methods rely on empirical system management, which involves making decisions based totally on commentary, experimentation, and comments. Teams constantly inspect and adapt their processes to enhance performance, and consumer satisfaction. Data-pushed decision-making and everyday retrospectives are important aspects of empirical process manipulation. Some famous agile strategies and frameworks are Scrum, Kanban, Extreme Programming, and Lean Agile. These methods provide specific hints, roles, and ceremonies to facilitate the implementation of agile standards in numerous contexts (Kuruppu and Egodawele, 2021).

10.3.8 Agile methodologies

There are several famous agile methodologies that corporations can pick from, depending on their unique desires and context. Following are a number of the most extensively used agile methodologies.

10.3.9 Scrum

Scrum is one of the most popular and extensively followed agile methodologies. It emphasizes iterative improvement in quick, time-boxed intervals referred to as sprints. Scrum work groups are cross-functional and feature specific roles like Scrum master, product owner, and development team. Daily stand-up meetings, dash planning, dash critiques, and retrospectives are key Scrum ceremonies.

10.3.10 Kanban

Kanban is a visible control device that focuses on visualizing work, restricting work-in-progress (WIP), and optimizing workflow. Teams that use Kanban have a Kanban board that shows the status of work items, along with those to-do, in development, and completed. It promotes non-stop flow and emphasizes pulling work as capability allows, in preference to placing fixed iterations (Munteanu and Dragos, 2021).

10.3.11 Lean Agile

Lean Agile, additionally known as Lean Software Development, combines agile concepts with Lean manufacturing standards. It emphasizes minimizing waste, optimizing flow, and constantly turning in fees to clients. Lean Agile methodologies focus on performance, non-stop improvement, and the removal of non-value-added activities (Concha et al., 2022).

10.3.12 Dynamic systems development method (DSDM)

DSDM is an agile methodology that provides a framework for delivering projects in a time-boxed and iterative manner. It emphasizes lively user involvement, frequent transport of enterprise price, and iterative improvement. DSDM includes ideas from both agile and traditional venture control approaches.

10.3.13 Feature-driven development (FDD)

FDD is an iterative and incremental agile methodology that makes a specialty of turning in functions incrementally. It emphasizes the introduction of a list of features, iterative design, and brief development cycles. FDD affords suggestions for domain item modeling, growing by using characteristics, and regular development reporting (Cizmecioglu et al., 2021).

10.3.14 Crystal

Crystal is a circle of relatives of agile methodologies with varying tiers of ritual, proper for distinctive project sizes and complexities. Crystal methodologies prioritize communication, simplicity, and close collaboration. They provide flexible hints and permit teams to tailor practices based on their specific context (Minor et al., 2009).

10.4 AGILE METHODOLOGY TOOLS

Agile methodologies are supported by a huge range of tools that help groups efficiently manipulate their tasks, collaborate, track progress, and facilitate agile practices.

10.4.1 Project management and collaboration tools

Jira: Jira is a broadly used venture control tool that allows teams to plan, track, and control agile projects with the use of capabilities like Scrum and Kanban boards, issue tracking, and dash-making plans.

Trello: Trello is a visual collaboration device that permits teams to organize tasks on boards, create cards for each undertaking, and circulate them throughout distinctive levels of workflow. It offers an easy and flexible manner to manipulate projects using agile methodologies.

Asana: Asana is a versatile assignment management device that helps agile practices with the aid of permitting teams to create goals, track development, assign duties, and collaborate on initiatives in a visual interface.

Monday.Com: Monday.Com is a group collaboration and task management platform that allows groups to plot, track, and manage projects with the use of customizable boards, mission assignments, and visual timelines.

10.4.2 Communication and collaboration tools

Slack: Slack is a famous communication tool that permits groups to collaborate in actual time through chat channels, direct messaging, and file sharing. It helps effective communication and collaboration among agile groups.

Microsoft Teams: Microsoft Teams is a complete collaboration device that combines chat, video conferencing, document sharing, and mission control capabilities. It permits agile teams to communicate and collaborate seamlessly.

Zoom: Zoom is a video conferencing device that facilitates remote meetings, stand-ups, and dash opinions. It lets teams maintain effective verbal exchange and collaboration, especially in distributed agile environments.

10.4.3 Version control and code management tools

Git: Git is a broadly used distributed version control gadget that supports agile development practices. It allows groups to collaborate on code, manage branches, track changes, and merge code seamlessly.

GitHub: GitHub is a web-primarily-based platform built on the pinnacle of Git that provides capabilities for version control, collaboration, and code overview. It permits agile teams to work together on code, tune problems, and manage challenge repositories.

Bitbucket: Bitbucket is another famous platform that supports Git-based model control, code collaboration, and non-stop integration and transport. It permits agile teams to manipulate code repositories and projects on code collaboratively.

10.4.4 Agile project tracking and metrics tools

Agilefant: Agilefant is an agile project control tool that enables groups to plan, music, and visualize assignment development with the use of features like backlog management, dash-making plans, and burndown charts.

VersionOne: VersionOne is an enterprise agile platform that provides comprehensive project management, tracking, and reporting abilities. It allows groups to manipulate backlogs, plan iterations, track progress, and generate agile metrics.

10.5 AGILE MANAGEMENT IN BANKING SECTOR PRACTICES

In the banking sector, agile management practices can be implemented in various areas.

10.5.1 Agile product development

Banks can utilize agile methodologies inclusive of Scrum or Kanban to broaden and decorate their virtual products and services. They can break down improvement responsibilities into small, practicable increments, set brief improvement cycles (sprints), and collaborate carefully with stakeholders to ensure the product meets purchaser needs. Regular comment loops and iterative improvement enable quicker delivery and continuous improvement (Rai, 2006).

10.5.2 Cross-functional teams

Banks can form cross-functional teams comprising individuals from different departments, such as IT, operations, marketing, and customer service (Dávila, 2002). These teams work together on specific projects or products, fostering collaboration and knowledge sharing. This approach eliminates silos and enables quicker decision-making and problem-solving (Mashhour et al., 2011).

10.5.3 Daily stand-up meetings

Daily stand-up conferences, additionally known as everyday scrums, can be held to make certain transparency, synchronization, and alignment within agile groups. Team members in short percentage their development, discuss any challenges or roadblocks, and plan their activities for the day. These quick, focused meetings preserve anyone at the equal web page and permit timely issue resolution (Frãþila et al., 2011).

10.5.4 Iterative releases

Instead of waiting for a complete product or feature to be developed, banks can adopt an iterative approach to release smaller increments more frequently (Ekweli, 2020). This allows for earlier customer feedback, rapid validation of ideas, and quicker adaptation to changing market conditions.

It also reduces the risk of investing significant resources in a product that may not meet customer expectations.

10.5.5 User story mapping

User story mapping is a technique used to capture and visualize user requirements and their journey within a product or service. Banks can employ this practice to create a visual representation of customer needs and prioritize development efforts accordingly. User story mapping helps ensure that the developed product aligns with user expectations and delivers maximum value (Fasnacht, 2020).

10.5.6 Continuous integration and continuous deployment (CI/CD)

CI/CD practices involve automating the integration, testing, and deployment of software changes. Banks can leverage CI/CD pipelines to streamline their development and deployment processes, ensuring that new features or updates are rapidly delivered to customers with minimal disruption. This practice facilitates a faster time to market and reduces the chances of errors or inconsistencies.

10.5.7 Retrospectives

Regular retrospectives offer a possibility for agile teams to mirror their overall performance, discover areas for development, and make important changes (Al-Nattar and Alazzawi, 2020). Banks can hold retrospective meetings at the quilt of each dash or assignment segment to gather remarks, compare effects, and put in force adjustments to improve team productiveness and efficiency.

10.5.8 Agile governance

Agile governance frameworks can be implemented in the banking sector to align agile practices with regulatory requirements. This ensures that the benefits of agility are achieved while adhering to compliance and risk management standards. Agile governance promotes transparency, accountability, and effective decision-making within the organization.

10.6 USAGE OF AGILE METHODS IN BANKING SECTOR

Agile methods have found practical application in the banking sector, enabling banks to enhance their operations, innovate products and services, and improve customer experience.

10.6.1 Digital transformation

Banks are undergoing digital transformation to meet evolving customer expectations and stay competitive (Chen & Zhang, 2024). Agile methods are utilized to develop and implement digital solutions, such as mobile banking apps, online payment platforms, and digital onboarding processes. Agile allows banks to deliver digital services in iterative releases, gather customer feedback, and quickly adapt to emerging technologies and market trends (Raj, 2018).

10.6.2 Product development

Agile methods are applied to the development of new banking products and services. Cross-functional teams collaborate to deliver incremental releases of products, allowing for earlier market feedback and validation. Agile practices enable banks to respond to changing customer needs and preferences, accelerate time to market, and iterate on products based on real-time insights (Gryzunova, 2020).

10.6.3 Customer experience

Agile methods support the improvement of customer experience in banking. By involving customers in the development process through regular feedback and usability testing, banks can ensure that their products and services are aligned with customer expectations. Agile also enables banks to quickly address customer issues or requests and implement necessary changes to enhance the overall customer experience (Singh and Lalita, 2018).

10.6.4 Risk management and compliance

Agile methods can be applied to risk management and compliance functions in banking. Agile practices facilitate a more iterative and collaborative approach to risk assessment, monitoring, and mitigation. Compliance requirements and regulatory changes can be addressed in shorter cycles, ensuring banks stay up-to-date and aligned with industry standards (Madusika and Dilshani, 2020).

10.6.5 Project management

Agile project management methodologies, such as Scrum or Kanban, are used to manage banking projects. These methods promote adaptive planning, frequent communication, and iterative delivery. Agile project management enables banks to respond quickly to changing project requirements, adapt to market conditions, and improve overall project transparency and efficiency (Quddus, 2020).

10.6.6 Operational efficiency

Agile methods can enhance operational efficiency within banks. By applying agile principles to internal processes and workflows, banks can identify and eliminate bottlenecks, streamline operations, and improve collaboration between different departments. Agile practices like daily stand-up meetings, visual task boards, and continuous improvement initiatives help banks optimize their operations.

10.6.7 Innovation and experimentation

Agile methods foster a culture of innovation and experimentation within banks. By encouraging cross-functional collaboration, creativity, and a willingness to take risks, banks can explore new ideas and rapidly test them in the market (Alhousary and Underwood, 2016). Agile practices support the development of innovative banking solutions, such as open banking initiatives, personalized digital experiences, and fintech partnerships.

10.6.8 Importance of agile methodology in banking

The importance of agile methodology in the banking sector cannot be understated. Agile methodologies offer several key benefits that are particularly relevant and valuable for banks. Following are some reasons why agile methodology is important in banking (Kamau et al., 2019).

10.6.9 Customer-centric approach

Agile methodologies prioritize customer collaboration and feedback. This customer-centric approach is crucial for banks as they strive to meet the evolving needs and expectations of their customers. By involving customers throughout the development process, banks can gather valuable insights, validate ideas, and ensure that their products and services align closely with customer preferences (Hinduja, 2015).

10.6.10 Speed and time to market

In today's fast-paced digital landscape, banks need to be able to innovate and deliver solutions quickly (Milosevic et al., 2021). Agile methodologies enable banks to shorten development cycles, release minimum viable products (MVPs) rapidly, and iterate based on customer feedback. This accelerates the time to market for new products and services, giving banks a competitive advantage in the industry.

10.6.11 Adaptability to changing requirements

Banks operate in a highly regulated and dynamic environment where requirements can change frequently. Agile methodologies promote adaptability, allowing banks to respond effectively to regulatory changes, market shifts, and customer demands. The iterative and incremental nature of agile enables banks to adjust and prioritize new requirements more efficiently (Owusu-Tucker and Stacey, 2018).

10.6.12 Risk mitigation and compliance

Compliance and risk management are critical functions in the banking sector. Agile methodologies can enhance risk management processes by enabling banks to identify and address risks earlier in the development cycle (Ceylan and Gürsev, 2020). Agile practices facilitate collaboration between risk management and development teams, ensuring that compliance requirements are integrated from the early stages of product development (Sagala et al., 2022).

10.6.13 Continuous improvement and innovation

Agile methodologies foster a culture of continuous improvement and innovation. Banks can leverage agile practices to encourage experimentation, learn from customer feedback, and continuously enhance their products and services (Dursun et al., 2020). By iterating and refining their offerings, banks can stay competitive, drive innovation, and differentiate themselves in the market.

10.6.14 Collaboration and team efficiency

Agile methodologies emphasize cross-functional collaboration and self-organizing teams. In the banking sector, where multiple departments and teams are involved in product development and service delivery, agile practices enhance collaboration, communication, and shared understanding. This leads to increased team efficiency, better coordination, and improved overall project outcomes (Kiburu and Mungai, 2022).

10.6.15 Enhanced customer experience

The banking industry is increasingly focused on providing exceptional customer experiences. Agile methodologies enable banks to gather customer feedback and iterate on their offerings, resulting in products and services that better align with customer needs and expectations. (Panda, 2017). This leads to higher customer satisfaction, increased loyalty, and improved customer retention.

10.7 AGILE BANKING IN THE REAL WORLD

Agile banking in the real world refers to the practical implementation of agile principles and methodologies within banking organizations.

10.7.1 Agile project management

Banks are adopting agile mission control methodologies, including Scrum or Kanban, to supply initiatives more successfully. This method involves breaking down tasks into smaller, potential tasks, creating shorter improvement cycles, and selling collaboration and frequent communication among cross-functional teams. Agile undertaking control enables banks to supply initiatives in incremental iterations, reply to changing requirements, and deliver costs to customers extra fast.

10.7.2 Digital transformation

Agile methods are widely used in the digital transformation initiatives of banks. Instead of undertaking large-scale transformations, banks adopt an iterative and incremental approach to digitize their operations and services. Agile practices enable banks to rapidly develop and deploy digital solutions, gather user feedback, and iterate on their offerings. This approach supports banks in keeping pace with evolving customer expectations and staying competitive in the digital landscape.

10.7.3 Customer-centric product development

Agile concepts are carried out in the improvement of patron-centric banking services and products. Banks engage in continuous customer remarks and usability reviews to recognize purchaser needs and options. Through agile techniques, banks can rapidly launch new capabilities, acquire patron remarks, and contain them in the next iterations. This iterative development technique allows banks to deliver products that align with customer expectations and offer value to clients.

10.7.4 Agile risk management and compliance

Agile methodologies are also being used in risk management and compliance functions within banks. By adopting an agile approach, banks can respond more effectively to changing regulatory requirements. Agile practices help banks adapt compliance processes, conduct risk assessments in shorter cycles, and quickly implement necessary changes. This agile approach to risk management and compliance ensures that banks remain compliant while maintaining their agility in a rapidly changing regulatory landscape.

10.7.5 Agile organizational culture

Banks are embracing an agile organizational culture that values transparency, collaboration, and continuous improvement. Agile principles, such as self-organizing teams and empowered decision-making, are being promoted to foster a culture of innovation, agility, and adaptability. Banks are encouraging employees to experiment, learn from failures, and continuously improve their processes and ways of working.

10.7.6 Agile customer service

Agile methodologies are being applied to customer service operations in banks to improve responsiveness and customer satisfaction. Banks utilize agile practices to streamline customer service processes, such as issue resolution and query handling. By adopting an agile approach, banks can respond to customer inquiries and requests more quickly, enhance the customer experience, and improve overall customer satisfaction.

10.8 BENEFITS OF AGILE METHODOLOGY

Agile methodologies offer numerous benefits to organizations that adopt them.

10.8.1 Flexibility and adaptability

Agile methodologies allow companies to be more conscious of change. The iterative and incremental nature of agile allows for flexibility in adjusting task requirements, priorities, and scope as wanted. Agile groups can speedily adapt to new statistics, market conditions, and patron comments, resulting in advanced effects and client pride.

10.8.2 Faster time to market

Agile methodologies emphasize turning in operating increments of a product or undertaking in shorter time frames. This permits groups to launch valuable features or functionalities in advance, permitting quicker time to the marketplace. By prioritizing the maximum important necessities, agile groups attention on turning in excessive-value increments quickly, gaining a competitive facet in the market.

10.8.3 Customer satisfaction

Agile methodologies place a robust emphasis on purchaser collaboration and involvement during the improvement method. By actively involving clients, accumulating their remarks, and adapting primarily based on their wishes, agile companies can make certain that the very last product meets patron expectations. This ends in higher purchaser satisfaction and loyalty.

10.8.4 Improved product quality

Agile methodologies promote non-stop integration, reviews, and comment loops, mainly to improve product satisfaction. By engaging in regular reviews and purchaser validation, issues and defects may be diagnosed and addressed early in the improvement technique. Agile teams additionally prioritize warranty practices, resulting in better first-class deliverables.

10.8.5 Enhanced team collaboration and morale

Agile methodologies encourage collaboration, transparency, and self-organizing groups. Cross-functional groups work intently together, fostering a collaborative and cohesive running environment. This results in advanced conversation, information sharing, and problem-solving. Agile also empowers team members, leading to increased activity satisfaction and morale.

10.8.6 Transparency and stakeholder engagement

Agile methodologies promote transparency by offering visibility into mission development, priorities, and challenges. Regular meetings, which include everyday stand-up and sprint reviews, permit stakeholders to be informed and provide remarks. This high stage of stakeholder engagement fosters consideration, collaboration, and alignment among the improvement crew and stakeholders.

10.8.7 Continuous improvement

Agile methodologies embrace a culture of continuous improvement. Through everyday retrospectives, groups replicate their work, perceive areas for development, and put in force changes in the next iterations. This consciousness of continuously gaining knowledge of and version drives innovation, manner optimization, and normal organizational boom.

10.8.8 Risk mitigation

Agile methodologies enable organizations to identify and address risks early in the development process. By breaking down work into smaller iterations, potential risks and challenges can be identified and managed promptly. The iterative nature of agile allows for frequent reassessment of risks and adjustments to mitigate them effectively.

10.9 CHALLENGES OF AGILE METHODOLOGY

10.9.1 Organizational culture

Shifting to an agile mindset requires a cultural change within the organization. Traditional hierarchical structures and command-and-control management styles can hinder the adoption of agile principles, which emphasize self-organizing teams, collaboration, and empowerment. Overcoming resistance to change and fostering a culture that supports agile values and practices can be a significant challenge.

10.9.2 Lack of experience and understanding

Agile methodologies may be new to the organization, and team members may have limited experience or understanding of agile principles and practices. Training and coaching may be required to help teams embrace and apply agile methods effectively. Building a shared understanding of agile concepts and practices is crucial for successful implementation.

10.9.3 Stakeholder engagement

Agile methodologies require active involvement and collaboration with stakeholders, including customers, product owners, and business representatives. However, stakeholders may have competing priorities, limited availability, or difficulty adapting to the iterative and incremental nature of agile development. Ensuring consistent and meaningful stakeholder engagement can be a challenge.

10.9.4 Managing changing requirements

Agile methods embrace changing requirements as a natural part of the development process. However, managing and prioritizing evolving requirements can be challenging, particularly when dealing with complex projects or large-scale initiatives. Clear communication, effective backlog management, and a robust prioritization process are essential for addressing changing requirements without compromising project objectives.

10.9.5 Team collaboration and communication

Agile methodologies emphasize close collaboration and effective communication within the team. However, geographical distribution, time zone differences, or cultural barriers can impede collaboration. Ensuring effective collaboration and communication, particularly in distributed or remote teams, requires the right tools, practices, and processes.

10.9.6 Scaling agile

Scaling agile methodologies to large and complex projects or organizations can pose challenges. Coordinating multiple teams, ensuring alignment, and maintaining consistency across teams can be complex. Adopting appropriate scaling frameworks, establishing shared practices, and fostering collaboration across teams are critical to successful scaling.

10.9.7 Balancing flexibility and predictability

Agile methodologies offer flexibility and adaptability, but balancing these with the need for predictability and planning can be challenging. Stakeholders often require visibility into project timelines, costs, and outcomes. Striking the right balance between responding to change and providing predictability is crucial for managing expectations and maintaining stakeholder confidence.

10.9.8 Agile maturity and continuous improvement

Achieving high levels of agile maturity and continuous improvement takes time and effort. Teams may encounter initial difficulties in implementing agile practices effectively or struggle to measure and demonstrate improvements. A commitment to ongoing learning, retrospectives, and adapting practices based on feedback is essential for maturing and continuously improving agile implementation.

IoT can be utilized in combination with agile techniques in the banking sector. Figure 10.1 presents different scenarios.

- Employ IoT sensors to gather up-to-the-minute data from diverse banking procedures, including transaction volumes, client interactions, and branch operations. Analyzing this data enables the making of well-informed decisions and adjustments throughout Agile planning and iterations (Chhaya et al., 2018).
- Deploy IoT devices (Chhaya et al., 2020) to collect data on client behavior and preferences. The provided data can be utilized to customize banking services, develop focused marketing initiatives, and enhance overall client contentment.

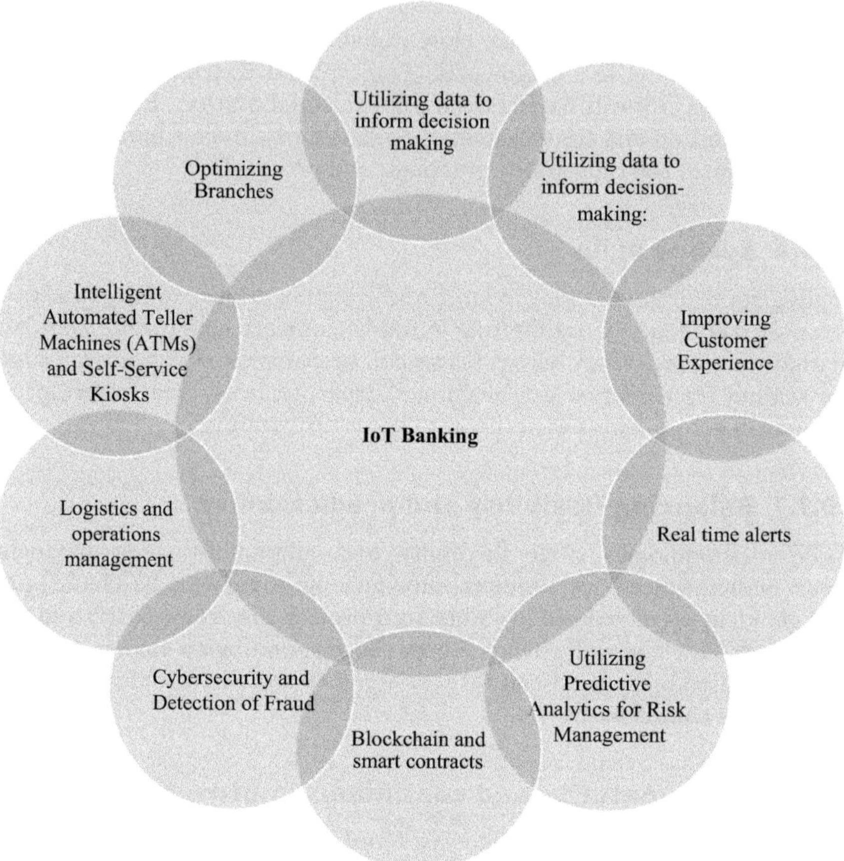

Figure 10.1 IoT in agile banking.

- Incorporate IoT devices to gather data on market trends, economic indicators, and other pertinent elements (Goel et al., 2023). By leveraging this data in conjunction with predictive analytics, the banking sector may optimize risk management tactics, empowering them to proactively tackle prospective difficulties.
- Utilize IoT sensors to track and assess the flow of people and examine client actions within physical branches. This data can be used to make decisions on the arrangement of branches, the number of staff members, and the range of services provided, thereby enhancing the overall branch experience (Gehlot et al., 2021).
- Integrate IoT technology into ATMs and self-service kiosks to facilitate proactive maintenance, track usage trends, and bolster security measures. This guarantees that these crucial customer interactions are consistently accessible and align with customer expectations.

- Utilize IoT devices in supply chain management to control the transportation and storage of banking items, such as cash, papers, and equipment. This facilitates the prompt accessibility of resources for banking operations.
- Cybersecurity and detection of fraud can be prevented by incorporating IoT devices (Gehlot et al., 2016), such as surveillance cameras and sensors, to augment security protocols. Implementing this not only aids in averting physical security breaches but also enhances fraud identification and mitigation by means of continuous real-time monitoring.
- Utilize IoT and blockchain technology to streamline and safeguard transactions. Implementing this technology can optimize diverse banking procedures, mitigate the likelihood of mistakes, and augment the clarity of financial transactions.
- Employ IoT asset tracking systems to monitor valuable assets, such as vehicles, equipment, or confidential documents. This improves security and guarantees that items are located correctly and promptly.
- Deploy IoT devices to oversee and enhance operating procedures within the banking infrastructure. This includes HVAC systems, lighting, and other elements of facilities management, which collectively contribute to enhancing energy efficiency and achieving cost savings.
- Incorporate IoT devices to deliver immediate notifications and alerts regarding system status, security problems, or other crucial occurrences. This facilitates prompt answers and mitigations, in accordance with agile principles of swift adjustment.

It is necessary to use data from the IoT to set up feedback loops so that banking processes are always getting better. During the agile retrospective, always look at success indicators, customer feedback, and operational data to find ways to make things better.

10.10 SUMMARY

The agile method has emerged as an effective method for undertaking control and software improvement, permitting businesses to thrive in hastily converting and consumer-pushed surroundings. By prioritizing flexibility, collaboration, and turning in cost, agile methodologies have revolutionized the way teams approach projects and engage with stakeholders. Throughout this chapter, we explored the central ideas, advantages, equipment, and roles related to agile methodologies. We discovered that agile methodologies, inclusive of Scrum, Kanban, Extreme Programming (XP), and Lean Agile, provide frameworks and practices that aid iterative and incremental development, patron collaboration, adaptability, and continuous improvement. These methodologies empower groups to respond fast to changing necessities, deliver first-rate products in shorter cycles, and increase purchaser satisfaction.

Furthermore, we explored a range of gear that facilitate agile practices, consisting of assignment control and collaboration tools, communication and collaboration systems, version manipulation and code management structures, and agile assignment monitoring and metrics tools. These tools provide essential help for agile teams to efficiently manage initiatives, collaborate, talk, and music development in agile surroundings. Lastly, we examined the diverse roles in agile methodologies, such as the product owner, Scrum master, development team, agile coach, and stakeholders. These roles emphasize collaboration, shared responsibility, and self-company within the group, allowing powerful conversation, collaboration, and delivery of price. The agile method offers corporations a dynamic and purchaser-centric technique for task control and software development. By embracing agile ideas, leveraging suitable equipment, and fostering a collaborative and adaptive tradition, businesses can effectively navigate the challenges of a brand-new business landscape, supply first-rate merchandise, and achieve customer satisfaction. Agile methodologies empower teams to innovate, reply fast to trade, and continuously improve in an evolving marketplace. Agile management and IoT can make banks more responsive and efficient. This connectivity helps banks adjust swiftly to market developments, improve customer experiences, and boost operational efficiency. When deploying IoT technologies in finance, security and privacy must be considered.

REFERENCES

Ahmed, N.S., & Elali, W. (2021). The Role of Agile Managers' Practices on Banks's Employees Performance in the Kingdom of Bahrain. *International Journal of Business Ethics and Governance*.

Ahmed, N.S., & Mohammed, S.R. (2019). Developing a Risk Management Framework in Construction Project Based on Agile Management Approach. *Civil Engineering Journal*.

Alhousary, T., & Underwood, J. (2016). Knowledge Sharing Culture: A Study on the Omani Commercial Banking Sector. *Eurasian Journal of Business and Management*, 4(3), 1–12.

Alipour, N., et al. (2022). Lean, Agile, Resilient, and Green Human Resource Management: The Impact on Organizational Innovation and Organizational Performance. *Environmental Science and Pollution Research International*, 29(55), 82812–82826.

Al-Nattar, B.A., & Alazzawi, A. (2020). Data Analytics of Strategic Agility and Competitiveness in Operation Performance: A Case of Banking Sector in Saudi Arabia. 2020 International Conference on Decision Aid Sciences and Application (DASA), 293–298.

Bardolet, D., & Dhanwai, R. (2020). *Agile Management in the Arts.*

Ceylan, Z., & Gürsev, S. (2020). AHP ve TOPSIS Yöntemleri Ile Bilgi Teknolojileri Projelerinde Scrum-Kanban-Şelale Uygulamaları Karşılaştırması. *Bilişim Teknolojileri Dergisi.*

Chen, Y., & Zhang, Y. (2024). The Impact of Digital Transformation on Firm's Financial Performance: Evidence from China. *Industrial Management & Data Systems,* https://doi.org/10.1108/IMDS-07-2023-0507.

Chhaya, L. et al. (2018). IoT-Based Implementation of Field Area Network Using Smart Grid Communication Infrastructure. *Smart Cities,* 1(1), 176–189.

Chhaya, L. et al. (2020). Cybersecurity for Smart Grid: Threats, Solutions and Standardization. *Advances in Greener Energy Technologies,* 17–29.

Cizmecioglu, H.S., et al. (2021). *Selection of the Best Software Project Management Model via Interval-Valued Neutrosophic AHP.*

Cojocaru, A.M., et al. (2022). The Impact of Agile Management and Technology in Teaching and Practicing Physical Education and Sports. *Sustainability.*

Concha, U.R., et al. (2022). *Integration of the Big Data Environment in a Financial Sector Entity to Optimize Products, Services and Decision-Making.*

Dávila, P.I. (2002). *La Contabilidad de Gestión en las Cajas Rurales. Una Perspectiva Empírica.*

de Borba, J.C.R., et al. (2019). Agile Management in Product Development. *Research-Technology Management,* 62(5), 63–67.

Dursun, M., et al. (2020). A Cognitive Map Integrated Intuitionistic Fuzzy Decision-Making Procedure for Provider Selection in Project Management. *Journal of Intelligent and Fuzzy Systems,* 39(5), 6645–6655.

Ekweli, F. (2020). Process Innovation and Organizational Agility in the Banking Sector of Nigerian Economy. *Strategic Journal of Business & Change Management,* 7(1).

Fasnacht, D. (2020). *Open Innovation in the Financial Services - The Magic Bullet. IRPN: Product Development Strategy & Organization (Topic).*

Fauska, P., et al. (2014). Agile Management of Complex Goods & Services Bundles for B2B E-commerce by Global Narrow-Specialized Companies. *Global Journal of Flexible Systems Management,* 15(1), 5–23.

Frāþila, L., et al. (2011). *Aspects Regarding to It Risk Management in Financial-Banking Sector.*

Gehlot, A. et al. (2016). IoT and Zigbee Based Street Light Monitoring System with LabVIEW. *International Journal of Sensor and its Applications for Control. Systems,* 4(2), 1–8.

Gehlot, A. et al. (2021). WPAN and Iot Enabled Automation to Authenticate Ignition of Vehicle in Perspective of Smart Cities. *Sensors,* 21(21), 7031.

Goel, A. et al. (2023). The Role of Artificial Neural Network and Machine Learning in Utilizing Spatial Information. *Spatial Information Research,* 31(3), 275–285.

Gromova, E.A. (2018). *Agile Management in the Context of Russian Industrial Sector.*

Gryzunova, N.V. (2020). *Adjustment of Prudential Norms, Determining Vector of the Banking Sector Development.* Upravlenie.

Hinduja, K.P. (2015). *Banking Sector: Emerging Challenges and Opportunities.* https://cberuk.com/cdn/conference_proceedings/2015iciee_india3.pdf.

Kamau, J.G., et al. (2019). Relationship Between Strategic Capabilities and Competitive Advantage in the Kenyan Banking Sector. *European Journal of Business and Strategic Management, 4*(2), 1–23. https://www.iprjb.org/journals/index.php/EJBSM/article/view/836.

Kiburu, L., & Mungai, E.M. (2022). *Equity Bank: Repositioning as a Fintech. Emerald Emerging Markets Case Studies.* https://www.emerald.com/insight/content/doi/10.1108/EEMCS-03-2022-0069/full/html.

Kuruppu, K.A.D.T.D., & Egodawele, M.H.A. (2021). The Relationship between Knowledge Management Processes and Workforce Agility in the Sri Lankan Banking Sector. *Wayamba Journal of Management, 12*(2), 400–427. https://doi.org/10.4038/wjm.v12i2.7546.

Labarrère, A., et al. (2017). *Hybrid Methods in Project Management: A Case Study in the Banking Sector [Structure Hybride de Management de Projet: Une Étude de Cas dans le Secteur Bancaire].* Pre-ICIS, Dublin.

Madusika, J.K.D., & Dilshani, A.K.D.N. (2020). *Big Five Personality and Employee Adaptability: A Study on Multigenerational Workforce in Banking Industry,* 11(2), 102–117. https://doi.org/10.4038/wjm.v11i2.7475.

Mashhour, A., et al. (2011). Identifying Internet Abuse by Analyzing User Behavior on the Internet. *The Journal of Internet Banking and Commerce, 16,* 1–15.

McDonald, A., & Welland, R. (2005). Agile Web Engineering (AWE) Process: Perceptions within a Fortune 500 Financial Services Company. *Journal of Web Engineering, 4,* 283–312.

Milosevic, N., et al. (2021). Managerial Perception of Human Capital, Innovations, and Performance: Evidence from Banking Industry. *Engineering Economics, 32*(5), 446–458.

Minor, M., et al. (2009). Demonstration of the Agile Workflow Management System CAKE II Based on Long-Term Office Workflows. International Conference on Business Process Management.

Muhammad, U., et al. (2021). Impact of Agile Management on Project Performance: Evidence from I.T Sector of Pakistan. *PLoS One, 16*(4).

Munteanu, V.P., & Dragos, P. (2021, December). A Theoretical View About Agile Management in Bank Sector. The Annals of The University of Oradea. *Economic of Sciences, 30*(2), 344–352.

Nomeer, M. (2021, March). *Intelligent Energy Platform.* Paper presented at the International Petroleum Technology Conference, Virtual. https://doi.org/10.2523/IPTC-21252-MS.

Nowotarski, P., & Pasławski, P. (2016). Lean and Agile Management Synergy in Construction of High-rise Office Building. *Archives of Civil Engineering, 62*(4), 133–148.

Olowe, A.E., et al. (2020). Strategic Agility and Organisational Performance Nexus: Evidence from Selected Deposit Money Banks in Lagos, Nigeria. *The International Journal of Business and Management, 8,* 191–203. http://www.internationaljournalcorner.com/index.php/theijbm/article/view/154970/107341.

Owusu-Tucker, E., & Stacey, P. (2018). *Assessing the Role of Cloud Computing in the Strategic Agility of Banking.* Loughborough University. Conference contribution. https://hdl.handle.net/2134/33184.

Panda, S. (2017). *Effects of Organizational Capabilities on Organizational Performance: Empirical Evidences from Indian Banking Industry*. PhD thesis. NIT Rourkela.

Philbin, S.P. (2015). *Exploring the Application of Agile Management Practices to Higher Education Institutions*, Proceedings of the American Society for Engineering Management 2015 International Annual Conference.

Pozhidaeva, N. (2022). Application of Project and Process Approaches on the Example of the Largest Banks of the Russian Federation. *Ideas and Ideals*, 14(1 pt 2), 278–290.

Quddus, A. (2020, November). Financial Technology and the Evolving Landscape of Banking and Financial Sector. *The IUP Journal of Bank Management*, XIX(4), 60–73.

Rai, S. (2006). Business Continuity Management in Banks – The Indian Experience. *Journal of Internet Banking of Commerce*, 11 (2).

Raj, R. (2018, November). Risk Management in Cloud Banking. *Journal of Emerging Technologies and Innovative Research*, 5(11), 80–82, http://www.jetir.org/papers/JETIRN006016.pdf.

Sagala, Y.P., et al. (2022). Digital Transformation Impact Analysis Towards Transition in the Role of Information Technology for Organization in New Digital Bank. 7th International Conference on Informatics and Computing (ICIC), 1–6.

Singh, S., & Lalita, D. (2018, November). Benefits of Cloud Computing in Transforming the Banking Sector and Precautions Needed While Making the Transformation. *Journal of Emerging Technologies and Innovative Research*, 5(11), 24–25. http://www.jetir.org/papers/JETIRN006007.pdf.

Tokel, A., et al. (2019). The Role of Learning Management in Agile Management for Consensus Culture. *The International Journal of Information and Learning Technology*, 36(4), 364–372.

Upadhyay, A., et al. (2020). A Review of Lean and Agile Management in Humanitarian Supply Chains: Analysing the Pre-disaster and Post-disaster Phases and Future Directions. *Production Planning and Control*, 33(6–7), 641–654.

Wang, R., et al. (2023, July). Analysis on Risk Awareness Model and Economic Growth of Finance Industry. *Annals of Operations Research*, 326(1), 143.

Wells, A. (2014). Agile Management. *IFLA Journal*, 40(1), 30–34.

Chapter 11

IoT-powered training for employee potential through smart embedded tools and technologies

Bishnu Paramguru Mahapatra,
K. Thomas Alwa Edison, L. Kavitha Nair,
Alagesan M., Prasanth P.S., Rasabihari Mishra,
and Pritish Bhanja

11.1 INTRODUCTION

In today's dynamic corporate environment, a dedicated training and development unit is needed. Training helps workers to increase their abilities and hone their current ones, leading to improved efficiency and output. Organizational success depends on the efforts of its employees, thus companies should spare no effort in encouraging peak performance in both favorable and unfavorable conditions.

Training refers to the methods an organization uses to educate its workers with new tools and technologies so that they can accomplish their jobs more effectively. The purpose of training employees is to provide them with the knowledge and skills they need to do their jobs effectively and efficiently in the modern workplace. Most businesses depend on carefully designed training programs to give workers the skills, knowledge, and character traits they need to do their jobs to the best of their ability, which would be possible through the usage of smart technologies. Workforce development is increasingly centered on the sum of an individual's characteristics—their intelligence, training, and the ease with which they can do a given activity (William 1982).

People feel training and growth are nearly synonymous with one another, yet they are not. Formal learning, such as that which takes place in a classroom, workshop, conference, seminar, online with instructor, etc., is the primary emphasis of training. On the other side, progress is seen as a more individual and casual approach to learning that may include self-paced study, tutoring, learning by observation, internships, etc.

Employee motivation via proper training is essential for achieving the goals of the organization. Keeping up with the competition in the business world necessitates that companies take the necessary steps to attract, retain, and persuade their employees in the face of new technologies, advances in

DOI: 10.1201/9781003510420-11

the field of e-business, and the globalization of business. In order to compete in the global and automotive markets with the best products and services, businesses must invest in training. Training prepares workers to make the most of new tools, to operate effectively in dynamic settings like remote teams (use of IoT technology), and to communicate and engage effectively with customers from a wide range of cultural backgrounds. Only by consistently investing in their employees' professional progress over time can companies ensure their own continued growth and development. Companies may get the most out of their workforces if they invest in their members via ongoing training and development programs run by the human resources department. The Internet of Things (IoT) (Chhaya et al., 2017) can play a vital role in the smart management of the system (Gupta and Kumar 2023; Rawat et al., 2018).

Training and development provided by external entities are crucial to an organization's bottom line. It is the obligation of all businesses to provide frequent opportunities for staff training and development. In order to maintain economic growth, stay ahead of the competition, and adapt to rapid changes in their environment and intensifying global rivalry, corporations and commercial organizations must continually extend their operations and establish themselves as industry leaders. Therefore, businesses need to regularly provide need-based and systematic training in order to enhance and inflate the knowledge, potential, and capacities of their workforce, enabling them to face the aforementioned problems and ultimately achieve their objectives. The goals, activities, and objectives of any business may be better understood by its staff, thanks to an excellent and comprehensive training program. Providing your staff with up-to-date training is a great way to boost their performance and productivity. When businesses invest in their employees' education and growth, they may cut down on turnover and boost their return on investment.

Training may be thought of as a planned, deliberate intervention to improve employee performance on the job. An employee's ability to acquire, enhance, and accomplish the organization's goals and objectives is directly related to their level of knowledge.

All workers, at some point in their careers, may have experienced the stress and frustration that may be alleviated via proper training. Effective training and development programs help businesses improve their performance in all of these areas. In order to maintain a competitive edge, businesses rely on training to consistently educate their staff. If a company wants to maintain its reputation and grow in today's complex and stagnant economy, it must invest in training and development to increase its workforce's skills and productivity (and reap a financial reward for doing so). Training occurs to be a key aspect of manager–staff interaction.

To survive in today's business environment, companies must take every precaution in managing their employees and achieving their goals. The concept of "employee empowerment" is gaining prominence in the modern business environment. As one of the key pillars of managerial and organizational performance, which rises when authority and control are shared in an organization, employee empowerment has garnered considerable recognition as an important issue in management circles. Empowerment has been identified as a key component of effective management in recent years. Kotter (1995) said that giving workers the authority to implement the company's vision is a crucial part of the transformation process. Leaders who instill confidence in their teams better prepare them—and their companies—to weather future market fluctuations and unanticipated demands (Lorsch, 1995). Increasing people's agency means giving them a voice in matters of consequence (Smith, 2002). The concept of empowerment is often discussed but seldom put into practice. Many individuals in business use the term "empowerment" without fully grasping its meaning (Dobbs, 1993; Randolph, 1995). There is a growing understanding that empowering employees may provide modern organizations with a competitive edge, as noted by Quinn and Spreitzer (1999). As a result, empowering employees should be a priority for every company. Methods and outcomes of employee empowerment are examined.

11.2 STUDY AIM

The primary purpose of this research is to delve into the topic of employee empowerment.

Based on the literature, we will:

- Identify the perspectives, methods, aspects, kinds, benefits, and practices of employee empowerment.
- Learn about the difficulties of empowering workers.
- Suggest ways of enhancing worker potential.

11.3 METHODOLOGY

The purpose of this study is to survey the literature on empowering workers. The information used in this analysis is secondary in nature. Books, journals, and scholarly articles are examined for secondary data. Newspaper stories and online resources have also been used here. However, the data has been interpreted by the authors in the context of the goals discussed earlier. Figure 11.1 presents the need for IoT in employee training.

Figure 11.1 IoT in training.

11.4 REVIEW OF RELATED LITERATURE

Cheng and Ho (2001) state that in any company, senior management will never stop stressing the importance of employee performance. Employees, on the other hand, are always worried about their own competence and output, and they are well aware of the rapid pace at which information and abilities become outdated in the workplace. Effective coaching may help employees become more self-motivated by increasing their awareness of opportunities for professional growth.

Anger, chaos, unhappiness, disrespect, and turbulence in the workplace may be mitigated by training and development of personnel. Training is the preferred medium of capital accumulation. Coaching and behavior should be moved to the organization's protection and mapped across the environment to increase worker productivity.

Ability may be developed via both formal and informal instruction. Improvements in performance are inevitable when talent is honed, fostered, and cared for. Teaching and development programs are what really cement an employee's sense of belonging and pride in their profession. Staff communication and competence in carrying out duties and activities are greatly enhanced, and the continuing duration is extended, as a result of the mandatory training. Instruction and guidance have been shown to positively affect employee engagement, job satisfaction, and conflict resolution in the workplace.

Employees are a company's most valuable resource, and treating them poorly at work may spell disaster for the business. Positive employee motivation for the organization's benefit may be fostered via ongoing training and education. Investment in employee training and development is seen by many companies as a catalyst for producing a competent workforce. In the current context, human resources are seen as an integral part of every successful business. Employee training and development may boost their competence and potential (Hardy et al. 1998). It may also help them diversify their skill sets to meet the requirements of an ever-evolving corporate culture. The profit of an organization and its workers' outlook on profit may both be influenced by training and education. It's fundamental to any industry's or business's ability to expand and thrive. Educating and training employees means providing them with the tools they need to perform well in a variety of roles within an organization. It's possible to see it as a company asset that helps workers learn new things, draw conclusions, and put those learnings to use.

Training closes the gap between current abilities and those needed to achieve one's goals. The goal of development is to strengthen an employee's abilities so that they can contribute to the company's success. Increasing productivity via higher employee performance is the goal of training and development programs. "Training" is described as "a learning activity that aids in the acquisition of specific knowledge and skills in performing a task effectively". Training is a systematic process wherein workers gain the information, skills, and competencies they need to do their jobs successfully.

Many individuals mistakenly assume that training and development are the same thing. Employees are more likely to succeed in their jobs if they are provided with the tools they need to do so, and training focuses on doing just that. However, development efforts are directed toward helping workers acquire skills that will be useful to the company as it strives to expand, compete, and succeed in today's dynamic business environment.

Workers' abilities, knowledge, and understanding are prioritized in development projects. It allows workforces to manufacture high-quality products, expeditious assistance, and productive organizations. Unlike training for a particular position, development is intended to help an individual in general. While it may be feasible to evaluate and assess educational and training programs, this is not an option when dealing with development. Even while progress can't be measured in a vacuum, companies should nonetheless invest in ongoing training and education for their staff. An employee's development comprises planned, sequential, and predictable alterations to the way they perform.

To succeed, businesses need employees who are fully invested in their jobs (Locke & Latham, 1990). A dedicated worker who enjoys their work greatly benefits the company. An employee who feels emotionally invested in his work is less likely to quit his position, takes higher satisfaction in his work, and contributes more to the success of his company (Kuo et al., 2009). Therefore, an organization's human resources team should work toward improving workers' skills and swaying their attitudes and actions in ways that bring about both employee happiness and the achievement of business objectives. Rather than relying only on the distribution of responsibility, proactive management practices foster an environment where workers feel valued, respected, and trusted while also receiving enough recognition for their achievements. Empowerment, as argued by Liden et al. (2000), is a crucial concept because of its capacity to affect beneficial results for both people and organizations.

Kim's (2002) research article "Participative Management and Job Satisfaction: Lessons for Management Leadership" found that encouraging employees to have a hand in making decisions at work increases their contentment with their jobs. From this, Kim concludes that "the essential implication is that executive leaders and managers should become aware of the importance of manager's use of participative management, employees' participation in strategic planning processes, and the role of effective channels of communication with supervisors." When agencies establish leadership development programs or other training for managers and supervisors, they should consider adding participatory management and employee empowerment approaches as essential components of the programs. Workforce agility has been documented as a consequence of employee participation programs, particularly power-sharing practices, according to research by Sumukadas and Sawhney (2004). Holden (2001) argued that encouraging employee participation and autonomy was key to boosting productivity.

Tzafrir (2004) argues that employee empowerment has two main components: (i) a more creative approach to working with people, and (ii) a redistribution of authority away from upper-level management and toward middle- and lower-level managers. For an organization to be deemed empowering, according to Lawler (1986), it must provide appropriate incentives for desirable contributions from workers. These views argue that organizational elements including values, working styles, and management systems are intertwined with empowerment.

The literature includes various definitions of employee empowerment. Researchers have come up with a variety of definitions of empowerment. Following are a few of the most common definitions.

According to Heathfield (2012), empowerment is "the process of enabling and authorising individuals to think, behave, take action and decision and control work autonomously." Vogt and Murrel (1990) defined empowerment as the process of gaining control over one's life by influencing one's actions and those of others via collaborative efforts (which they called "interactive empowerment") and independent decision-making (which they called "self-empowerment"). Conger and Kanungo (1988) argued that empowerment is the process of increasing people's perceptions of their competence inside an organization by identifying and eliminating the causes of their own helplessness. Smith and Mouly (1998) defined employee empowerment as the delegation of decision-making authority to workers.

Brymer (1991) characterized empowerment as a process of decentralizing decision-making in an organization, wherein management offers greater discretion and authority to the front-line workers. According to Kinlaw's "The Practise of Empowerment: Making the Most of Human Competence" (1995), empowerment is the process of achieving continuous improvement in organizational performance by increasing the influence of individuals and teams who are competent and have authority over most aspects of their work. Empowerment, as defined by Nonaka and Takeuchi (1991), is the process of facilitating people's independent cognition, behavior, and action. It encourages workers to take pride in their efforts and accept responsibility for the outcomes. Automation and technical advancements have made it such that businesses rely heavily on their workers' ability to think creatively and take initiative in order to succeed.

Various scholars have made attempts to define employee empowerment from a variety of angles. Although there are some overlaps in their descriptions of what it means to empower workers, their definitions are not identical. Here are some of them:

To begin, empowering workers gives them a voice in decision-making and the authority to act on it.

It allows workers to assume responsibility for their own jobs and it emphasizes employees' actual capabilities.

An employee is considered empowered when they have been given the authority to make decisions on their own.

When people are empowered, they are held accountable for the consequences of their actions.

Three distinct forms of employee empowerment were identified by Suminen (2005):

i) Verbal empowerment: Verbal empowerment refers to the capacity to declare one's viewpoint and dispute one's beliefs in various sorts of organizations. One of the key components of having a voice in decision-making is having that voice heard. Employees' dedication to the company, their sense of autonomy (the opportunity to put their talents and expertise to use), and their happiness in the work have all been shown to rise when given more say in important decisions.

ii) Behavioral empowerment: Behavioral empowerment is the capacity to acquire new skills and take on a more difficult job, to work in groups to solve issues, to recognize problems that need to be addressed, to gather data about work difficulties, and to make recommendations based on that data. Reporting and teamwork are two other components of behavioral empowerment.

iii) Outcome empowerment: Outcome empowerment comprises the capacity to discover the sources of issues and to address them, as well as the ability to make adjustments and changes to the way the job is done with a view to boosting the effectiveness of the organization. Employees are expected to develop under the guidance of empowering leaders, in an empowering culture, and with the support of empowering management practices.

11.5 METHODS OF EMPLOYEE INVOLVEMENT AND EMPOWERMENT

Workers play a crucial role in the prosperity of any company. Employee productivity is crucial to the achievement of these goals. Therefore, it is the responsibility of every company to create a working atmosphere that encourages workers to take initiatives that further the company's goals. A total of 983 Israeli secondary school educators participated in research by Bogler and Somech (2004). They came up with a list of six characteristics of empowered people: (i) the ability to make decisions, (ii) the chance to advance in one's career, (iii) social standing, (iv) confidence in one's own abilities, (v) freedom to act, and (vi) the ability to have an impact. Successful empowerments require the following information sharing, as identified by Brower (1995): (i) the organization's financial trends, current situation, and projections; (ii) the current and potential changes in markets, customers, and competition; (iii) the trends in the technology of production and products; (iv) the cost per unit on a daily or hourly basis, if possible; (v) the strategic direction of the organization and its priorities for the coming year; and (vi) key measures.

Today's world is dynamic and presents complex problems for businesses. Every company has to go beyond command-and-control management if it wants to thrive and remain competitive in the future. Workplace empowerment is a major concern. The methods of employee empowerment have been the subject of several books and studies. The following is an explanation of these methods:

- Informal communication: Information is shared between managers and their staff at all levels. The goals of the company may be better communicated to staff with this information. The organization's objective, vision, and goals; new legislation and technology; and any other relevant developments will all be disseminated.
- Delegating authority: Delegating authority is another strategy of empowerment. It's the act of giving power to others under one's command. Leaders should assess their team members' skills before asking them to take on new responsibilities.
- Recognizing and rewarding achievement: Motivating workers to do their best work is easy when they are rewarded for it. Management and leaders may show their gratitude to employees by holding award ceremonies and sending out letters of thanks.
- Response and commentary: By pointing out workers' strengths and helping them build on their weaknesses, feedback helps boost performance. Employees are more inspired and equipped to take on difficulties when they work in an atmosphere that is both supportive and encouraging. Another method of empowering workers is to have faith in their employees' judgment and allow them latitude to use it. Honesty and openness in communication are keys to establishing mutual trust.
- Training methodologies: To guarantee training is efficacious it should concentrate on essentials that staff learn by watching it with their eyes, hearing it with their ears, telling with their mouths, and doing it with their hands. On-the-job and off-the-job training, computer-based testing (CBT), internet-based testing (IBT), mentoring, e-books, and other forms of non-traditional education and development are all viable options.

There is value in giving workers more say in their work when done well. Empowering employees may be a powerful tool for increasing morale in the workplace. In this way, employee empowerment improves organizational performance by encouraging more employee input and initiative (Greasley, 2005). Ziyakashany (2009) argues that empowerment is a valuable technique for improving the caliber of human resources and the efficiency of organizations. When workers are given more responsibility, they feel more invested in the success of the business.

Empowering workers is a concept with tangible advantages for businesses. With these advantages, a company might get a leg up on the competition. Businesses that want to boost productivity and morale among workers often turn to employee empowerment as a means to that end.

Employee empowerment has several benefits, one of which is increased job satisfaction. When workers are happy in their jobs, they are more committed to the company. Workers who have more autonomy at work are more invested in their jobs, less likely to report high levels of stress, and less likely to consider quitting the company. Employees who have been with the firm for a while often serve as role models for new hires and are invaluable resources for upper management. Employees who are given more autonomy report higher levels of satisfaction with their jobs and are more likely to take pride in their work as a consequence. Involving staff in the goal-setting process allows them to share their perspectives, beliefs, and expertise. Results from empowered workers may have a positive effect on both top and bottom lines.

Competing in the modern corporate environment requires a workforce that is empowered to make choices and act autonomously. When workers are given responsibility and respect, they develop faith in their skills, which has a ripple effect across the company. Employees who have a good sense of self-worth are more likely to be cooperative. These behaviors encourage team collaboration. Employees who have agency over their work environment are more likely to take ownership of their work and strive for excellence. Therefore, businesses gain from empowered workers since they produce superior goods and services.

Empowering staff to make their own choices provides additional advantages, such as improved response times and fewer service interruptions. Employee autonomy has been linked to increased creativity. An employee who is given more autonomy may be able to perceive a problem in a new light and come up with an original solution. Costs related to customer service difficulties may be reduced in part because of empowered staff. Manager–worker relationships may be strengthened via workers' increased autonomy (Gresley, 2005).

These personnel shifts are crucial to an organization's success because they:

Improve workers' abilities to do complex jobs.
Instruct new hires on the tried-and-true methods of doing things.
Enhance efficiency and quality of execution; conform to legal requirements.

Development includes a variety of training and education programs designed to help employees mature. There is less of an emphasis on specific talents and more on broad education and attitude, both of which will serve workers well in their future roles. Individual motivation and desire are very important in any growth effort.

Among the many reasons businesses invest in their employees' professional growth and development are the following:

To address weaknesses
To improve performance
To reassure and appease employees
To boost productivity
To share vital information
To enable and upgrade skills (including adaptability, dexterity with technology, the ability to convey meaning, the capacity to read emotions, and the capacity to avoid or deal with collusion)
To increase productivity

Finding your way through the digital age: Because of the flashy nature of the business world and technological advancements, it is important to invest in the training and education of your team in order to successfully adopt and use new technologies.

11.6 RESEARCH RESULTS

From what has been discussed, it is clear that employee involvement, feedback and information sharing, delegating authority, training, rewards, recognition, and appreciation are the most important employee empowerment practices. As a consequence of these practices, employees report higher levels of satisfaction with their jobs, performance, motivation, creativity, innovation, cost savings, loyalty, commitment, enthusiasm, stress relief for managers, and a stronger sense of teamwork and camaraderie.

To summarize the previous discussions of literature studies, it can be claimed that in recent years, employee empowerment has been presented as an influential tool to increase performance. It's got some good points and some bad ones. The various difficulties of employee empowerment are in line with findings from a variety of other studies. Here are several examples:

- Managers don't truly comprehend what employee empowerment entails.
- Managers don't set limits on employees' autonomy.
- Employees' carelessness might damage the company in a number of ways.
- Employees may have disagreements with their co-workers or bosses from time to time.

- When managers lack the expertise to properly delegate tasks to their subordinates, it may lead to mismanagement of staff.
- The existence of empowerment may put a lot of strain on workers.
- Lack of training might make workers hesitant to complete assignments.
- Managers may be reluctant to disclose problems at work out of concern for their own jobs, while workers may be reluctant to report problems out of concern that their feedback may be misconstrued.

11.7 STUDY RECOMMENDATIONS

In today's fast-paced and competitive business environment, it is essential for companies to have policies that encourage employee autonomy and initiative. Building employee agency requires concerted efforts from all relevant parties. In light of the aforementioned, the following steps might be considered for the efficient growth of employee empowerment:

- Organizations can increase employee motivation and loyalty by giving them interesting, difficult jobs to do. Innovation and creativity can only happen if employees are given the opportunity to contribute.
- The values, norms, and perspectives of the staff should be respected (Ethica, 2013).
- Businesses should foster an approachable setting where workers may feel comfortable talking to one another and where employees get the attention they need to develop mutually beneficial bonds.
- Adequate training opportunities must be guaranteed in all businesses.
- Workers need to be held to a higher standard of accountability and independence.
- Creating awareness for empowerment may make employee empowerment more acceptable in the culture.

11.8 CONCLUSION AND RECOMMENDATIONS FOR FURTHER STUDY

The top employee training statistics of Peck Delvin reveal the following:

- Organizations with effective employee training programs have 218% higher income per employee than Organizations without effective training.
- 17% of employees of organizations are more productive when they undergo effective training every year.
- 59% of workforces realize training directly improves their productivity.

- Less than one-third of the workforce is satisfied with the existing opportunities for promotion.
- A maximum of 68% of employees show their interest to learn and train at workplaces.
- Around 45% of the workforce is more likely to continue in their roles and responsibilities if they get training.
- More than 90% of employees believe they won't leave the job if they get career advancement opportunities.
- 92% of the workforce are of the opinion that workplace training affects their job engagement positively.
- To remain competitive, 40% of Fortune 500 organizations utilize learning management systems.
- 10% of the skills of the workforce can be enhanced by regular training given to the employees in organizations on a yearly basis.

It is essential to equip the workforce of an organization to expand and thrive in the future with training. These days, workers know their rights better than ever before. Therefore, in order for businesses to keep up with the times, their leaders must grasp the essence of the empowerment ideology and implement it across the company. Both managers and workers may benefit from the suggestions made here in order to realize the full potential of employee empowerment. Managers will be motivated to use a wide range of employee empowerment strategies after reading this article. However, field study validation would have made for a more robust study. Creating surveys and collecting answers from staff members is a necessary step in this validation process.

ACKNOWLEDGMENTS

The authors thank the book editors and anonymous reviewers for their valuable feedback on the manuscript.

REFERENCES

Blank, William E., *Handbook for Developing Competency- Based Training Programmes*, New York, NJ: Prentice-Hall (1982).

Bogler, R. and Somech, A., "Influence of teacher empowerment on teachers' organizational commitment, professional commitment and organizational citizenship behavior in schools". *Teaching and Teacher Education* (2004), Vol. 20(3), pp. 277–289.

Brower, M., "Empowering teams: What, why, and how". *Empowerment in Organizations* (1995), Vol. 3(1), pp. 13–25.

Brymer, R.A., "Employee empowerment, a guest-driven leadership strategy". *The Cornell Quarterly* (1991), Vol. 32, pp. 56–58.

Cheng, E.W.L. and Ho, D.C.K., "The influence of job and career attitudes on learning motivation and transfer". *Career Development International* (2001), Vol. 6, pp. 20–27.

Chhaya, L., Sharma, P., Bhagwatikar, G. and Kumar, A., Wireless sensor network based smart grid communications: Cyber-attacks, intrusion detection system, and topology control. *Electronics* (2017), Vol. 6(1), p. 5.

Conger, J. and Kanungo, R., "The empowerment process: Integrating theory and practice". *Academy of Management Journal* (1988), Vol. 13, pp. 471–482.

Dobbs, J., "The empowerment environment". *Training and Development* (1993), pp. 55–57.

Ethica, T., Employee empowerment: A critical review, Dhaka university. *Journal of Management* (2013), Vol. 5(1), pp. 1–16.

Goel, A., Goel, A.K. and Kumar, A., "Performance analysis of multiple input single layer neural network hardware chip". *Multimedia Tools and Applications* (2023), Vol. 82 (18), pp. 1–22.

Goel, A., Goel, A.K. and Kumar, A., "The role of artificial neural network and machine learning in utilizing spatial information". *Spatial Information Research* (2023), Vol. 31(3), pp. 275–285.

Gresley, K., "Employee empowerment". *Employee Relation* (2005), pp. 364–368.

Gupta, S. and Kumar, A., Study on early accurate diagnosis and treatment of COVID-19 with smartphone tracking using bionics. *Security and Privacy* (2023), Vol. 6 (5), p. e303.

Hardy, C. and Leiba-O'Sullivan, S., "The power behind empowerment: Implications for research and practice". *Human Relations* (1998), Vol. 51(4), pp. 451–483.

Heathfield, S.M., "Empowerment definition and examples of empowerment" (2012). Retrieved on December 2, 2013. http://humanresources.about.com/od/glossarye/a/empowerment_def.htm.

Holden, L., "Employee involvement and empowerment" (2001). Retrieved on December 12, 2013. https://ulib.derby.ac.uk/ecdu/CourseRes/dbs/manpeopl/hold.pdf.

Kinlaw, D.C., *The Practice of Empowerment: Making the Most of Human Competence*, Hampshire: Gower Publishing Ltd., 1995.

Kim, S., "Participative management and job satisfaction: Lessons for management leadership". *Public Administration Review* (2002), Vol. 62(2), pp. 231–243.

Kotter, J.P., "Leading change: Why transformation efforts fail". *Harvard Business Review* (1995), pp. 59–67.

Kuo, Tsung-Hsien, Ho, L.A., Lin, C. and Lai, K.K., "Employee empowerment in a technology advanced environment". *Industrial Management and Data Systems* (2009), Vol. 110(1), pp. 24–42.

Lawler, E.E., *High Involvement Management*, San Francisco: Jossey-Bass, 1986.

Locke, E.A. and Latham, G.P., *A Theory of Goal Setting and Task Performance*, Englewood Cliffs, NJ: Prentice Hall, 1990.

Lorsch, J.W., "Empowering the board". *Harvard Business Review* (1995), Vol. 73 (1), pp.107–117.

Liden, R., Wayne, S. and Sparrowe, R., "An examination of the mediating role of psychological empowerment on the relations between the job, interpersonal relationships and work outcomes". *Journal of Applied Psychology* (2000), Vol. 85(3), pp. 407–416.

Nonaka, I. and Takeuchi, H., *The Knowledge Creating Company*, New York: Oxford University Press, 1991.

Quinn, R. and Spreitzer, G., "The road to empowerment: Seven Questions every leader should consider". *Journal of Organizational Dynamics* (1999), Vol. 26(2), pp. 37–49.

Randolph, W.A., "Navigating the journey to empowerment". *Journal of Organizational Dynamics* (1995), Vol. 23(4), pp. 19–32.

Rawat, A.S., Rana, A., Kumar, A. and Bagwari, A., Application of multi layer artificial neural network in the diagnosis system: A systematic review. *IAES International Journal of Artificial Intelligence* (2018), Vol. 7(3), p. 138.

Smith, C. and Mouly, V., "Empowerment in New Zealand firms: Sights two cases". *Journal of Empowerment and Organizations* (1998), Vol. 6, pp. 69–80.

Smith, Jane, *Empowerment People, Translating* Persian By Said Bagherian, Tehran: Khorram Publications, 2002.

Suminen, T, "Work empowerment as experienced by head nurses". *Journal of Nursing Management* (2005), Vol. 13(2), pp. 147–153.

Sumukadas, N. and Sawhney, R., "Workforce agility through employee involvement". *IIE Transactions* (2004), Vol. 36(10), pp. 1011–1021.

Tzafirir, S., "The consequences of emerging HRM practices, trust in their managers", *Journal of Personnel Review* (2004), Vol. 33, pp. 624–647.

Vogt, J.F. and Murrel, K.L., *Empowerment in Organizations: How to Spark Exceptional Performance*, San Diego, CA: University Associates (1990).

Ziyakashany, L., "The role of human resource empowerment on organizational effectiveness". *Sanate Lastike Iran* (2009), Vol. 53, pp. 94–101.

Chapter 12

IoT and blockchain technology as transformative technology

Vijai, Worakamol Wisetsri, and Deval Verma

12.1 BLOCKCHAIN TECHNOLOGY

Blockchain technology has had a profound effect on the economic sector, disrupting conventional structures and introducing new opportunities for performance, security, and transparency (Nakamoto, 2008; Tapscott & Tapscott, 2016; Swan, 2015). Its precise features, along with decentralization, immutability, and cryptographic safety, have paved the way for several modern packages in finance (Raval, 2016). Some of the important ways wherein blockchain is revolutionizing the economic zone include:

1. **Decentralized transactions:** Blockchain allows peer-to-peer transactions without the need for intermediaries like banks (Antonopoulos, 2017). This has given rise to cryptocurrencies, inclusive of Bitcoin and Ethereum, that permit human beings to conduct direct economic transactions globally.
2. **Faster and cheaper cross-border payments:** Traditional cross-border transactions are frequently sluggish and steeply priced due to intermediaries and agreement techniques. Blockchain streamlines these payments, reducing prices and agreement instances considerably.
3. **Smart contracts:** Smart contracts are self-executing contracts with terms and conditions written into the code (Pilkington, 2015; World Economic Forum, 2016). These contracts automatically execute at the same time as predefined conditions are met, removing the need for intermediaries and decreasing the threat of fraud.
4. **Improved security and fraud prevention:** Blockchain's immutable nature guarantees that after transactions are recorded, they cannot be altered or tampered with, offering robust protection and fraud prevention.
5. **Enhanced identity management:** The blockchain era may be leveraged to create decentralized and relaxed identity control systems, allowing people to curate their private profiles and control how they share data.

DOI: 10.1201/9781003510420-12

6. **Efficient supply chain management:** Blockchain facilitates real-time tracking of goods and financial transactions throughout the supply chain, enhancing transparency and lowering fraud.
7. **Tokenization of assets:** Blockchain permits the tokenization of belongings, changing real-world assets like real estate or artwork into digital tokens. This allows fractional possession and less complicated transferability, growing liquidity, and accessibility.

12.2 KEY CHARACTERISTICS OF BLOCKCHAIN

Blockchain possesses several key characteristics that make it precise and special. These traits are foundational to its functioning and have contributed to its widespread adoption throughout numerous industries (European Parliament, 2017). The key characteristics of blockchain are:

1. **Decentralization:** Blockchain operates as a decentralized community of nodes, without the use of a government controlling the system. Transactions and statistics are distributed across the network, improving safety, transparency, and resilience.
2. **Immutability:** Once statistics are recorded on a blockchain, they cannot be altered or deleted retroactively. Each block includes a cryptographic hash of the preceding block, developing a chain of blocks that can be linked collectively. This immutability ensures the integrity of the records and makes blockchain tamper-resistant (Gartner, 2020).
3. **Transparency:** Blockchain is a transparent machine, wherein all transactions and statistics are visible to all contributors on the network. This transparency complements agreement with and duty among customers and gets rid of the need for third-party intermediaries to validate transactions.
4. **Security:** Blockchain's safety is completed through superior cryptographic algorithms that verify the statistics and save you from unauthorized entry or tampering. Consensus mechanisms, inclusive of proof of work (PoW) and proof of stake (PoS), further enhance the protection and integrity of the community.
5. **Consensus mechanism:** Blockchain is primarily based on a consensus mechanism to agree on the validity of transactions and add them to the blockchain. Various consensus algorithms, which incorporate PoW, PoS, and delegated proof of stake (DPoS), ensure agreement among network users.
6. **Privacy:** While blockchain is on public databases, it may additionally offer varying levels of privacy. Some blockchains hire non-public or permissioned setups by which the right of entry is restricted to authorized participants.

7. **Trustless system:** Blockchain allows members to transact and interact without worry as to whether the other party can be trusted. The trust is placed within the system and the consensus mechanisms that ensure the validity of transactions.

12.3 OPPORTUNITIES FOR ADOPTION OF BLOCKCHAIN TECHNOLOGY IN FINANCE SECTOR

The adoption of the blockchain era within the finance area affords several opportunities that can result in transformative changes and great benefits for various stakeholders. Some of the important possibilities encompass:

1. **Enhanced security and data integrity:** Blockchain's inherent cryptographic mechanisms and decentralized nature offer reliable safety, making it extremely difficult for malicious actors to tamper with records or launch cyberattacks (Financial Stability Board, 2019). This heightened protection is valuable in a sector coping with sensitive financial facts and transactions (KPMG, 2019).
2. **Increased transparency:** Blockchain's obvious and immutable nature permits real-time monitoring and auditing of monetary transactions (Böhme et al., 2015). This level of transparency can assist in fostering agreement among clients, consumers, and regulators, fostering additional open and accountable economic surroundings.
3. **Streamlined cross-border payments:** Blockchain technology can facilitate quicker and much less high-priced cross-border transactions by removing the need for intermediaries and complicated agreement approaches. This can motivate value savings, faster remittances, and superior monetary inclusion for underserved populations.
4. **Cost efficiency:** Implementing blockchain solutions could lessen operational fees for economic institutions (Benos et al., 2017). By automating techniques and removing intermediaries, blockchain streamlines numerous functions, together with settlement, clearing, and reconciliation.
5. **Improved accessibility and financial inclusion:** The blockchain era can provide financial offerings to unbanked and underbanked populations in remote and underserved regions. DeFi structures offer several financial services without the need for classic intermediaries.
6. **Smart contracts and automation:** Smart contracts allow the automated execution of predefined conditions without human intervention. This feature can considerably streamline diverse financial processes, such as lending, insurance, and compliance, leading to quicker and greater green operations.

7. **Tokenization of assets:** Blockchain permits the fractional ownership and tokenization of real-world assets, making it less complicated for traders to take part in previously illiquid markets, such as real estate and fine art (Chiu & Koeppl, 2017). This democratization of asset possession can release new investment opportunities.

12.4 BLOCKCHAIN APPLICATIONS IN FINANCIAL SECTOR

Blockchain applications inside the economic sector have received huge traction, remodeling the manner monetary transactions are performed and reshaping diverse aspects of the enterprise. Some of the blockchain applications used within the financial arena include:

1. **Cryptocurrencies:** The famous blockchain utility is the introduction and use of cryptocurrencies like Bitcoin and Ethereum (Gipp et al., 2018). These digital currencies perform on blockchain networks, permitting ease and decentralized transactions without the need for classic economic intermediaries.
2. **Cross-border payments:** Blockchain-based cross-border payments offer quicker and secured remittances. By avoiding intermediaries and leveraging cryptocurrency or stablecoins, these structures facilitate real-time international transactions.
3. **Smart contracts:** Smart contracts are self-executing agreements with predefined conditions written into code. Blockchain enables the execution of smart contracts without human intervention, automating strategies like insurance claims, trade settlements, and loan agreements (Accenture, 2017).
4. **Identity verification:** Blockchain presents a decentralized and unrestrained identification verification system. Individuals can prevent access to their private information and selectively share it with trusted parties, reducing the chance of identity fraud.
5. **Supply chain management:** Blockchain complements supply chain transparency by recording every step of a product's movement on an immutable ledger (National Institute of Standards and Technology, 2018). This allows higher traceability, reduces counterfeiting, and guarantees compliance with regulations.
6. **Digital asset tokenization:** Blockchain permits real assets, including real estate, art, and commodities, to be tokenized. This fractional ownership of belongings will increase liquidity and accessibility to a broader range of traders.
7. **Decentralized finance:** DeFi systems leverage blockchain to provide decentralized economic offerings, together with lending, borrowing,

staking, and decentralized exchanges. DeFi removes the need for classic intermediaries, providing more financial inclusion.

8. **Trade finance:** Blockchain streamlines trade finance with the aid of digitized information, which includes letters of credit and bills of lading. This complements efficiency, reduces paperwork, and minimizes the chance of fraud.

9. **Regulatory compliance and auditing:** Blockchain's transparency simplifies regulatory compliance and auditing strategies. Regulators can access real-time transaction facts, making sure financial institutions adhere to compliance requirements.

10. **Central bank digital currencies (CBDCs):** Some international locations are exploring the issuance of CBDCs in the use of blockchain. CBDCs offer the advantages of real-time agreement, economic inclusion, and easier implementation of economic rules (Chen et al., 2018). The adoption of blockchain packages in the economic sector is progressively growing, with both conventional economic establishments and innovative startups exploring their capacity (Yli-Huumo et al., 2016). However, challenges, which include scalability, interoperability, regulatory compliance, and public notion stay as areas that need to be addressed to ensure the successful integration of blockchain technology into mainstream economic services.

12.5 BLOCKCHAIN WITHIN THE FINANCIAL SECTOR

Blockchain use has had a profound effect on the economic sector, disrupting conventional structures and introducing new possibilities for performance, security, and transparency. (Hayes, 2017). At its core, blockchain is a decentralized and immutable ledger that keeps a continuously developing listing of information (blocks) connected through cryptography. Each block includes a timestamp and a connection with the previous block, developing a secure and tamper-resistant chain of data.

1. **Cryptocurrencies:** One of the most famous packages of blockchain in finance is the introduction and use of cryptocurrencies like Bitcoin and Ethereum (Saxena et al., 2017). These virtual currencies permit peer-to-peer transactions without the need for intermediaries like banks, revolutionizing the manner cash is transferred and stored.

2. **Cross-border payments:** Traditional cross-border transactions may be sluggish and costly due to multiple intermediaries and complicated settlement tactics. Blockchain-based totally solutions streamline these bills, substantially reducing charges and settlement instances.

3. **Smart contracts:** Smart contracts are self-executing contracts with the terms and conditions automatically written into code (Wood, 2014).

These contracts automatically execute while predefined situations are met, removing the need for intermediaries and decreasing the danger of fraud.

4. **Identity management**: Blockchain can be leveraged to create decentralized and secure identity management structures. Individuals can curate their non-public information and selectively share it with relied-on parties, enhancing protection and reducing the hazard of identification theft (McKinsey & Company, 2017).

5. **Supply chain management**: Blockchain permits real-time monitoring and auditing of products and financial transactions at every point in the supply chain. This stronger transparency facilitates less fraud and ensures compliance with regulations.

6. **Tokenization of assets**: Blockchain permits the share ownership and tokenization of real-world assets, making it less complicated for investors to participate in formerly illiquid markets like real estate and fine artwork.

7. **Decentralized finance**: DeFi structures leverage blockchain to offer diverse economic services without conventional intermediaries. These include lending, borrowing, staking, and decentralized exchanges, supplying extra economic inclusion and accessibility.

12.6 USE CASES OF BLOCKCHAIN TECHNOLOGY IN FINANCE

Blockchain has several compelling use instances within the finance sector. These use instances leverage the specific capabilities of blockchain, which include decentralization, transparency, and immutability, to revolutionize various aspects of economic services. Here are some extraordinary use cases of blockchain technology in finance (Zohar, 2015; Buterin, 2013). Blockchain generation gives numerous use cases inside the financial quarter, presenting modern answers to longstanding challenges and improving the performance and safety of monetary offerings. Some of the outstanding use instances of blockchain in finance include:

1. **Cross-border payments and remittances**: Blockchain permits quicker, cheaper, and greater ease of cross-border transactions by removing the need for intermediaries and reducing agreement times (Ethereum Foundation, 2019).

2. **Smart contracts**: Smart contracts automate the execution of contractual terms and conditions without intermediaries, permitting self-executing agreements for various economic methods which include lending, insurance, and exchange settlements.

3. **Tokenization of assets:** Blockchain permits the share ownership and tokenization of real-world assets, making it less complicated to exchange and transfer ownership of belongings like actual property, art, and commodities.

4. **Decentralized finance:** DeFi systems make use of blockchain to provide a wide range of economic offerings, consisting of lending, borrowing, staking, and decentralized exchanges, without the need for classic intermediaries.

5. **Supply chain management and trade finance:** Blockchain enhances supply chain transparency and efficiency, facilitating trade finance by digitizing trade documents and decreasing paperwork.

6. **Identity verification and KYC:** Blockchain-based identity control systems offer secure and decentralized identity verification, improving KYC (know your customer) approaches and reducing the risk of identity fraud.

7. **Regulatory compliance and auditing:** Blockchain's transparency and immutability simplify regulatory compliance and auditing tactics, imparting real-time access to transaction statistics.

8. **Central bank digital currencies:** Some principal banks are exploring using blockchain to transact CBDCs, developing digital representations of countrywide fiat currencies for secure and green transactions.

9. **Fraud prevention:** Blockchain's immutability and traceability abilities assist in preventing fraud and enhancing the safety of economic transactions.

12.7 ETHICAL ISSUES FOR BLOCKCHAIN

Blockchain technology introduces several ethical issues that need careful consideration and resolution as it becomes more prevalent in various industries. Some of the key ethical issues for blockchain include:

1. **Cross-border payments and remittances:** Blockchain permits quicker, cheaper, and easier cross-border transactions by removing the need for intermediaries and reducing settlement instances. (Ethereum Foundation, 2019).

2. **Smart contracts:** Smart contracts automate the execution of contractual terms and conditions without intermediaries, allowing self-executing agreements for numerous financial methods, which consist of lending, insurance, and change settlements.

3. **Tokenization of assets:** Blockchain allows the shared ownership and tokenization of real-world belongings, making it much less complicated to exchange and switch ownership of assets like actual property, artwork, and commodities.

4. **Decentralized finance:** DeFi systems employ blockchain to provide a huge range of economic services, inclusive of lending, borrowing, staking, and decentralized exchanges, without the need for classic intermediaries.
5. **Supply chain management and trade finance:** Blockchain enhances delivery chain transparency and performance, facilitating trade finance through digitizing trade files and decreasing paperwork.
6. **Identity verification and KYC:** Blockchain-based identity control structures provide comfy and decentralized identification verification, improving KYC methods and reducing the hazard of identification fraud.
7. **Regulatory compliance and auditing:** Blockchain's transparency and immutability simplify regulatory compliance and auditing tactics, supplying real-time access to transaction records.
8. **Central bank digital currencies:** Some main banks are exploring the use of blockchain to CBDCs, growing digital representations of countrywide fiat currencies for secure and transparent transactions.
9. **Digital identity and security:** Blockchain-based identity systems hold personal information on-chain, and protecting this data from unauthorized access or misuse is a critical ethical consideration.

12.8 DEVELOPMENT OF BLOCKCHAIN IN THE FINANCIAL SECTOR

Blockchain development in the financial industry has been transformative, supplying revolutionary solutions to lengthy status-demanding situations and reshaping diverse components of finance (Littman & Pitts, 2016; PwC, 2018). Since the advent of Bitcoin in 2009 as the first blockchain utility, the financial sector has been one of the earliest adopters of this technology. Here are a few key milestones and traits of blockchain within the economic area:

1. **Emergence of cryptocurrencies:** Bitcoin's introduction marked the birth of cryptocurrencies, which perform on blockchain technology. Bitcoin's success as a digital forex and its consistency paved the way for the improvement of different cryptocurrencies, which include Ethereum, Ripple, and Litecoin.
2. **Initial coin offerings (ICOs):** ICOs have become a popular method for startups and initiatives to elevate finances by issuing their very own tokens or cash on blockchain structures (Huckle et al., 2016). While ICOs faced regulatory challenges, they demonstrated the ability for blockchain-based crowdfunding.

3. **Decentralized finance:** DeFi emerged as a massive improvement in the financial area, imparting numerous financial services without traditional intermediaries. (Agboh, 2019). DeFi systems offer lending, borrowing, staking, yield farming, and decentralized exchanges, among others, using clever contracts on blockchain networks.

4. **Enterprise blockchain solutions:** Financial establishments and businesses started exploring non-public and permissioned blockchain networks for unique use instances like supply chain control, alternative finance, and cross-border transactions (Capgemini Research Institute, 2017). These networks offer scalability and privacy for company applications.

5. **Central bank digital currencies:** Several vital banks around the sector began studying and experimenting with CBDCs, exploring the opportunity of issuing virtual variations of their countrywide currencies in the blockchain.

6. **Interoperability and collaboration:** Various blockchain consortia and alliances have been fashioned to sell interoperability between one-of-a-kind blockchain networks and to collaborate on enterprise standards and high-quality practices.

7. **Regulatory frameworks:** Governments and regulators worldwide have been operating on growing regulatory frameworks for cryptocurrencies, ICOs, and other blockchain-based economic offerings to protect customers and foster innovation.

12.9 INFLUENCE OF BLOCKCHAIN ON THE FINANCIAL INDUSTRY

Blockchain has had a profound impact on the financial industry, disrupting conventional structures and introducing transformative changes. Some of the key effects of blockchain in the monetary area consist of:

1. **Decentralization and disintermediation:** Blockchain allows peer-to-peer transactions without the need for intermediaries like banks, decreasing the dependency on conventional monetary institutions and potentially lowering transaction costs.

2. **Faster and cheaper cross-border payments:** Blockchain-based cross-border transactions offer faster agreement instances and decreased charges as compared to traditional global cash transfers, reaping rewards for businesses and individuals alike.

3. **Enhanced security and fraud prevention:** Blockchain's immutable and transparent nature provides protection and reduces the danger of fraud, as every transaction is recorded and verified via a couple of contributors in the network.

4. **Smart contracts and automation:** Smart contracts automate contract execution based on predefined situations, streamlining diverse economic contracts including insurance claims, change settlements, and mortgage agreements.

5. **Tokenization of assets:** Blockchain enables the shared possession and tokenization of real-world property, consisting of real estate and commodities, increasing liquidity and accessibility to a broader range of buyers.

6. **Decentralized finance:** DeFi systems leverage blockchain to offer a wide range of financial offerings without conventional intermediaries, inclusive of lending, borrowing, decentralized exchanges, and yield farming.

7. **Improved identity management:** Blockchain-based identity verification systems provide secure and decentralized control over non-public records, decreasing the risk of identification fraud.

8. **Supply chain transparency:** Blockchain complements delivery chain control with the aid of presenting real-time monitoring and verification of products and financial transactions, improving transparency and decreasing fraud.

9. **Regulatory compliance and auditing:** Blockchain's transparent and auditable nature simplifies regulatory compliance and auditing strategies, allowing real-time access to transaction records.

12.10 BLOCKCHAIN BENEFITS AND OPPORTUNITIES IN GLOBAL FINANCIAL INDUSTRY

Blockchain technology offers a wide range of benefits and opportunities for the global financial industry, revolutionizing traditional financial processes and creating new possibilities. Some of the key benefits and opportunities include:

1. **Enhanced security:** Blockchain's decentralized and cryptographic nature offers strong protection in opposition to fact tampering and unauthorized access, reducing the hazard of fraud and cyberattacks.

2. **Transparency and accountability:** Blockchain's obvious and immutable ledger ensures that each transaction is visible to network individuals, promoting belief and accountability among stakeholders.

3. **Faster and cheaper transactions:** Blockchain-based structures allow faster and greater cost-effective cross-border bills and settlements by disposing of intermediaries and lowering transaction times.

4. **Smart contracts and automation:** Smart contracts automate agreement execution based on predefined conditions, streamlining techniques like trade settlements, coverage claims, and compliance assessments.

5. **Financial inclusion:** Blockchain can make more financial services available to unbanked and underbanked populations, imparting them with access to banking, lending, and other economic services.
6. **Regulated financial products:** Blockchain-based systems can provide regulated monetary merchandise like protection tokens, presenting a greater efficient and transparent approach for traders to take part in conventional markets.
7. **Improved identity verification:** Blockchain's decentralized identity systems provide people with control over their private statistics, enhancing the security and performance of identification verification strategies.
8. **Interoperability and collaboration:** Blockchain fosters collaboration between economic establishments and innovators, permitting interoperability between exclusive systems and systems.
9. **Data privacy and consent:** Blockchain can offer records privacy by providing users with extra control over their information and the capability to provide consent for information sharing.

12.11 CHALLENGES AND RISKS OF BLOCKCHAIN IN FINANCIAL SECTOR

While blockchain gives numerous advantages, it also comes with several challenges and risks, especially when applied within the monetary area (Xue et al., 2020; Crosby et al., 2016). Some of the important challenges and dangers of blockchain within the financial enterprise consist of:

1. **Scalability:** Public blockchain networks like Bitcoin and Ethereum face scalability troubles, limiting the variety of transactions they could method consistent with 2d. This ought to lead to slower transaction instances and higher fees at some stage in peak durations.
2. **Regulatory compliance:** The evolving regulatory scene for blockchain and cryptocurrencies can be complicated and uncertain (Deloitte, 2018). Financial establishments should navigate exclusive regulations throughout jurisdictions to ensure compliance, which may be difficult and time-consuming.
3. **Data privacy and GDPR compliance:** While blockchain offers transparency, a few applications may also require records privacy and compliance with General Data Protection Regulation (GDPR) laws. Striking a balance between transparency and privacy can be a challenge.
4. **Interoperability:** Ensuring seamless data sharing and collaboration between different blockchain networks and conventional financial structures remains a challenge (Iansiti & Lakhani,

2017). Interoperability standards should be established to facilitate integration.

5. **Security concerns:** While blockchain is known for its robust security, it isn't always foolproof against attacks. Smart contracts can be prone to coding errors and exploitation, leading to monetary losses.
6. **Energy consumption:** Some blockchain networks, especially those using proof of work consensus, consume massive amounts of electricity for mining and validation, elevating concerns about environmental impact.
7. **User experience:** Blockchain interfaces and pockets management can be complicated and intimidating for non-technical customers. Improving consumer satisfaction and presenting user-friendly interfaces can be vital for wider adoption.
8. **Legal and regulatory risks:** Financial products and services presented on blockchain structures might not usually align with existing monetary policies, leading to illegal practices and capability liabilities.
9. **Centralization risks:** While the idea of blockchain is to be decentralized, in practice, some blockchains can grow to be centralized due to a small number of powerful nodes or mining pools, raising issues about control and censorship.
10. **Token value volatility:** Cryptocurrencies and tokens can experience volatility, posing dangers for traders and groups coping with digital assets.
11. **Recovery of lost private keys:** If users lose access to their non-public keys or virtual wallets, it could bring about permanent loss of funds, as there's no centralized authority to help them recover their funds.

12.12 SUMMARY

Blockchain creates a decentralized record that can't be changed, which makes it hard for hackers to change transaction data. On the blockchain, smart contracts can automatically follow the rules and lower the risk of theft. By cutting out middlemen and shortening settlement times, blockchain makes cross-border trades faster and cheaper. It makes the payment process clearer, which cuts down on mistakes and disagreements. KYC is applied by keeping a safe record of customer information that can't be changed, blockchain can speed up and improve KYC processes. This can make the hiring process run more smoothly and help companies follow the rules better. Blockchain can be used to make supply lines clear and easy to track. This makes it safer for banks to offer supply chain financing. Lending choices are better when there is better visibility into activities in the supply chain. Smart contracts can handle many banking tasks, like approving loans, settling payments, and making sure that rules are followed. This makes things run more smoothly, saves money, and cuts down on mistakes.

Putting IoT and blockchain together promotes data security and integrity (Chhaya et al., 2018, 2020). When IoT is combined with blockchain, the data received by IoT devices is safer and more reliable. Blockchain's independent and open nature makes sure that data is safe and can't be changed. Blockchain can be used to make a record of IoT device data that can't be changed. This makes sure that different banking processes are open and can be tracked. AI can play an important role in this (Goel et al., 2023). When IoT and blockchain are combined, processes can be simplified and automated, which means that middlemen and manual steps are not needed as much. When used in IoT apps, smart contracts can be used to make transactions and agreements between devices more efficient and safer.

Blockchain technology has emerged as a revolutionary force in various industries, including the financial sector. Its decentralized and immutable nature has the potential to transform traditional financial systems, introducing greater efficiency, transparency, and inclusivity. Throughout this chapter, we explored the applications, benefits, and opportunities of blockchain in finance, as well as the challenges and risks that accompany its adoption. The financial industry has witnessed significant changes driven by blockchain technology. Cryptocurrencies have disrupted traditional payment systems, providing faster and cheaper cross-border transactions. Smart contracts have automated complex financial processes, reducing administrative costs and the risk of errors. DeFi platforms have opened a new realm of decentralized financial services, offering lending, borrowing, and yield farming without intermediaries. Blockchain technology holds tremendous promise for the financial industry. Its potential to enhance security, transparency, and accessibility is reshaping traditional financial systems. As the technology matures and addresses existing challenges, it is crucial for stakeholders, including financial institutions, technology innovators, regulators, and consumers, to collaborate in creating a balanced and sustainable blockchain ecosystem.

With responsible development, robust regulatory frameworks, and a focus on user experience, blockchain has the potential to foster financial inclusion, enable new business models, and revolutionize global financial services. As we move forward, the ethical considerations of blockchain's implementation must be at the forefront to ensure a fair and equitable financial landscape that benefits all participants. The future of blockchain in finance holds immense promise, and its transformative power is set to redefine the financial industry for years to come. However, even though the Internet of Things and blockchain have the potential to bring about major improvements in the banking business, it is imperative that the industry address difficulties such as regulatory compliance, scalability, and interoperability. For these technologies to be adopted, there will need for cooperative efforts among the various stakeholders, investments in infrastructure, and continual attempts to establish trust in them.

REFERENCES

Accenture. (2017). *Banking on Blockchain: Charting the Progress of Distributed Ledger Technology in Financial Services.*

Agboh, K. S. (2019). Blockchain and artificial intelligence as tools for fintech development in emerging economies. *Procedia Manufacturing*, 35, 1096–1102.

Antonopoulos, A. M. (2017). *Mastering Bitcoin: Unlocking Digital Cryptocurrencies.* O'Reilly Media.

Benos, E. et al. (2017). *The Economics of Distributed Ledger Technology for Securities Settlement.* Bank of England Staff Working Paper No. 670. Available at SSRN: https://ssrn.com/abstract=3023272 or http://dx.doi.org/10.2139/ssrn.3023272

Böhme, R., Christin, N., Edelman, B., & Moore, T., (2015). Bitcoin: Economics, technology, and governance. *Journal of Economic Perspectives*, 29(2), 213–238.

Buterin, V. (2013). *Ethereum White Paper.* Ethereum Project. https://github.com/ethereum/wiki/wiki/White-Paper.

CapGemini Research Institute. (2017). *Blockchain in Financial Services: A Reality Check and Five Building Blocks for Success.* CapGemini Research Institute.

Chen, C., et al., (2018). Blockchain based trust management in Vehicular Networks. *IEEE Transactions on Vehicular Technology*, 67(12), 11688–11698.

Chhaya, L., Sharma, P., Kumar, A., & Bhagwatikar, G., (2018). IoT-based implementation of field area network using smart grid communication infrastructure. *Smart Cities*, 1(1), 176–189.

Chhaya, L., Sharma, P., Kumar, A., & Bhagwatikar, G. (2020). Cybersecurity for smart grid: Threats, solutions and standardization. In Bhoi, A., Sherpa, K., Kalam, A., Chae, G. S. (eds) *Advances in Greener Energy Technologies*, Green Energy and Technology. Springer: Singapore.17–29. https://doi.org/10.1007/978-981-15-4246-6_2

Chiu, J., & Koeppl, T., (2017). *The Economics of Cryptocurrencies—Bitcoin and Beyond.* Bank of Canada Staff Working Paper No. 2014-33. Available at SSRN: https://ssrn.com/abstract=3048124 or http://dx.doi.org/10.2139/ssrn.3048124.

Crosby, M.A., et al., (2016). Blockchain technology: Beyond bitcoin. *Applied Innovation*, 2, 6–10.

Deloitte. (2018). *Breaking Blockchain Open: A Global Blockchain Survey.* Deloitte.

Ethereum Foundation. (2019). The state of Ethereum 2.0. https://ethresear.ch/t/beacon-chain-ethereum-2-0-finalized/6263.

European Parliament. (2017). *Virtual Currencies and Central Banks' Monetary Policy: Challenges Ahead.* European Parliament.

Financial Stability Board. (2019). *Decentralised Financial Technologies.* Financial Stability Board.

Gartner. (2020). *The Hype Cycle for Blockchain Technologies and Other Distributed.* LED.

Gipp, B., et al. (2018). Decentralized trusted timestamping using the crypto currency bitcoin. Proceedings of the International Conference on Theory and Practice of Digital Libraries (TPDL), 378–389.

Goel, A., Goel, A. K., & Kumar, A., (2023). The role of artificial neural network and machine learning in utilizing spatial information. *Spatial Information Research*, 31(3), 275–285.

Hayes, A. S., (2017). A brief survey of cryptocurrency systems. *PLoS One*, 12(1), e0169556.

Huckle, s., et al., (2016). Internet of things, blockchain and shared economy applications. *Procedia Computer Science*, 98, 461–466.

Iansiti, & Lakhani, K. (2017). The truth about blockchain. *Harvard Business Review*, 95(1), 118–127.

KPMG. (2019). *The Pulse of Fintech*, Q4 2018.

Littman, J., & Pitts, L. (2016). *Ethereum for Dummies: Everything You Need to Know about Ethereum, How to Mine Ethereum, How to Exchange Ethereum, and How to Buy ETH.*

McKinsey & Company. (2017). *Blockchain beyond the Hype: What Is the Strategic Business Value?*

Nakamoto, S. (2008). Bitcoin: A peer-to-peer electronic cash system. https://bitcoin.org/bitcoin.pdf.

National Institute of Standards and Technology (NIST). (2018). *Blockchain Technology Overview.*

Pilkington, M. (2015). *Blockchain Technology: Principles and Applications. Research Handbook on Digital Transformations*, 225.

PwC. (2018). *Blockchain Is Here. What's Your Next Move?*. PwC.

Raval, S. (2016). *Decentralized Applications: Harnessing Bitcoin's Blockchain Technology.* O'Reilly Media.

Saxena, S., et al., (2017). A survey of consensus protocols on blockchain applications. *Journal of Network and Computer Applications*, 107, 42–60.

Swan, M. (2015). *Blockchain: Blueprint for a New Economy.* O'Reilly Media.

Tapscott, D., & Tapscott, A. (2016). *Blockchain Revolution: How the Technology Behind Bitcoin Is Changing Money, Business, and the World.* Portfolio.

Wood, G., (2014). Ethereum: A secure decentralised generalised transaction ledger. *Ethereum Project Yellow Paper*, 151. 1–32

World Economic Forum. (2016). *The Future of Financial Infrastructure: An Ambitious Look at How Blockchain Can Reshape Financial Services.* World Economic Forum.

Xue, X., et al. (2020). A blockchain-based secure and transparent data sharing scheme for smart city applications. *Future Generation Computer Systems*, 108, 996–1008.

Yli-Huumo, J., et al. (2016). Where is current research on blockchain technology?—A systematic review. *PloS One*, 11(10), e0163477.

Zohar, A. (2015). Bitcoin: Under the hood. *Communications of the ACM*, 58(9), 104–113.

Chapter 13

IoT in schools

Revolutionizing education through smart technology

Priyanka Kaushik

13.1 INTRODUCTION

About 29% of India's organizations have adopted the Internet of Things (IoT) in the education sector for student and institution monitoring purposes. In 2019, Europe had the second largest IoT in education market share of 29.8%. By 2023, the IoT in education market in Asia–Pacific (APAC) is expected to reach $7.01 billion.

The advent of IoT has ushered in a new era of connected devices and intelligent data analytics. In the education sector, IoT offers the potential to revolutionize teaching and learning by integrating smart devices and applications into classrooms, thereby creating an interconnected and efficient learning environment. This chapter aims to elucidate the various applications of IoT technology in the education sector, highlighting its potential to enhance educational processes and outcomes.

The Internet of Things, a network of interconnected devices beyond standard computers and smartphones, has significantly impacted various aspects of modern life. While its applications in education might not be immediately apparent, the education sector is undergoing a transformation owing to IoT technology. The integration of IoT in education has brought about advancements, ranging from enhanced safety measures and efficient resource management to improved access to information within the learning environment.

IoT-driven devices have swiftly supplanted traditional tools like pencils, chalkboards, and paper, ushering in a transformative era for instructional methods in education. In recent years, educational institutions, once reliant on conventional teaching approaches, have enthusiastically embraced technology. Recognizing its potential to significantly enhance learning outcomes, industry leaders have embraced IoT with zeal. From flipped classrooms and SMART boards integrating augmented and virtual reality capabilities to the provision of secure learning platforms, IoT has gained traction among eager adopters in the education sector.

DOI: 10.1201/9781003510420-13

Amidst the COVID pandemic, educational institutions worldwide had to pivot from in-person instruction to curb the spread of the novel coronavirus. Caught mid-academic year, both teachers and students were compelled to adapt to a crisis-mode continuation of their studies. The education system shifted to remote learning, and IoT played a pivotal role in aiding this transition for teachers and students alike. However, the journey through e-learning is ongoing, with IoT poised to further bridge gaps and lay the groundwork for innovative companies focusing on digital education.

13.2 DEFINITION OF IOT

The Internet of Things, or IoT, is a system of interconnected (networked) devices that can communicate with each other, sharing information. It can include sensors, cameras, robots, and user interfaces such as mobile phones. This integrated system of devices can either be local or global.

In IoT, the term "things" encompasses a broad spectrum of devices equipped with diverse sensors, capable of autonomously gathering and transmitting data over networks without human intervention. The embedded technology in these objects facilitates interaction with both internal states and the external environment, aiding in decision-making processes.

The IoT is essentially a framework wherein every entity possesses a representation and presence on the internet. It strives to introduce novel applications and services that bridge the physical and virtual realms. Machine-to-machine (M2M) communications form the fundamental mode of communication, enabling seamless interactions between "things" and cloud-based applications. This perspective is corroborated by *IEEE Communication Magazine*.

Oxford Dictionaries provides a concise definition, considering the internet as a critical component of IoT: "Internet of things (noun): The interconnection via the Internet of computing devices embedded in everyday objects, enabling them to transmit and receive data."

In practice, the IoT establishes a comprehensive information system, composed of interconnected smaller information systems. Smart devices are interconnected within smart home systems, which in turn are linked to broader smart city systems. However, it's essential to recognize that the reality of IoT is considerably more intricate than this simplified representation.

For example, in an office building, the lighting, CCTV, printers, desktop computers, HVAC system, and biometric attendance system can all be connected with one another. The various devices in the network can be automated to adjust to the needs and activities of the employees.

It's possible to customize the coffee maker based on employee preferences. The system could also be programmed to dim or switch off the lights based on the presence or absence of people in a room.

In a more personal setting, many homes are now using virtual assistant technology such as Alexa, not only for entertainment purposes but also to integrate household functions, such as video doorbells and lights.

13.3 IOT IN THE EDUCATION INDUSTRY: PIONEERING A NEW ERA

The transformative impact of IoT on education warrants a thorough examination, considering its potential to augment existing e-learning and SMART board infrastructure through connected smart devices utilized by numerous educational institutions.

The IoT sector within education is projected to experience substantial growth, with a remarkable compound annual growth rate (CAGR) of 14.22%, propelling it from $6.01 billion in 2021 to an anticipated $17.42 billion by 2028. This surge represents a nearly threefold increase, signifying a substantial leap forward. Several factors contribute to this astounding adoption:

Popularity of IoT-enabled devices: The simplicity and cost-effectiveness of these IoT-enabled gadgets have fueled their widespread adoption.

Advancements in cloud services: With unlimited storage capacities and flexible payment models, IoT facilitates convenient and affordable data access, analysis, and storage.

Tailored specialized applications: The ability to swiftly develop highly specialized apps to cater to the unique needs of educational institutions has further propelled IoT adoption.

IoT possesses the transformative potential to streamline operations for all stakeholders, including teachers, administrators, parents, and students. Through efficient resource utilization, better results can be achieved with equivalent effort. IoT applications encompass various facets of the education sector, delivering solutions that:

- Optimize application management to maximize cost-effectiveness and viability.
- Facilitate classroom management for an enhanced learning experience.
- Enhance engagement in academics by fostering stronger connections among stakeholders.

The emergence and integration of IoT within the education sector have fundamentally altered the approach to learning. By minimizing idle time and maximizing operational efficiency, this technology paves the way for an automated learning environment. IoT seamlessly connects diverse devices to the internet, enabling convenient data sharing and processing. A simple touch now transmits information swiftly and seamlessly, without requiring human intervention. Consequently, IoT in the education sector is steering the creation of automated learning environments, revolutionizing the landscape of education.

13.4 KEY CHARACTERISTICS OF IOT

- Interconnectivity: In the realm of IoT, any entity can seamlessly connect to the global information and communication infrastructure.
- Thing-related services: IoT has the capability to offer services tailored to specific things, ensuring privacy protection and maintaining semantic consistency between physical and virtual representations of these entities. Achieving this requires adjustments in both the technologies of the physical and information worlds.
- Heterogeneity: IoT devices come in diverse forms, utilizing various hardware platforms and networks. They can effectively interact with other devices or service platforms through different network types.
- Dynamic changes: Devices within IoT experience dynamic changes in their states, transitioning between phases like sleep and wakefulness, connected and disconnected, and altering context elements such as location and speed. Additionally, the number of devices can vary dynamically.
- Enormous scale: The IoT ecosystem involves a vast number of devices that necessitate efficient management and communication. This scale is significantly larger than the devices currently connected to the traditional internet. Managing the immense volume of generated data and deriving meaningful insights for applications is a critical challenge. This entails addressing data semantics and ensuring efficient data handling.
- Safety: As we reap the benefits of IoT, ensuring safety is paramount. Both creators and users of IoT must prioritize safety, encompassing the security of personal data and physical well-being. Establishing a robust security paradigm that encompasses endpoints, networks, and data is crucial and must be designed to scale effectively.
- Connectivity: Connectivity forms the backbone of IoT, enabling network accessibility and compatibility. Accessibility involves joining a network, while compatibility ensures a shared ability to consume and produce data.

13.5 IOT IN EDUCATION

The Internet of Things is a crucial tool for making education more accessible, interactive, and collaborative. It can facilitate online, real-time interaction between students and teachers. It has expanded the classroom into cyberspace. Physical distance has become less of a hindrance to learning. Even now that face-to-face classes have returned post-pandemic, the role of IoT in education remains crucial. It can make classroom instruction more interesting and interactive and collaborative group projects are a lot easier, from the design to the prototype phase.

But more than this, IoT matters in education because we're giving students the skills, knowledge, and tools they'll need in the future. IoT is becoming such an integrated part of our lives that students will be missing out if they don't learn about it from as early an age as possible.

13.6 IMPORTANT ROLES OF TECHNOLOGY IN EDUCATION

Since the beginning of the first formal school more than 4,000 years ago, the use of contemporary technology has always been part of the education system.

From papyrus in ancient Egypt to touchscreen tablets in modern classrooms, the use of the most current technologies available to the public is a must for education. Although there are several roles that technology plays in education, most of these can be categorized into three categories: accessibility, interactivity, and collaboration.

- Accessibility: Technology in the form of computers (both handheld and desktop devices), computer peripherals, electronic kits, audio-visual devices, mechanical devices, and the Internet of Things makes education more accessible to a wider range of students. Educational materials and classes can also be easily accessed via online platforms.
- Interactivity: Technology is used not only to access the vast store of information online but also as a means of communication and interaction between students and teachers. Learning materials can be made more dynamic and interactive. For example, those who are just starting to learn coding and programming can conduct test runs of their programs on online platforms. Virtual experimentation on structural designs can also be done using certain applications before prototyping the design. Some schools have 3D printers that can be used by students to prototype designs or create scaled-down models.
- Collaboration: technology makes it possible to collaborate at various levels, both online and face to face. For example, students can collaborate on a research paper using cloud storage such as Google Drive. Documents can be edited in real time among the various members of a group. They can also easily share information.

13.7 IMPLEMENTATION OF IOT IN EDUCATION SECTOR

1. Ensuring secure and seamless remote classroom access: Enabling secure and remote access to classes has been significantly bolstered by IoT, ensuring the safety and well-being of students. Across the globe, educational institutions swiftly transitioned to remote modes of education, a transformation made feasible through the implementation of IoT technologies. The deployment of sensors throughout the campus plays a pivotal role in monitoring student activities and swiftly addressing instances of bullying or damage to school property. Moreover, the utilization of RFID tags embedded in student ID cards has streamlined attendance tracking, providing an efficient and automated method to record daily student attendance. This holistic approach enhances the overall safety and security of the educational environment.

2. Energy management: Efficient energy management is now imperative in the education sector due to escalating global warming concerns. Large-scale printing of notebooks and textbooks leads to substantial paper wastage, urging the need for change. IoT offers a sustainable solution by providing textbooks in electronic formats on devices, promoting a paperless approach and facilitating easier updates and annotations by students. Moreover, the integration of sensors and actuators in lighting and water systems allows for automated operations, minimizing energy consumption and reducing harmful emissions like CO_2 and CO, contributing to a greener environment.

3. Innovative applications for enhanced interactive learning: Leveraging IoT technology can significantly elevate the educational experience, making it enjoyable and engaging for students. IoT not only captivates students' interest in acquiring knowledge but also aids in their overall personality development and mental acuity. Even during periods of illness, learning can persist through virtual classrooms accessible via televisions and mobile devices. The incorporation of IoT into educational settings introduces innovative graphic classrooms, breathing life into textbooks through vivid 3D graphics, vibrant colors, and interactive sounds, transforming the way students absorb information. Additionally, applications like voice-to-text converters prove instrumental in enhancing note-taking efficiency and ease for students, further optimizing the learning process.

4. Educators: Educators, including teachers, tutors, and professors, hold a significant place in the education system, complementing the importance of students. IoT technology has emerged as a valuable tool to streamline and enhance the role of educators. IoT applications aid in automating attendance tracking and monitoring students' progress,

categorizing them based on various criteria such as grades, sports, and more. By doing so, IoT bridges the gap between students and teachers, providing valuable insights to educators about the unique needs of their students and enabling a more personalized approach to teaching.

5. Health monitoring for students and staff: Maintaining the well-being of both students and staff is paramount for an educational institution to function at its best. It's essential to ensure that students prioritize their health by getting adequate sleep and taking care of their overall well-being, as this directly impacts their active engagement and participation in class. Utilizing physiological signals allows for continuous monitoring of the health of students, with the collected data stored in databases for easy access and analysis by healthcare professionals. Furthermore, IoT technology plays a vital role in securely storing essential health information such as blood pressure, medical history, allergies, and prescription details. This stored data becomes invaluable in promptly administering appropriate treatment in case of unforeseen accidents or health emergencies.

6. Exam surveillance: In the realm of exam surveillance, where the integrity of examinations is pivotal to a student's success, IoT plays a critical role. It helps institutions maintain discipline and fairness during exams by leveraging IoT-embedded LED screens. These devices are equipped to detect even faint noises during exams and promptly send alerts to examiners. Simultaneously, warning messages can be displayed to students, ensuring a controlled and fair examination environment. The integration of IoT in exam surveillance contributes to upholding the integrity and credibility of the examination process.

7. Electronic board or smart board: An electronic or smart board stands as a prime illustration of how IoT significantly simplifies the tasks of educators while enhancing the learning experience. These boards are designed with simplicity, reliability, and student-friendliness in mind. They facilitate the display of images, encouraging interaction and engagement during lessons. Utilizing a touchscreen interface, teachers can easily manipulate the displayed content, moving images, drawing graphics, and personalizing the learning process. Smart pens further augment the functionality, enabling the creation of 3D objects with vivid colors and accompanying audio. Additionally, voice-to-text sensors play a crucial role, seamlessly converting the teacher's spoken words during a lesson into digital notes, promptly emailed to the students for easy access and review.

8. Supervising disability through IoT in education: In the realm of educational tool development, children with disabilities, like dyslexia, are sometimes overlooked. However, with the integration of IoT, specialized applications can be crafted to facilitate enhanced learning experiences for these students. Transcripts, for instance, can effectively

convert voice into text, enabling students with hearing disabilities to read and comprehend the notes. Conversely, sound-to-voice applications prove invaluable for students with visual impairments, audibly conveying written content. Tailored applications on tablets are transforming the landscape of education for disabled students, undergoing thorough analysis to ensure optimal assistance in their learning journey.

9. Improving attendance tracking and exam assessment: Certain academic institutions enforce attendance requirements, like maintaining an 80% attendance rate, as a prerequisite for passing a semester. IoT applications prove instrumental in monitoring student attendance and generating comprehensive reports for staff members. Students with lower attendance receive automated messages and can monitor their performance through IoT-powered applications. Moreover, IoT technology enables remote exam-taking, where sensors in devices monitor student movement to prevent cheating.

10. Enhancing safety measures: In the educational sphere, the integration of technology, particularly IoT, plays a critical role in bolstering safety for students. IoT applications in schools encompass a range of safety features:

    • Integrated fire detection sensors promptly alert the school premises and can even notify the fire department in case of a fire hazard.
    • IoT-enabled online activity monitoring ensures students do not access inappropriate content while within the school premises.
    • Video monitoring allows for comprehensive oversight of student activities, both within and outside classrooms, including during lunch breaks.
    • Utilizing IoT for digital signage, access control, and visitor management contributes to ensuring a safe school environment.
    • The collaborative use of these IoT applications significantly elevates the safety and security measures in educational institutions, promoting a conducive and secure learning environment for all.

11. Real-time data collection for educational enhancement: A predominant concern within the present education system is the outdated nature of syllabi and teaching methodologies, often not aligning with the needs and preferences of today's students. This is where IoT proves to be a game-changer by aiding in real-time data collection and analysis. IoT technology allows for the accumulation of valuable data on student engagement, enabling the creation of topics that emphasize interaction, enhance focus, and promote more effective learning on a larger scale. Furthermore, IoT facilitates the collection of data pertaining to both staff and students, providing insights into their preferences and behaviors. This data-driven approach is instrumental in improving the production of school gadgets, accessories, and

educational content, tailoring them to the specific needs and preferences of the educational community.

12. Enhancing school bus safety through IoT: Sending children to school via school buses often raises safety concerns among parents due to potential reckless driving and the frequent occurrence of accidents involving school buses worldwide. Fortunately, IoT presents a viable solution to address this pressing issue. By leveraging IoT applications, drivers can actively monitor the well-being of students aboard the school bus. The school administration can simultaneously track both the driver's behavior and the bus's location in real time directly from the school premises. Additionally, parents can use IoT-enabled platforms to stay informed about their children's whereabouts, ensuring they reach school safely and punctually. Furthermore, IoT can monitor critical safety aspects such as speed limits and tire punctures, promptly triggering alerts in case of potential dangers, thus significantly enhancing overall school bus safety.

13. Ensuring safe arrival home for children through IoT: Parents often grapple with concerns about their children reaching home on time and safely after school. IoT technology provides a reassuring solution by leveraging various devices to track and monitor children's journey back home. For instance, sensors integrated into school bags or shoes relay location updates and whereabouts to parents in real time. This allows parents to remotely keep track of their children and promptly address any worries. Additionally, IoT applications are designed to raise immediate alerts if a child is in a potentially dangerous situation. They not only share precise location details with parents but also notify nearby police departments if the child deviates from the usual and safe route leading home, collectively contributing to the safety and peace of mind of both parents and children.

14. Augmented reality (AR) technology: Leveraging AR technology in education has revolutionized the learning experience by enabling live interactions and breathing life into audio, videos, and graphics. This dynamic approach significantly enhances students' comprehension and grasp of various concepts. One notable application of AR is in project mapping, where it projects concepts and distances in AR, allowing students to perceive and understand the topics more effectively. AR facilitates interactive learning within the classroom environment, extending beyond traditional learning methods confined to books and textbooks.

15. Efficient temperature control within premises: In the realm of maintaining appropriate temperature levels within educational premises, IoT applications prove to be invaluable. The temperature requirements in classrooms and gymnasiums can vary, and IoT applications efficiently regulate temperature and lighting conditions based on the

prevailing environment. Moreover, these applications have the capability to detect human presence, intelligently adjusting electricity usage, accordingly, thus contributing to energy efficiency and ensuring comfort for occupants.

13.8 ADVANTAGES OF IOT IN EDUCATION

- Streamlined school management: IoT streamlines administrative tasks, reducing manual paperwork, managing funds, and tracking supplies, ultimately saving time and costs.
- Real-time data collection: IoT processes vast amounts of data, predicting and analyzing student and staff activities, contributing to the development of an efficient school system.
- Resource management improvements: IoT aids in the effective management of resources like water and electricity, significantly reducing paper usage within the education system.
- Enhanced global connectivity: IoT's reliance on the internet facilitates global connections, offering a platform for interaction with schools and universities worldwide, thereby improving the quality of education through shared insights.
- Addressing safety concerns: In rural areas, parents often worry about their children's safety while accessing education. IoT addresses these concerns by ensuring a secure learning environment, especially for children who may face resource constraints.

13.9 DRAWBACKS OF IMPLEMENTING IOT IN EDUCATION

- High implementation costs: IoT systems entail significant upfront expenses, including costly software, hardware, and the need for a skilled IT team to manage and maintain the system.
- Academic integrity concerns: IoT-based assessments rely on grading knowledge-based tests, raising concerns about preventing cheating, plagiarism, and other fraudulent means during evaluations.
- Data storage challenges: The substantial amount of data processed by IoT necessitates efficient cloud-based storage solutions. Traditional data storage methods may not efficiently handle the data generated within IoT ecosystems.
- Security and privacy risks: Implementing IoT exposes private information to a wide online user base, potentially putting sensitive student and staff data at risk. Security measures must be robust to safeguard against unauthorized access and potential exploitation by hackers.

13.10 FUTURE PROSPECTS

The future of IoT in education is promising, especially with integration alongside artificial intelligence (AI), virtual reality, and blockchain. This amalgamation can greatly enhance personalized learning, virtual classrooms, and secure educational records. IoT's evolution, combined with emerging tech like AI and augmented reality, will further enrich tailored learning experiences. Anticipated advancements encompass virtual classrooms, intelligent tutoring systems, and adaptive learning platforms, propelling education into an exciting IoT-powered future.

However, integrating IoT devices in classrooms isn't without hurdles. Education providers must navigate challenges such as limited network bandwidth, reliable Wi-Fi, security, and teacher training. Despite these obstacles, IoT is set to revolutionize teaching and learning processes, making education more accessible and efficient for both students and teachers.

In the upcoming years, IoT is expected to redefine the role of educators. Modern educators will utilize IoT frameworks to engage with students, monitor their progress, and enhance their overall performance. IoT data will provide insights into students' attitudes toward training topics. AI techniques like face and facial expression recognition will become crucial in understanding students' behavioral parameters.

IoT tools will create a flexible, engaging, and quantifiable education system that caters to diverse student needs. Automation through IoT can reduce time spent on administrative tasks, allowing teachers to focus more on engaging with students. Automated attendance, cognitive brain activity assessment, and discreet alerts to students via wearables are among the exciting possibilities IoT can offer. The vision of a technologically enriched learning environment is on the horizon, even though many educational institutions have yet to fully embrace IoT.

13.11 SUMMARY

The Internet of Things presents a promising future for the education sector. With the integration of IoT devices alongside emerging technologies like artificial intelligence, virtual reality, and blockchain, education is poised for significant enhancements. From personalized learning experiences to virtual classrooms and secure educational records, IoT has the potential to revolutionize education. Despite challenges such as infrastructure costs, security concerns, and data management, the transformative impact of IoT on education is undeniable.

As IoT technology continues to evolve, addressing these challenges and leveraging its benefits is crucial for educational institutions. The future will likely see a seamless integration of IoT devices in classrooms, offering engaging, efficient, and personalized learning experiences. Educators will be empowered to monitor student progress and tailor teaching methodologies. The IoT-powered education landscape is on the verge of transforming traditional teaching and learning paradigms into a technologically enriched and globally connected educational experience.

BIBLIOGRAPHY

Achilovich, Q. O. (2021). Efficiency of using smart technologies in teaching technical sciences in higher educational institutions. *Middle European Scientific Bulletin*, 17, 133–137.

Al-Emran, M., Malik, S. I., & Al-Kabi, M. N. (2020). *A Survey of Internet of Things (IoT) in Education: Opportunities and Challenges. Toward Social Internet of Things (SIoT): Enabling Technologies, Architectures and Applications: Emerging Technologies for Connected and Smart Social Objects*, 197–209. Springer, Cham.

Badshah, Afzal, Ghani, Anwar, Daud, Ali, Jalal, Ateeqa, Bilal, Muhammad, & Crowcroft, Jon (2023). Towards smart education through internet of things: A survey. *ACM Computing Surveys*, 56(2), 1–33.

Maksimović, M. (2018). IOT concept application in educational sector using collaboration. *Facta Universitatis, Series: Teaching, Learning and Teacher Education*, 1(2), 137–150.

Tripathy, H. K., Mishra, S., & Dash, K. (2021). Significance of IoT in education domain. In Santosh Kumar Pani & Manjusha Pande (Eds.), *Internet of Things: Enabling Technologies, Security and Social Implications*, 59–83. Springer, Singapore.

Zeeshan, K., Hämäläinen, T., & Neittaanmäki, P. (2022). Internet of things for sustainable smart education: An overview. *Sustainability*, 14(7), 4293.

Chapter 14

The efficiency of blending AI technology to enhance behavior intention and critical thinking in higher education

Ahmad Al Yakin, Muthmainnah, Ahmed J. Obaid, Eka Apriani, Souvic Ganguli, and Abdul Latief

14.1 INTRODUCTION

In recent years, the integration of artificial intelligence (AI) technology has revolutionized various sectors, including education. In particular, the realm of higher education has witnessed a paradigm shift with the adoption of AI-driven tools and systems. As traditional teaching methods face challenges in meeting the demands of a fast-paced world, educators and institutions are exploring innovative ways to enhance the learning experience and cultivate important skills among students (Devagiri et al., 2022; Leal Filho et al., 2023). This exploration has led to increasing interest in blending AI technologies with behavioral intention and critical thinking to achieve greater efficiency in higher education. Rapid advances in AI technology have transformed various industries, and the education sector is no exception.

It cannot be denied that the rise of automation in several industries has increasingly driven the advancement of AI (Lund et al., 2023). AI has great potential to revolutionize the way we think and analyze in education. However, research on AI in education has largely ignored the factors influencing the use of AI on behavioral intention and critical thinking in university settings, focusing instead on the technical aspects of systems development. The development of research for a new generation of AI demands the use of AI to accelerate the transformation of educational practices, including in higher education. In order to build better models of AI-assisted education in the future, it is clear that more work needs to be done to understand educators' perspectives on AI and the aspects that affect it when implemented in hybrid classrooms. The integration of AI into higher education has sparked a wave of innovation and potential, propelling the traditional classroom experience into an era of personalized and efficient learning (Joshi and Pramod, 2023; Alhazmi et al., 2023). One of the most promising areas where AI can significantly impact higher education is in supporting behavioral intentions and critical thinking among students.

DOI: 10.1201/9781003510420-14

Although some studies are unsure about the effectiveness of AI, as shown by Chen et al. (2023), there may be several obstacles for educators looking to adapt to AI and maximize the potential of AI in the classroom. Acceptance and innovation on the part of educators are necessary for AI-supported learning to become widespread (Park and Kwon, 2023). Therefore, while the majority of educators in universities agreed to incorporate technological tools into their lessons, many factors, both external (such as lack of funding or technical support) and internal (such as lack of expertise or fear of seeing their pedagogical authority eroded), caused them to stop and eventually return to their previous model of teaching (Tokareva et al., 2019). Therefore, it is important to examine the adoption of AI by teachers and the relevant external and internal aspects that teachers consider when deciding whether to incorporate AI as a new technology in their classroom learning process, considering the conditions and needs of undergraduate students. We believe that the co-development of pedagogy and technology can benefit from an appreciation of the dynamics between internal pedagogical elements' usability criteria.

Many researchers have used AI to analyze how people's impression of a technology influences their intention to use it, and the impact on learning motivation, engagement in learning, and critical skills (Muthmainnah et al., 2022). The acceptance and design of teachers' teaching models are interrelated dimensions that need to be addressed to facilitate the development of their skills for the pedagogical use of new technologies, as has been noted in previous research (Chocarro et al., 2023). The usefulness of AI isn't just a result of technological design; it also depends on how a lecturer internalizes the material and views the ability of the teaching model and AI to influence each other. Many educators, for example, view AI as a valuable resource for delivering content (Muthmainnah et al., 2023) and, therefore, support its use in instructor-led classes. AI can be used as a knowledge-creation tool that supports interactive engagement by teachers with student-centered pedagogical tendencies, enabling students to develop an understanding of a topic based on information and communication technology (ICT) (Khang et al., 2023). Using technology in any of these ways is an indication that teachers are comfortable with it, but teachers with a deeper understanding of traditional, student-centered pedagogy may be more comfortable with it because they are better able to find pedagogical applications for the technology and can integrate it in the classroom to its full potential. That is, educators' perceptions of the utility of technology and ease of use can be influenced by the level of technological and/or pedagogical expertise, high teaching motivation, and the lecturer's intention to use Web 2.0 technologies in the classroom.

Behavior intention refers to the conscious decision-making process that leads to planned behavior, while critical thinking entails the ability to analyze, evaluate, and synthesize information to make well-informed decisions and solve complex problems. Both behavior intention and critical thinking are vital skills for students to excel academically and thrive in the dynamic and evolving professional landscape.

The efficiency of blending AI technology with behavior intention and critical thinking in higher education can yield transformative outcomes, enhancing student engagement, performance, and overall academic success. This chapter delves into the various aspects of this amalgamation, exploring its potential benefits, challenges, and implications for educators, students, and institutions. The aim of this research is to investigate the efficiency of blending AI technology-supported behavior intention and critical thinking in higher education and its potential impact on students' learning outcomes, problem-solving abilities, and overall academic success. By examining this emerging approach, we hope to shed light on the benefits and challenges of integrating AI into educational settings, providing valuable insights for educators, policymakers, and stakeholders.

Research questions:

1. How does blending AI technology with behavior intention influence students' engagement in learning?
2. What is the impact of integrating AI-supported critical thinking skills in learning?

14.2 BEHAVIORAL INTENTION THEORY

The concept of behavior intention, derived from the theory of planned behavior (TPB) proposed by Ajzen (2020), plays a pivotal role in understanding how individuals' attitudes, subjective norms, and perceived behavioral control influence their intention to engage in specific behaviors. In the context of higher education, students' behavioral intentions are crucial factors that affect their level of commitment, motivation, and perseverance toward academic tasks (Bargmann et al., 2022; Kim et al., 2021). By leveraging AI technology to support behavior intention, educators can identify and address students' individual learning needs, providing personalized feedback and adaptive learning pathways to enhance the learning experience.

Technology has become an integral part of modern education, significantly influencing students' learning experiences and behaviors (Al Breiki et al., 2023). The behavioral intention theory (BIT), a widely accepted psychological model, plays an important role in understanding how students' intentions to engage in certain behaviors are shaped by their attitudes, perceived norms, and perceived behavioral controls (Siragusa and Dixon, 2008). This literature review explores the application of BIT in the context of technology adoption and its impact on students' behavioral intentions.

BIT, originally developed by Ajzen and Fishbein in the theory of reasoned action (TRA) (Alqasa et al., 2014), argues that behavioral intention is the main determinant of actual behavior. Intention is influenced by three main factors: attitudes toward the behavior, subjective norms (social influence), and perceived behavioral control (individual beliefs about their ability to perform the behavior). BIT has been widely applied in various fields to understand and predict human behavior, including technology adoption in education. Technology in education and behavioral intentions have an impact on academic engagement and influence learning outcomes and socio-emotional aspects. According to Ahmad (2020), the integration of technology in educational settings has revolutionized approaches to teaching and learning. Technology offers a variety of learning opportunities, personalized instruction, and real-time feedback. This section examines how exposure to technology affects students' behavioral intentions, considering factors such as motivation, engagement, and academic achievement.

Some studies have investigated BIT's impact on academic engagement (Zahedi et al., 2021; Aydınlıyurt et al., 2021). Xu et al. (2023) show how various technology tools, such as online learning platforms, gamification, and educational applications, impact students' academic engagement. Research shows that well-designed and interactive technology can increase students' motivation, engagement, and active participation in the learning process. The relationship between technology and academic performance is a matter of ongoing debate. We review research exploring the impact of technology on student learning outcomes, including test scores, knowledge retention, and critical thinking skills. Next, we discuss how technology can provide immediate feedback and facilitate formative assessment to improve learning outcomes.

Technology integration also has implications for the socio-emotional development of students (Bucci, 2023; Yondler and Blau, 2023; Saleme et al., 2023). This section reviews studies that examine the effects of technology on students' social interactions, communication skills, and emotional well-being. It also explores concerns regarding screen time, digital addiction, and the potential erosion of face-to-face communication.

Despite the many benefits of technology in education, certain challenges need to be overcome. Issues such as the digital divide, access to technology, and teacher readiness require careful consideration. This section identifies potential solutions and strategies to ensure fair and effective integration of technology in education.

The integration of technology in education has enormous potential to positively shape students' behavioral intentions. Understanding the influence of technology through the lens of BIT is critical for educators and policymakers to harness its benefits effectively while mitigating potential challenges. This provides valuable insights into the multifaceted relationship between technology and student behavior in educational contexts.

14.3 CRITICAL THINKING IN HIGHER EDUCATION

Critical thinking is a cognitive process that involves the objective analysis and evaluation of information to form well-reasoned judgments and decisions. In the context of higher education, critical thinking skills are fundamental for students to navigate complex academic challenges, become effective problem-solvers, and make informed decisions in their future careers. The integration of AI technology into the critical thinking process can offer innovative tools, such as natural language processing and data analytics, to assist students in information synthesis and evidence-based reasoning.

Technology has emerged as a powerful aid to critical thinking, revolutionizing how we approach and solve complex problems (Sweet and Michaelsen, 2023). This has opened new avenues for information access, data analysis, communication, and collaboration, enabling individuals to make informed decisions and draw logical conclusions based on evidence. Here are some ways in which technology serves as a catalyst for critical thinking:

1. Access to vast information (Xiang et al., 2021). The internet has given us an unprecedented amount of information at our fingertips. With just a few clicks, we can access diverse perspectives, research studies, historical data, and expert opinions on almost any topic. This abundance of information prompts critical thinkers to evaluate sources, double-check facts, and consider multiple points of view before drawing their own conclusions.

2. Data analysis and visualization (Shrivastava et al., 2023). Technology enables us to collect and analyze data more efficiently and accurately. Data visualization tools allow complex datasets to be presented in an understandable and visually appealing format, making it easier to identify patterns, trends, and correlations. This data-driven approach enhances critical thinking by promoting evidence-based decision-making.

3. Simulation and modeling (Cooper, 2023). Technology provides the means to create simulations and models for real-world scenarios. Critical thinkers can test different hypotheses, evaluate potential outcomes, and assess the consequences of various actions without having to carry them out. This empowers individuals to make better-informed choices and anticipate potential challenges.

4. Collaborative problem-solving (Chan et al., 2023). Online collaboration tools and platforms facilitate teamwork and knowledge sharing across geographic boundaries. By engaging people with diverse backgrounds and expertise, critical thinkers can gain unique insights and challenge their own assumptions. This collaborative process encourages vigorous debate and helps refine ideas.

5. Fact-checking and critical analysis tools (Wangdi and Savski, 2023). Technology offers fact-checking tools and software that help verify the authenticity of information and identify misinformation or fake news. These tools help users develop a keen eye and teach them to question the credibility of sources before accepting information as truth.
6. Online courses and educational platforms (Shihab et al., 2023). The digital age has given rise to many online learning platforms, providing access to educational resources and courses from prestigious institutions around the world. Through this platform, individuals can develop critical thinking skills, learn new subjects, and engage in thought-provoking discussions with instructors and peers.
7. Artificial intelligence assistance (Berson et al., 2023). AI-powered tools can augment critical thinking by aiding problem-solving, suggesting relevant information, and identifying potential problems in complex data sets. Although AI can be a valuable aid, it is essential for critical thinkers to understand its limitations and interpret the results with human judgment.
8. Adaptive learning systems (Martin et al., 2020). Technology has paved the way for adaptive learning systems that personalize educational content based on individual strengths and weaknesses. This system challenges learners at an appropriate level, encouraging them to think critically and solve problems in their area of interest.

Understanding what motivates students to embrace innovative approaches to education, such as those that heavily incorporate information and communication technologies, is essential before introducing new models. The first stage in this endeavor is to correctly identify the factors that will increase the likelihood of success. Therefore, this study examines the influence of modern learning approaches and resources on the inculcation and evaluation of students' important skills such as critical thinking. Students will work on a variety of tasks to build their critical thinking skills in a group setting, which is made possible by the use of AI cases in this teaching technique.

We started by observing when lecturers introduced AI-Chatbot, AI-Andy, and AI-My Virtual Dream Friends, giving an overview of the benefits of AI in the class, for example, Chatbot is a completely online technology that is highly dependent on the internet network. We also provide a definition of critical thinking skills and an explanation of why AI aims to build and implement a teaching innovation model that leverages "AI cases" to complement student learning through the application of new technological resources.

However, despite the many benefits of technology in assisting critical thinking, it is important to be aware of the potential pitfalls. Overreliance on technology without cultivating fundamental critical thinking skills can lead to a superficial understanding of complex problems. Therefore, it is very important to strike a balance between leveraging technology and cultivating the cognitive abilities necessary for effective critical thinking. By harnessing the potential of technology while maintaining an inquisitive and analytical mindset, individuals can enhance their critical thinking abilities and become more informed and intelligent decision-makers in an increasingly complex world.

The influence of technology on critical thinking and students' behavioral intentions in learning can be positive and negative (Chai et al., 2022; Park, 2009; Songkram et al., 2023; Humida et al., 2022). Next are some important points to consider.

To optimize the positive effects of technology while reducing the negative effects, educators and students need to strike a balance. Encouraging conscious use of technology, teaching digital literacy and critical evaluation skills, and promoting face-to-face interactions alongside digital learning can help students develop strong critical thinking skills and positive behavioral intentions in their educational journey. Figure 14.1 shows the diagram of the blending of AI technology to enhance behavioral intention and critical thinking and the following list explains the diagram:

1. Higher education institutions: These are educational centers where the process of increasing behavioral intention and critical thinking takes place. They consist of students and the teaching and learning process.
2. Students (behavioral intention and critical thinking): Student behavioral intention and critical thinking are the main outcomes that we aim to improve through the integration of AI technologies.
3. AI models and data: These represent the various AI models, algorithms, and datasets used to develop and support AI-based tools and solutions.
4. Blending AI technologies: This is where AI technologies are integrated into the context of higher education. AI tools can be designed to provide personalized learning experiences, offer real-time feedback, analyze student performance, and recommend resources to improve behavioral intention (motivation, engagement) and critical thinking skills.
5. Learning and teaching process: This involves the core educational activities of lectures, assignments, assessments, discussions, and interactions between students and instructors. AI technology integration complements this process, providing insights to educators and helping students on their learning journey.
6. Impact of AI: AI models and data interact with students, helping to understand their behavioral intentions and critical thinking patterns.

Table 14.1 Technology, behavioral intention, and critical thinking

Positive effect	Access to information	Technology gives students easy access to vast amounts of information, allowing them to explore multiple perspectives and deepen their understanding of subjects. It can improve critical thinking by encouraging students to analyze, evaluate, and synthesize information from various sources.
	Interactive learning	Technology offers interactive learning tools and platforms that engage students actively in the learning process. This interactive activity can stimulate critical thinking skills as students solve problems, make decisions, and think critically to progress their studies.
	Collaborative learning	Technology facilitates collaboration between students, even when they are physically far apart. Collaborative learning experiences can expose students to different points of view, encouraging critical thinking by promoting discussion, debate, and teamwork.
	Personalized learning	Technology enables adaptive learning experiences that are tailored to each student's needs and learning style. This personalized approach can enhance critical thinking as students receive content that challenges them appropriately and supports their growth.
	Analytical skills	Technologies often include data analysis and simulation tools that allow students to engage with real-world scenarios. This experience can develop critical thinking by asking students to interpret data, draw conclusions, and make decisions.
Negative effect	Information overload	The abundance of information available through technology can lead to information overload. Students may struggle to distinguish credible sources from unreliable ones, affecting their critical thinking abilities as they may rely on superficial or biased information.
	Reduced deep thinking	Constant exposure to digital distractions, such as social media and entertainment, can shorten attention spans and inhibit deep thinking. This can affect students' ability to focus and engage in complex critical thinking tasks.
	Limited face-to-face interaction	Overreliance on technology for learning can reduce face-to-face interaction, which is important for developing social and emotional intelligence. A lack of interpersonal skills can affect behavioral intentions and create difficulties in resolving conflicts or empathizing with others.
	Reliance on solutions	With technology providing quick answers, some students may become too dependent on these solutions instead of developing problem-solving skills independently. This can hinder their ability to think critically and devise creative solutions.

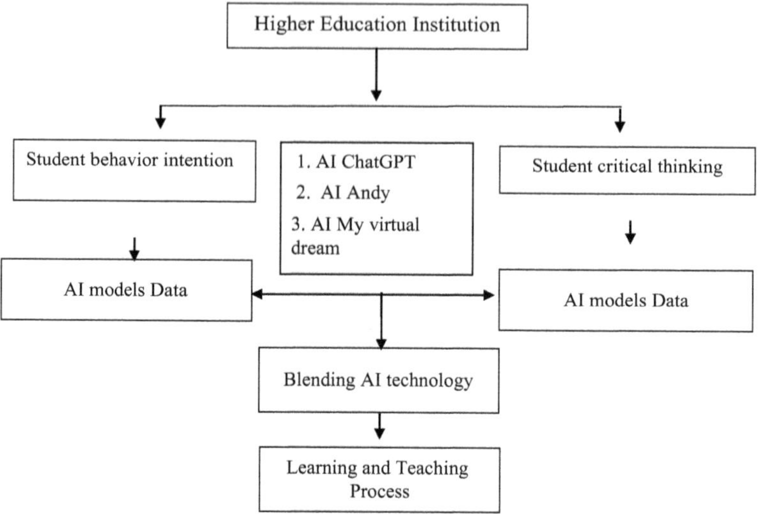

Figure 14.1 Diagram of the blending of AI technology to enhance behavioral intention and critical thinking.

By analyzing student performance, engagement, and interaction, AI can provide personalized suggestions and interventions to improve these aspects. Within this blended AI approach, technology acts as a driver to create more personalized, data-driven, and effective educational experiences, ultimately enhancing students' behavioral intentions and critical thinking skills in higher education.

14.4 METHOD

We designed a survey to test our theory using a quantitative approach (Gall et al., 2003). Students at the Universitas Al Asyariah Mandar in Indonesia were surveyed after being exposed to artificial intelligence in four different faculties: computer science, public health, government, and teaching and education science. Since January 2023, the English course they were taking encouraged the use of AI tools in the classroom. So, they were deliberately selected to participate in this study.

There were 350 total undergraduate students who were included in this study, and a sample of 64 respondents was randomly selected and answered the survey. Participants in this study were 32 undergraduate students representing the faculty of computer science, 12 undergraduate students from the teaching and education faculty, 9 undergraduate students from the faculty of public health, and 11 undergraduate students from the faculty of government science. Sixty-four valid replies were collected from surveys

Figure 14.2 AI-ChatGPT.

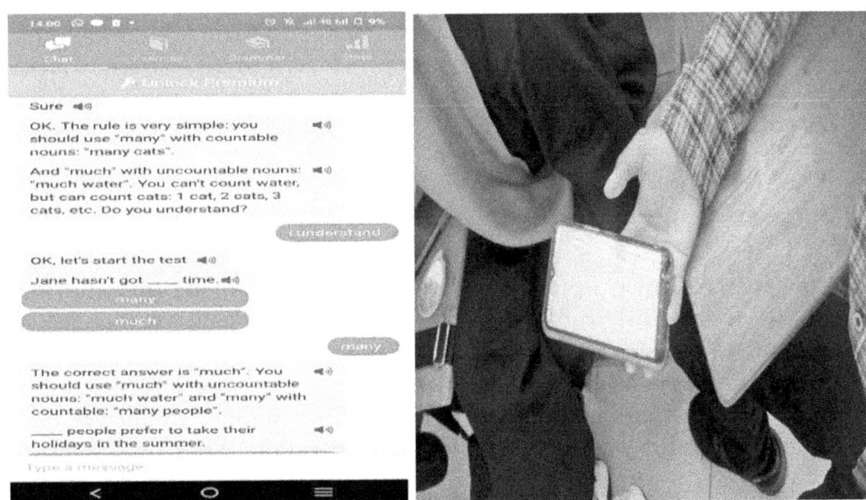

Figure 14.3 AI-Andy.

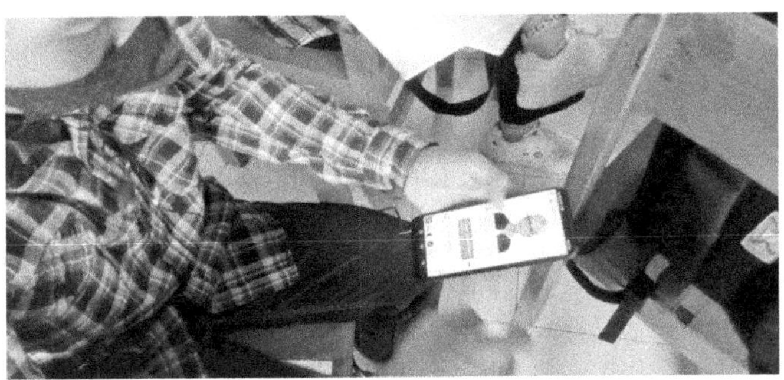

Figure 14.4 AI-My Virtual Dream Friends.

administered between March and May 2023. The gender breakdown of respondents (49% male, 51%) reflects the current gender ratio among freshmen.

This study was conducted with consideration of the unique nature of human–AI collaboration, studying undergraduate students' behavioral intentions toward using AI to enhance critical thinking skills that require a human-centered approach. To be successful as a team, humans and AI must not be seen in isolation from one another. Through observation, it is important for our proposed study model to consider the human perspective after interacting with machine learning.

Due to the potential for positive and adverse outcomes, the attitudes and behavioral intentions to learn to accept or avoid adopting AI to enhance critical thinking skills can be influenced by factors such as those listed earlier. Factors relating to personal issues, such as the possibility that using AI during the learning process can give undergraduate students emotional growth and well-being when engaged in learning while interacting with learning machines, can have a significant impact on their behavioral intentions and critical thinking abilities to adopt AI. As a result, the study model that we suggest combines behavioral intention and critical thinking due to the use of AI during learning.

14.5 DATA ANALYSIS

The purpose of this chapter is to find out the effect of AI on behavioral intention and critical thinking skills, especially AI-Chatbot technology, AI-Andy, and AI-My Virtual Dream Friends, which are used by lecturers when teaching English courses. A group of undergraduate students participated in the research, and they were given three minutes to answer questions that had been posted on the WhatsApp group. After one month of data collection, 64 undergraduate students were representative of the population. After the data is obtained, it is analyzed. In this study, the results from the questionnaire had a Likert scale range of 1 to 5 and SPSS version 26.0 was used for data analysis.

14.6 RESULTS

To understand the elements that influence behavioral intention and critical thinking that utilize AI during the learning process, we rely on AI robots, namely, AI-My Virtual Dream Friends, AI-Andy, and AI-Chatbot. We observed lecturers using these three AI technologies during English learning, which can be categorized as a combination of technology and learning media. First, our study contributes to the growing canon of AI in education;

theoretical models are still needed to facilitate the integration of conceptual development and empirical research around AI phenomena. A more nuanced discussion of the benefits and drawbacks of AI could help guide educational judgments about the use of AI-based learning media and technologies if made. Additionally, by providing new empirical evidence in the context of AI use, we contribute to the development of a more comprehensive innovation-class framework for understanding IT adoption in general that expands on the underresearched empirical literature on the application of AI in such contexts of behavioral intention and critical thinking skills.

In this study, we observed English lecturers who integrated foreign language learning theory and the use of smartphone-based AI technology, which influenced behavioral intentions and critical thinking skills toward AI use. Existing work suggests that individuals tend to have positive attitudes and intentions to use AI for certain interactions and instructional materials when they experience its usefulness and ease of use or are motivated to use AI as a daily practice to develop their own language skills. However, little is known about the relationship between the two because current AI adoption models do not yet represent such behavior; the need to incorporate elements of AI effects on behavioral intention and critical thinking skills is often overlooked in this research and the IT literature. This study helps fill this gap by evaluating the application of controversial AI technologies in contexts where people tend to view them with skepticism. Table 14.2 describes the results of the behavioral intention survey data collected statistically from undergraduate reactions after the learning process with AI.

Table 14.2 displays the responses to the questionnaire survey, all of which show that AI users show a positive response, which can be seen from the mean score for each questionnaire answer; most of them belong to the good category. Scholars who use machine learning AI (AI-Chatbot, AI-Andy, and AI-My Virtual Dream Friends) report that they are motivated to learn when using AI technology in the learning process in the classroom, have an increased interest in learning English, actively participate, are confident in learning, have easier-to-understand material, have better social interaction, and that the use of AI in learning raises awareness of progress and personal development in learning. Respondents also strongly believe AI improves time efficiency in completing learning tasks. Respondents also feel more engaged and excited about the learning process when AI technology is involved, that AI can provide a personalized and customized learning experience, and that using AI in the learning process helps develop digital skills and better access to information.

However, a score of 3 is found in the statements "I feel more motivated to seek additional information outside of classroom learning thanks to the application of AI in the learning approach," "The quality of the feedback that AI provides during the learning process is very good," and "AI

Table 14.2 Students' behavior intention influences students' engagement in learning

	N	Minimum	Maximum	Mean	Std. deviation
Descriptive statistics					
1. I feel motivated to learn when using AI technology in the learning process in the classroom.	64	4.00	5.00	4.5313	.50297
2. The use of AI in learning increases my interest in actively participating in class discussions.	64	3.00	5.00	4.4844	.56322
3. When using AI technology, I feel more confident approaching learning tasks.	64	3.00	5.00	4.5156	.53429
4. I believe using AI can help you understand learning materials better.	64	4.00	5.00	4.3594	.48361
5. The use of AI can increase your involvement in the teaching and learning process in the classroom.	64	3.00	5.00	4.6250	.51946
6. I feel more compelled to complete learning tasks when there is interaction with AI in the process.	64	4.00	5.00	4.5781	.49776
7. I use AI to help me be more focused and organized when managing my study time.	64	4.00	5.00	4.8750	.33333
8. The role of AI is very large in helping to understand learning material.	64	3.00	5.00	4.3906	.72631
9. The use of AI affects the level of cooperation and social interaction between students during the learning process in the classroom.	64	3.00	5.00	4.4844	.71252
10. I feel more motivated to seek additional information outside of classroom learning thanks to the application of AI in the learning approach.	64	3.00	5.00	3.9531	.51731
11. AI helps me to find relevant and useful learning resources.	64	4.00	5.00	4.5313	.50297
12. The quality of the feedback that AI provides during the learning process is very good.	64	3.00	5.00	3.9375	.87060

(Continued)

Table 14.2 (Continued) Students' behavior intention influences students' engagement in learning

	Descriptive statistics				
	N	Minimum	Maximum	Mean	Std. deviation
13. The use of AI in learning raises awareness of progress and personal development in learning.	64	3.00	5.00	4.1719	.76749
14. AI helps a lot in improving critical and analytical thinking skills during the learning process.	64	3.00	5.00	4.1094	.89296
15. The comparison between interactions with AI and interactions with teachers in terms of learning motivation is the same.	64	3.00	5.00	4.1250	.89974
16. AI helps in better coping with learning challenges, such as difficult or complex assignments.	64	3.00	5.00	4.0469	.88065
17. I feel AI improves time efficiency in completing learning tasks.	64	3.00	5.00	3.9375	.61399
18. I feel more engaged and excited about the learning process when AI technology is involved.	64	3.00	5.00	4.1250	.70147
19. AI can provide a personalized and customized learning experience.	64	3.00	5.00	4.4531	.53243
20. Using AI in the learning process helps develop digital skills and better access to information.	64	3.00	5.00	4.4219	.55791
Valid N (listwise)	64				

improves time efficiency in completing learning tasks." Based on the observation results, it is known that the behavioral intention of undergraduate students as a result of AI is that they ask questions, give feedback, do assignments on time, participate actively, use smartphones to study and use AI-Chatbot as a learning resource, and actively participate in English lectures, which initially make them feel sluggish and unenthusiastic. The presence of AI inspires enthusiasm for classes and enhances digital literacy skills to the fullest. Learning activities that combine this technology emphasize that they can adapt to AI technology so that they can remain motivated to learn even though they are not in the classroom.

Table 14.3 shows the respondents' answers to a survey regarding the effect of AI on their critical thinking skills. In general, statistical results show a value in the good category (score 4). The positive response given by undergraduate students was seen in class when the lecturer teaching English asked undergraduate students to access AI-Chatbot and use it as teaching material (e-material). Also, students reacted positively when asked to solve complex problems and challenges related to grammar with AI that may be difficult to solve conventionally. Through the use of AI technology, students can be faced with problems that require analytical, creative, and innovative thinking to find effective solutions during the learning process.

Although AI is capable of massively collecting, presenting, and analyzing data, by using this technology, students can acquire data analysis skills that are important in making evidence-based and rational decisions, but some of the undergraduate students still have doubts, as shown in statements 1, 2, and 3. They still have doubts even though information is easy to obtain from various sources. Information evaluation skills become very important and require lecturers to direct and guide them. This can be seen when they discuss and provide feedback. AI can help students identify credible sources of information and verify the truth of the information they receive, thus developing critical abilities for consuming information.

In addition, students' critical thinking skills have also increased due to AI according to statements 4, 7, 8, 19, and 20. The results of this survey suggest that AI technology can adapt learning methods and materials according to each student's pace and learning style. This helps improve critical thinking skills, as students are encouraged to understand the material in greater depth and ask relevant questions.

Through AI technologies such as virtual assistants or chatbots that encourage discussion, students can practice developing good arguments and learn how to respond rationally and logically to other people's arguments. In addition, AI enables the development of realistic simulations and interactive learning games. Students can be involved in situations that are similar to real life, forcing them to think critically and make smart decisions to overcome the challenges they face. AI technology can facilitate collaboration and communication between students and machines, with students using AI-Andy and AI-My Virtual Dream Friend as practice and training partners.

The results in Table 14.4 show that the average value is in the good category. Having good intentions and critical thinking skills helps students develop independence in their learning. They can manage time, manage resources, and adapt to various learning methods, which contributes to their future academic and professional success. In addition, positive behavioral intentions and thinking skills help students communicate more effectively and openly. They can present their arguments logically, support their opinions with strong evidence, actively participate in academic discussions,

Table 14.3 Integrating AI-supported critical thinking skills in learning

		Descriptive statistics			
	N	Minimum	Maximum	Mean	Std. deviation
1. Does using AI in classroom learning help improve your critical thinking skills?	64	3.00	5.00	4.6250	.51946
2. Do you feel that AI can help you tackle more complex problems during the learning process?	64	3.00	5.00	4.5938	.55546
3. To what extent has AI affected your ability to critically analyze information?	64	3.00	5.00	4.5625	.58757
4. Does AI help you recognize and understand different points of view in learning materials?	64	4.00	5.00	4.5781	.49776
5. How has the use of AI affected your ability to make evidence- or data-based judgments and decisions?	64	3.00	5.00	4.3281	.53614
6. Does AI help you be more critical in evaluating sources of information or references used in learning?	64	3.00	5.00	4.6094	.55255
7. To what extent does AI affect your ability to solve problems innovatively?	64	4.00	5.00	4.6719	.47324
8. Has the use of AI inspired you to seek more creative and different solutions in understanding learning materials?	64	4.00	5.00	4.8125	.39340
9. How has AI influenced the way you think about complex problems in certain learning areas?	64	3.00	5.00	4.3437	.59678
10. Has using AI helped you become more skilled at finding patterns or relationships between different concepts?	64	3.00	5.00	4.0469	.86244
11. How often do you feel that AI provides appropriate solutions to the learning challenges you are facing?	64	3.00	5.00	4.0625	.46718
12. How has AI impacted your ability to search for additional, in-depth information on a learning topic?	64	3.00	5.00	4.3438	.56957

(Continued)

Table 14.3 (Continued) Integrating AI-supported critical thinking skills in learning

	Descriptive statistics				
	N	Minimum	Maximum	Mean	Std. deviation
13. Has the use of AI in learning made you more confident in voicing your opinions and arguments?	64	3.00	5.00	3.8281	.82721
14. To what extent does AI affect your ability to identify weaknesses in proposed learning arguments or approaches?	64	3.00	5.00	4.3281	.69132
15. How can AI help you understand the implications and impact of certain decisions or actions in learning contexts?	64	3.00	5.00	4.2344	.83080
16. Does AI trigger further questions in you about the learning topic being studied?	64	3.00	5.00	4.1719	.84618
17. To what extent does AI improve your ability to formulate analytical questions in learning?	64	3.00	5.00	4.1562	.83986
18. How has the use of AI affected your level of involvement in the process of problem solving and critical reflection?	64	3.00	5.00	4.0156	.57714
19. Does AI help you identify and overcome biases or assumptions that may arise in the learning process?	64	3.00	5.00	4.1406	.66350
20. How effective do you think the integration of AI in classroom learning is for improving overall critical thinking skills?	64	4.00	5.00	4.4531	.50173
Valid N (listwise)	64				

Table 14.4 Mean score of behavioral intentions and critical thinking skill survey

	Descriptive statistics		
	Mean	Std. deviation	N
Behavior intention	86.6563	6.69095	64
Critical thinking skill	86.9063	6.42594	64

and encourage students to think outside conventional boundaries. They are more likely to find innovative and creative solutions to problems at hand, which are qualities that are highly valued in both the academic and professional worlds.

Table 14.5 Students' behavioral intention and critical thinking skills in learning

Interval	Category
90–100	Very good
70–89	Good
50–69	Fairly
30–49	Low
0–29	Very low

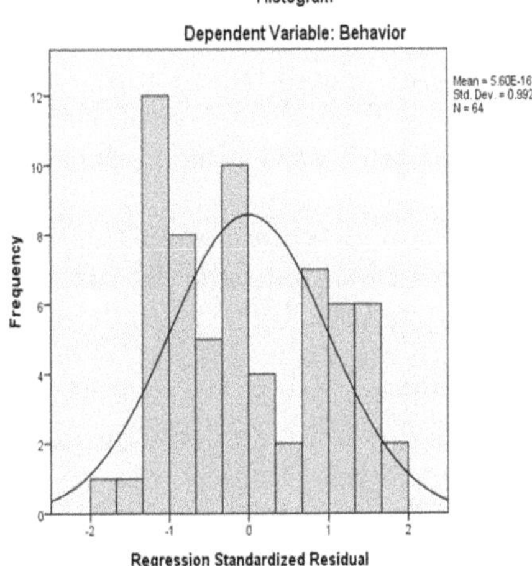

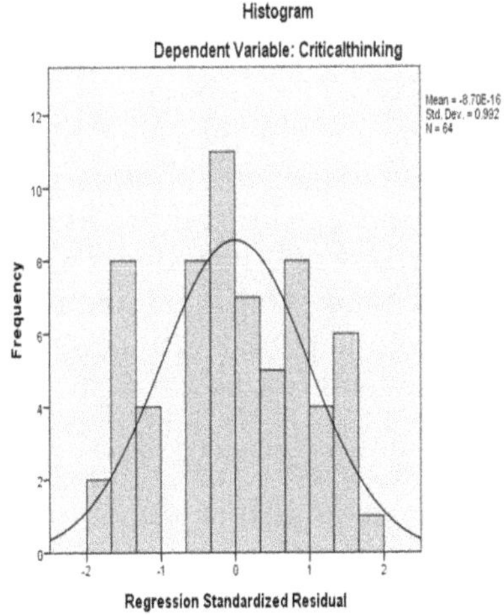

Figure 14.5 Histogram of AI to BI and CT.

Table 14.5 and Figure 14.5 show the statistical results of the effect of using AI on the behavioral intention and critical thinking skills of the students, which achieved an average score using an interval scale with a length of 29 and a maximum value of 100; the positive response with a score of 86 is in the good category.

The histogram images (Figure 14.5) describe the results of a survey of the effects of AI on behavioral intention and critical thinking skills showing the range of scores obtained by students from 76 to 100. The histogram bar shows the highest value is 99, so it is assumed that the average value of behavioral intention and critical thinking skills is in the good category.

14.7 DISCUSSIONS

This research refers to Weber (Adler, 2012), who shows that the rationality method takes a social view to explain behavior. The findings of this study illustrate that students' behavioral intentions and critical thinking skills increase due to the application of AI in the classroom, as described in Tables 14.2 and 14.3 It states that the influence of AI technology (My Virtual Dream Friends, Andy, and Chatbot) in teamwork has a positive effect on the intention to use. Many previous studies in the context of IT adoption for students found that facilitating conditions had a negligible effect on behavioral intention. Information technology that supports artificial intelligence technology as a medium and learning technology is the focus of our research, which is different from other research because our technology class is not limited to AI as a learning technology but also AI-Chatbot as a learning resource, material, and smart teaching in class (Figure 14.1, Figure 14.2, and Figure 14.3).

The results of current research in the context of AI are consistent in improving behavioral intentions and critical thinking skills (see Muthmainnah et al., 2022). The AI-based teaching model and its related factors fully represent the context for using AI for innovative and creative classroom management. Students' positive responses to the use of AI affect behavioral intentions and critical thinking skills.

Our research introduces and empirically supports the notion that perceived behavioral intention and perceived critical thinking skills can help explain and predict attitudes and intentions to use AI. To better understand how individuals feel and plan to behave with respect to IT, the literature on IT adoption should benefit from the addition of these new concepts (Hensel, 2023). This study found that undergraduate students' attitudes and behavioral intentions toward AI were positively influenced with our findings showing that attitudes were positively influenced by the benefits and experiences of learning with AI that were interesting, enjoyable, and self-motivating. These results support our research proposition that motivational problems and IT-based classroom management for undergraduate students can play an important role in positively influencing their attitudes and behavioral intentions toward the impact of AI (for example, Cabrera-Sánchez et al., 2021), as well as previous research which emphasizes the importance of understanding the affective dimensions of human interaction with IT (Li et al., 2022). In addition, these results indicate that behavioral intention tends to be an important component that influences undergraduate students' attitudes toward the use of AI in everyday life.

Many studies have examined learning performance from a cognitive perspective. To improve their spoken English skills, students in an immersive environment can practice communicating with AI robots. My Virtual Friend and Andy are in a natural setting and relive daily activities, whereas using the Chatbot application develops existing teaching materials. It is important to note that this research addresses issues related to high-level cognitive skills, teamwork, or communication resulting from AI.

Behavioral intention, described in Table 14.2, is the most frequently discussed subject. Chang and Chen (2023), for example, observed that incorporating robot construction into transdisciplinary activities increased students' interest in learning. Students are assisted by a combination of artificial intelligence robots and personal digital assistants (Horvath et al., 2023). According to the findings, students' participation in class and their ability to solve problems increased because of using this strategy (Table 14.3).

This study also looked at the effect of educational robots on students' cognitive load during classroom learning and found that these robots can increase students' enjoyment, interest, confidence, and motivation. Moreover, both models of technology acceptance and the prospect of users to implement their models show that AI robots help them to be more active and have a role in the classroom.

The research results in Table 14.2 show that learning performance for higher-order thinking skills and collaborative or communicative abilities are just a few examples of good cognitive research topics. Other topics including affection (such as models of technology acceptance or intention to use) can be enhanced by AI. Apart from the behavioral intention and critical thinking factors, there are many other factors that can support the success of AI apart from the results of this study, such as skills, learning behaviors, discussing learning effectiveness, the nature of learning, the extent of student and robot interaction, the learning environment, the attitude of the lecturer in class, and maximum IT-based support.

The AI robot workload in this study is described in Tables 14.2 and 14.3. This is the most common use of AI robots demonstrated by students. For example, students were assigned to learn and given partners, but this time they interacted with AI. Thus, they could improve their skills in constructing and interpreting English. Having the AI robot act as a conversation partner shows that the AI robot offers customized instructions and feedback to students on the topics of introducing oneself, giving directions, favorite food, likes and dislikes, and weather. The term "tool" refers to the use of artificial intelligence robots in education. For example, Khang et al. (2023) and Muthmainnah et al. (2022) found that introducing undergraduate students to virtual AI robots using smartphones helped foster the students' natural curiosity.

Our results have important implications for educators who want to use AI to improve students' behavior intentions and critical thinking skills. Because the results show that undergraduate students' attitudes and critical thinking skills about the use of AI are indirectly affected by student conditions, universities should create good facilitation conditions and permit lecturers to design AI-based teaching models and provide appropriate learning instructions. Lecturers must first ensure that the necessary supporting technology and infrastructure are available and that they have been able to apply the right technology and support to make effective use of AI.

This study provides output regarding the effect of technology on behavioral intention and critical thinking skills. It is important to design effective systems in which the problems of self-learning and weak self-determination by undergraduate students can be corrected because personal concerns about the negative effects of technology can negatively impact their behavioral intentions and critical thinking skills. Our findings suggest that undergraduate students' subjective evaluation of the role of AI in learning is influenced by their own biases, which can dramatically reduce the effectiveness of their use of AI-based media and technologies. This shows that classes that wish to integrate AI into the learning process need to provide socialization on the use of AI and the availability of resources, such as training for lecturers. The results of this study contribute to the need to address undergraduate students' concerns about AI. The decision to implement an AI-based system can lead to the adoption of AI that is not limited to use in the classroom but can also be adequately utilized outside the classroom. To minimize the fundamental problem of the dangers of AI, the reader said that AI as a learning tool has been sufficiently addressed.

Ultimately, the research results reveal that the use of AI to improve behavioral intention and critical thinking skills is not without its drawbacks. Using the pros and cons of AI to decide whether to use it can lead to bad AI adoption decisions, inefficient and/or ineffective AI-based decision systems, and/or missed chances to improve learning outcomes in a big way. AI can be used to weigh the benefits of AI against its drawbacks, with the aim of improving the former while reducing the negative effects of its use in the classroom. The survey findings are not without limitations, some of which may inspire future research. Our study is intended to include the behavior of AI users as learning media and technology, which of course influences behavioral intentions and critical thinking skills in maximally adopting AI and being able to select the right AI tools according to the learning material. This is because the actual use of AI for behavioral intention and critical thinking skills has not been widely studied. Evaluation of validity in the future may be useful, considering that the behavior and critical thinking skills of technology users are very important factors in its application in the classroom. Likewise, future research may wish to examine the influence of AI on other aspects such as character development, morals, and values.

14.8 SUMMARY

The blending of AI technology with behavioral intention and critical thinking in higher education holds immense potential for enhancing students' learning experiences and academic outcomes. Personalized learning, intelligent tutoring systems, automated assessment, and AI-supported critical thinking tools offer innovative ways to cater to diverse learning styles, improve problem-solving skills, and foster critical thinking abilities. However, it is essential to address the ethical considerations and challenges associated with AI implementation to ensure that its integration remains ethical, unbiased, and beneficial for all students. As the education landscape continues to evolve, educators and institutions must embrace AI as a complementary tool to empower students in their pursuit of knowledge and intellectual growth. This research concludes that an AI-based learning culture is very important for education and the use of digital technology can create a fun learning ecosystem. Researchers may want to see the impact of an AI-enabled learning culture in the future.

REFERENCES

Adler, P. S. (2012). Perspective—The sociological ambivalence of bureaucracy: From Weber via Gouldner to Marx. *Organization Science*, 23(1), 244–266.

Ahmad, T. (2020). Student perceptions on using cell phones as learning tools: Implications for mobile technology usage in Caribbean higher education institutions. *PSU Research Review*, 4(1), 25–43.

Ajzen, I. (2020). The theory of planned behavior: Frequently asked questions. *Human Behavior and Emerging Technologies*, 2(4), 314–324.

Al Breiki, M., Al Abri, A., Al Moosawi, A. M., & Alburaiki, A. (2023). Investigating science teachers' intention to adopt virtual reality through the integration of diffusion of innovation theory and theory of planned behaviour: The moderating role of perceived skills readiness. *Education and Information Technologies*, 28(5), 6165–6187.

Alhazmi, A. K., Alhammadi, F., Zain, A. A., Kaed, E., & Ahmed, B. (2023). AI's role and application in education: Systematic review. In: Nagar, A.K., Singh Jat, D., Mishra, D.K., and Joshi, A. (Eds.). *Intelligent Sustainable Systems*. Lecture Notes in Networks and Systems, Springer, Singapore. vol 578. https://doi.org/10.1007/978-981-19-7660-5_1

Alqasa, K. M., Mohd Isa, F., Othman, S. N., & Zolait, A. H. S. (2014). The impact of students' attitude and subjective norm on the behavioural intention to use services of banking system. *International Journal of Business Information Systems*, 15(1), 105–122.

Aydınlıyurt, E. T., Taşkın, N., Scahill, S., & Toker, A. (2021). Continuance intention in gamified mobile applications: A study of behavioral inhibition and activation systems. *International Journal of Information Management*, 61, 102414.

Bargmann, C., Thiele, L., & Kauffeld, S. (2022). Motivation matters: Predicting students' career decidedness and intention to drop out after the first year in higher education. *Higher Education*, vol 83 1–17.

Berson, I. R., Berson, M. J., Luo, W., & He, H. (2023, June). Intelligence augmentation in early childhood education: A multimodal creative inquiry approach. In: *International Conference on Artificial Intelligence in Education* (pp. 756–763). Cham: Springer Nature Switzerland.

Bucci, N. P. (2023). The evolution of educational technology in inclusive learning spaces from Pre to post pandemic. *Integration of Instructional Design and Technology*, 3. pp. 8–57.

Cabrera-Sánchez, J. P., Villarejo-Ramos, Á. F., Liébana-Cabanillas, F., & Shaikh, A. A. (2021). Identifying relevant segments of AI applications adopters– Expanding the UTAUT2's variables. *Telematics and Informatics*, *58*, 101529.

Chai, C. S., Chiu, T. K., Wang, X., Jiang, F., & Lin, X. F. (2022). Modeling Chinese secondary school students' behavioral intentions to learn artificial intelligence with the theory of planned behavior and self-determination theory. *Sustainability*, *15*(1), 605.

Chan, C. L., Shroff, R. H., Tsang, W. K., Ting, F. S., & Garcia, R. C. (2023). Assessing the effects of a collaborative problem-based learning and peer assessment method on junior secondary students' learning approaches in mathematics using interactive online whiteboards during the COVID-19 pandemic. *International Journal of Mobile Learning and Organisation*, *17*(1–2), 6–31.

Chang, C. C., & Chen, Y. K. (2023). A transdisciplinary STEM course integrated through project-based learning on robotics: Perspective from teacher and student feedback. *Asia Pacific Journal of Education*, 1–16. https://doi.org/10.1080/02188791.2023.2209698

Chen, Y., Jensen, S., Albert, L. J., Gupta, S., & Lee, T. (2023). Artificial intelligence (AI) student assistants in the classroom: Designing chatbots to support student success. *Information Systems Frontiers*, *25*(1), 161–182.

Chocarro, R., Cortiñas, M., & Marcos-Matás, G. (2023). Teachers' attitudes towards chatbots in education: A technology acceptance model approach considering the effect of social language, bot proactiveness, and users' characteristics. *Educational Studies*, *49*(2), 295–313.

Cooper, G. (2023). Examining science education in ChatGPT: An exploratory study of generative artificial intelligence. *Journal of Science Education and Technology*, *32*(3), 444–452.

Devagiri, J. S., Paheding, S., Niyaz, Q., Yang, X., & Smith, S. (2022). Augmented reality and artificial intelligence in industry: Trends, tools, and future challenges. *Expert Systems with Applications*, vol 207. 118002.

Gall, M. D., Gall, J. P., & Borg, R. W. (2003). *Educational Research: An Introduction*. New York: Longman.

Hensel, N. H. (Ed.). (2023). *Course-Based Undergraduate Research: Educational Equity and High-Impact Practice*. Taylor & Francis, Washington.

Horvath, A. S., Jochum, E., Löchtefeld, M., Vissonova, K., & Merritt, T. (2023). Soft robotics workshops: Supporting experiential learning about design, movement, and sustainability. In: *Cultural Robotics: Social Robots and Their Emergent Cultural Ecologies* (pp. 189–218). Cham: Springer International Publishing.

Humida, T., Al Mamun, M. H., & Keikhosrokiani, P. (2022). Predicting behavioral intention to use e-learning system: A case-study in Begum Rokeya University, Rangpur, Bangladesh. *Education and Information Technologies*, *27*(2), 2241–2265.

Joshi, S., & Pramod, P. J. (2023). A collaborative metaverse based A-La-Carte framework for tertiary education (CO-MATE). *Heliyon*, *9*(2), pp.e13424, https://doi.org/10.1016/j.heliyon.2023.e13424.

Khang, A., Muthmainnah, M., Seraj, P. M. I., Al Yakin, A., & Obaid, A. J. (2023). AI-aided teaching model. In Education 5.0. In *Handbook of Research on AI-Based Technologies and Applications in the Era of the Metaverse* (pp. 83–104). IGI Global.

Kim, D., Jung, E., Yoon, M., Chang, Y., Park, S., Kim, D., & Demir, F. (2021). Exploring the structural relationships between course design factors, learner commitment, self-directed learning, and intentions for further learning in a self-paced MOOC. *Computers and Education, 166,* 104171.

Leal Filho, W., Yang, P., Eustachio, J. H. P. P., Azul, A. M., Gellers, J. C., Gielczyk, A., ... Kozlova, V. (2023). Deploying digitalisation and artificial intelligence in sustainable development research. *Environment, Development and Sustainability, 25*(6), 4957–4988.

Lee, H. (2023). The rise of ChatGPT: Exploring its potential in medical education. *Anatomical Sciences Education.* 1–11

Li, X., Jiang, M. Y. C., Jong, M. S. Y., Zhang, X., & Chai, C. S. (2022). Understanding medical students' perceptions of and behavioral intentions toward learning artificial intelligence: A survey study. *International Journal of Environmental Research and Public Health, 19*(14), 8733.

Lund, B. D., Wang, T., Mannuru, N. R., Nie, B., Shimray, S., & Wang, Z. (2023). ChatGPT and a new academic reality: Artificial Intelligence-written research papers and the ethics of the large language models in scholarly publishing. *Journal of the Association for Information Science and Technology, 74*(5), 570–581.

Martin, F., Dennen, V. P., & Bonk, C. J. (2020). A synthesis of systematic review research on emerging learning environments and technologies. *Educational Technology Research and Development: ETR and D, 68*(4), 1613–1633.

Muthmainnah, M., Khang, A., Al Yakin, A., Oteir, I., & Alotaibi, A. N. (2023). An innovative teaching model: The potential of metaverse for English learning. In: Alex Khang, Vrushank Shah, and Sita Rani. *Handbook of Research on AI-Based Technologies and Applications in the Era of the Metaverse* (pp. 105–126). IGI Global.

Muthmainnah, P. M. I., Ibna Seraj, P. M., & Oteir, I. (2022). Seraj, Ibrahim Oteir, Playing with AI to investigate human-computer interaction technology and improving critical thinking skills to pursue 21st century age. *Education Research International, 2022,* Article ID 6468995, 17 pages. https://doi.org/10.1155/2022/6468995.

Park, S. Y. (2009). An analysis of the technology acceptance model in understanding university students' behavioral intention to use e-learning. *Journal of Educational Technology and Society, 12*(3), 150–162.

Park, W., & Kwon, H. (2023). Implementing artificial intelligence education for middle school technology education in Republic of Korea. *International Journal of Technology and Design Education,* vol 34, 109–135. https://doi.org/10.1007/s10798-023-09812-2

Saleme, P., Dietrich, T., Pang, B., & Parkinson, J. (2023). Design of a digital game intervention to promote socio-emotional skills and prosocial behavior in children. *Multimodal Technologies and Interaction, 5*(10), 58. *Co-Creating Gamified Social Marketing Programs Using Living Labs, 72,* 2021.

Shihab, S. R., Sultana, N., Samad, A., & Hamza, M. (2023). Educational technology in teaching community: Reviewing the dimension of integrating Ed-Tech tools and ideas in classrooms. *Eduvest-Journal of Universal Studies, 3*(6), 1028–1039.

Shrivastava, A., Krishna, K. M., Rinawa, M. L., Soni, M., Ramkumar, G., & Jaiswal, S. (2023). Inclusion of IoT, ML, and blockchain technologies in next generation industry 4.0 environment. *Materials Today: Proceedings, 80,* 3471–3475.

Siragusa, L., & Dixon, K. (2008). Planned behaviour: Student attitudes towards the use of ICT interactions in higher education. Hello! Where are you in the landscape of educational technology? *Proceedings Ascilite Melbourne, 2008*, 942–953.

Songkram, N., Chootongchai, S., Osuwan, H., Chuppunnarat, Y., & Songkram, N. (2023). Students' adoption towards behavioral intention of digital learning platform. *Education and Information Technologies*, vol 28, 11655–11677.

Sweet, M., & Michaelsen, L. K. (Eds.). (2023). *Team-Based Learning in the Social Sciences and Humanities: Group Work That Works to Generate Critical Thinking and Engagement*. Taylor & Francis, New York.

Tokareva, E. A., Smirnova, Y. V., & Orchakova, L. G. (2019). Innovation and communication technologies: Analysis of the effectiveness of their use and implementation in higher education. *Education and Information Technologies*, 24(5), 3219–3234.

Wangdi, J., & Savski, K. (2023). Linguistic landscape, critical language awareness and critical thinking: Promoting learner agency in discourses about language. *Language Awareness*, 32(3), 443–464.

Xiang, Z., Fesenmaier, D. R., & Werthner, H. (2021). Knowledge creation in information technology and tourism: A critical reflection and an outlook for the future. *Journal of Travel Research*, 60(6), 1371–1376.

Xu, W., Zhang, N., & Wang, M. (2023). The impact of interaction on continuous use in online learning platforms: A metaverse perspective. *Internet Research*, 34(1), 79–106.

Yondler, Y., & Blau, I. (2023). What is the degree of teacher centrality in optimal teaching of digital literacy in a technology-enhanced environment? Typology of teacher prototypes. *Journal of Research on Technology in Education*, 55(2), 230–251.

Zahedi, L., Batten, J., Ross, M., Potvin, G., Damas, S., Clarke, P., & Davis, D. (2021). Gamification in education: A mixed-methods study of gender on computer science students' academic performance and identity development. *Journal of Computing in Higher Education*, 33(2), 441–474.

Chapter 15

Embedded machine learning

Shweta Gupta and Abhigna Ragala

15.1 INTRODUCTION

The significance of embedded machine learning (EML) extends beyond a variety of applications, including driverless cars, healthcare systems, and intelligent Internet of Things gadgets.

The importance of EML manifests across diverse applications. In healthcare, for example, wearable devices equipped with EML algorithms can continuously monitor vital signs, swiftly detecting anomalies in real time and delivering timely alerts to both patients and healthcare professionals. In the automotive industry, EML empowers vehicles to recognize and respond to intricate traffic scenarios, thereby enhancing safety and efficiency. Additionally, within industrial settings, embedded systems incorporating machine learning capabilities optimize processes, predict equipment failures, and enhance overall operational efficiency.

The evolution of embedded systems and machine learning has been interwoven, each influencing the trajectory of the other. Initially designed for specific tasks with limited computational capabilities, embedded systems saw an expansion in scope with the advent of more powerful and energy-efficient processors. Concurrently, machine learning algorithms evolved, becoming increasingly sophisticated and adaptable.

This convergence led to the emergence of EML. Traditional machine learning models, predominantly executed on powerful servers, seamlessly transitioned into embedded devices, transforming them into intelligent entities capable of learning and adapting. This evolution signifies a paradigm shift from the conventional reliance on external servers for processing to a decentralized model, empowering devices to independently analyze and respond to data.

This chapter's primary objectives encompass exploring the foundational principles of EML, investigating its potential applications, and comprehending the challenges associated with its implementation. Additionally, it seeks to underscore the significance of EML (Chen et al. 2018) in augmenting the efficiency, autonomy, and intelligence of embedded systems.

DOI: 10.1201/9781003510420-15

The scope of this chapter spans a comprehensive review of existing EML techniques, a presentation of case studies showcasing successful implementations, and a discussion of the future directions and challenges within the field. By addressing these objectives, this chapter endeavors to contribute to the expanding body of knowledge surrounding EML, fostering further research and innovation in this dynamic and transformative domain. EML represents a pivotal advancement in integrating intelligent capabilities into embedded systems (Johnson and Smith 2020). Its significance traverses various applications, promising to reshape how devices operate and interact in our interconnected world. This chapter serves as an illumination on the critical aspects of EML, laying the foundation for ongoing exploration and development in this exciting and evolving field.

15.2 FOUNDATIONS OF EMBEDDED SYSTEMS

Embedded systems form the backbone of modern technological advancements, seamlessly integrating into our daily lives to enhance efficiency and functionality. At its core, an embedded system is a specialized computing device dedicated to performing specific functions within a larger system. These systems are designed to be embedded into a larger device or product, contributing to its overall intelligence and functionality.

The evolution of embedded systems traces back to the early days of computing when mainframe computers were the size of rooms. As technology progressed, the need arose for compact, dedicated computing systems that could perform specific tasks independently. This led to the development of embedded systems, which have since become ubiquitous in various industries.

The defining characteristic of embedded systems is their focus on performing predefined tasks efficiently and reliably. Unlike general-purpose computers, embedded systems are designed for specific applications, such as controlling a car's engine, managing the temperature in a smart thermostat, or processing data from sensors in medical devices (Dhyani et al. 2023). This specialization allows embedded systems to deliver high performance in a compact form, often with low power consumption. Microcontrollers and microprocessors (Zhang and Lee 2019) play pivotal roles in the functionality of embedded systems. Microcontrollers are compact integrated circuits that include a processor, memory, and input/output peripherals, specifically tailored for embedded applications. They excel in real-time operations and are commonly found in devices like washing machines, microwave ovens, and automotive control systems. On the other hand, microprocessors, though similar, are more powerful and versatile, suitable for applications requiring advanced computing capabilities. The evolution of embedded systems, driven by advancements in microcontroller and microprocessor technologies, has led to a diverse array of applications. From the simplicity of early embedded systems to the complexity of modern devices, these systems continue to shape the technological landscape, making them indispensable in today's interconnected world.

Embedded systems frequently encounter resource constraints, including limitations in processor capacity, memory capacity, and energy availability. Thus, machine learning models intended for these systems must be compact, effective, and tailored to the existing hardware. Embedded systems require models that are small and customized to the application. Methods such as quantization, pruning, and model compression are frequently employed to diminish the dimensions and intricacy of machine learning models. Numerous embedded applications necessitate immediate or minimal delay processing. Machine learning models must possess the ability to deliver prompt and precise outcomes while adhering to the limitations of the embedded system. Embedded systems frequently engage with sensors to gather data. Machine learning models can be incorporated to analyze and comprehend sensor data, allowing the system to make informed decisions depending on the input. Power consumption is a crucial factor to consider for embedded systems, particularly for applications operating on battery power. Efficient algorithms and hardware optimizations are essential for minimizing energy usage throughout both the training and inference processes. Embedded machine learning is intricately linked to the concept of edge computing, which involves performing data processing locally on the device instead of depending on servers housed in the cloud. Edge computing minimizes the delay and data transfer needs, making it well-suited for numerous embedded applications. Machine learning frameworks and libraries help in installing models on embedded systems. TensorFlow Lite, PyTorch for Mobile, and Arm's CMSIS-NN are frameworks specifically designed for situations with limited resources.

15.3 MACHINE LEARNING FUNDAMENTALS

In this pivotal chapter, we embark on a journey into the heart of machine learning, unraveling its fundamental concepts and exploring the critical elements that underpin its functionality. Machine learning, a subfield of artificial intelligence (AI), empowers systems to learn and improve from experience without being explicitly programmed (Goel et al. 2023a). It leverages algorithms that enable computers to recognize patterns, make predictions, and optimize decisions based on data. This chapter delves into the essence of machine learning, illustrating its transformative role in the evolution of computational intelligence (Rawat et al. 2018). The foundation of machine learning lies in diverse algorithms, each catering to specific tasks. This chapter illuminates the three primary types:

- **Supervised learning:** Where models learn from labeled data, making predictions or classifications.
- **Unsupervised learning:** Involves unlabeled data, allowing the algorithm to find patterns and relationships independently.
- **Reinforcement learning:** Focused on decision-making, with agents learning through interactions with an environment and receiving feedback in the form of rewards or penalties.

15.3.1 Importance of training data

At the core of any successful machine learning endeavor is the quality and quantity of training data. This chapter accentuates the pivotal role of training data in shaping the accuracy and robustness of machine learning models. The importance of diverse, representative datasets is underscored, as they enable models to generalize well beyond their training samples. Additionally, considerations for data preprocessing and cleaning are explored, emphasizing the need for meticulous attention to ensure the reliability of the learning process (Gupta and Kumar 2023).

As we navigate through the intricacies of machine learning fundamentals, readers will gain a solid grasp of the terminology, principles, and foundational elements essential for comprehending the subsequent chapters. The journey into the world of machine learning has just begun, promising insights into its applications, challenges, and transformative potential.

15.4 INTEGRATION OF MACHINE LEARNING IN EMBEDDED SYSTEMS

Integration of machine learning in embedded systems represents a transformative paradigm shift, offering a unique blend of challenges and opportunities. This chapter delves into the intricacies of merging machine learning capabilities with embedded systems, exploring real-world applications, and dissecting the associated benefits and limitations.

15.4.1 Challenges and opportunities

The fusion of machine learning and embedded systems presents a set of challenges that demand innovative solutions. One primary hurdle is the constrained computational resources inherent in embedded devices. Balancing the computational demands of machine learning algorithms with the limited processing power of these devices requires careful optimization and algorithmic efficiency. Additionally, real-time processing requirements pose a challenge, as certain machine learning algorithms may introduce latency, impacting the responsiveness of embedded systems (Kumar et al. 2021).

However, these challenges open avenues for innovation and problem-solving. Researchers and engineers are actively working on developing lightweight machine-learning algorithms tailored for embedded systems (Kumar 2023). This optimization involves exploring techniques like model quantization, pruning, and compression to reduce the computational burden while maintaining acceptable performance.

15.4.2 Real-world examples of embedded machine learning applications

The real-world applications of embedded machine learning are diverse and impactful. In healthcare, wearable devices (Gupta and Kumar 2018) equipped with embedded machine-learning algorithms can monitor vital signs and provide early detection of health issues. In automotive systems, embedded machine learning enables advanced driver-assistance systems (ADAS), enhancing safety through features like object recognition and adaptive cruise control. Smart homes leverage embedded machine learning for energy management, security, and personalized user experiences.

15.4.3 Benefits and limitations

The integration of machine learning in embedded systems offers several advantages. Improved decision-making, adaptability, and the ability to learn from data make embedded systems more intelligent and responsive. This leads to enhanced user experiences and increased efficiency in various applications.

However, limitations exist, such as the potential for overfitting due to limited training data and the need for continuous updates to adapt to evolving patterns. Power consumption is another concern, as energy-efficient algorithms are crucial for embedded devices with limited battery life. In navigating the integration of machine learning into embedded systems, understanding and addressing these challenges while capitalizing on the opportunities is paramount for unlocking the full potential of this dynamic synergy.

15.5 HARDWARE FOR EMBEDDED MACHINE LEARNING

In the realm of EML, the hardware components play a pivotal role in shaping the capabilities and efficiency of the system. This chapter delves into the essential hardware aspects, exploring the key components, the choice between custom and off-the-shelf solutions, and the critical consideration of energy efficiency.

15.5.1 Overview of hardware components

Embedded machine learning systems rely on a synergistic integration of various hardware components to function effectively. Sensors serve as the sensory input, capturing real-world data that becomes the foundation for decision-making. Processors, the brains of the system, execute the machine learning algorithms to derive meaningful insights from the input data. Memory, both volatile and non-volatile, is crucial for storing data and model parameters. Understanding the interplay between these components is vital for optimizing the performance of embedded machine-learning applications.

15.5.2 Custom versus off-the-shelf solutions

One of the fundamental decisions in developing embedded machine learning systems is whether to opt for custom-designed hardware or leverage off-the-shelf solutions. Custom solutions offer the advantage of tailored specifications, optimizing the hardware for specific applications. However, they often come with higher development costs and longer time-to-market. On the other hand, off-the-shelf solutions provide a more cost-effective and time-efficient approach but may lack the fine-tuned optimization required for certain applications. Striking the right balance between customization and practicality is a key consideration for developers and engineers.

15.5.3 Energy efficiency considerations

Energy efficiency is a critical factor in the success of embedded machine learning applications, particularly in scenarios where power resources are constrained. The choice of hardware components significantly influences the energy consumption of the system. Developers need to explore low-power sensors, energy-efficient processors, and optimized memory solutions to ensure sustainable operation. Balancing performance requirements with energy constraints is a delicate task, requiring a nuanced understanding of the application's needs and the available power resources.

In conclusion, this chapter underscores the importance of carefully selecting and optimizing hardware components for embedded machine-learning systems. The intricate dance between sensors, processors, and memory, coupled with the strategic choice between custom and off-the-shelf solutions, defines the efficacy of EML applications. Moreover, the imperative consideration of energy efficiency ensures the sustainability of these systems in diverse operational environments.

15.6 SOFTWARE AND PROGRAMMING MODELS

In the realm of EML, the foundation lies in the software and programming models that empower the integration of intelligence into constrained devices. This chapter delves into the crucial aspects of programming languages, development environments, and the integration of machine learning libraries within the embedded systems landscape.

15.6.1 Programming languages for embedded systems

Choosing the right programming language is pivotal for successful embedded systems development, considering the resource constraints inherent in these environments. C and C++ stand out as the dominant languages for embedded programming due to their low-level control, efficiency, and proximity to hardware. These languages provide a level of abstraction that facilitates direct manipulation of memory and hardware registers, crucial for optimizing performance in resource-constrained environments.

Python, despite being known for its ease of use and versatility in machine learning, has gained traction in certain embedded applications. Micro Python and Circuit Python, for instance, cater to microcontrollers, offering a higher-level language for embedded development. However, the trade-off often involves sacrificing some degree of performance compared to lower-level languages.

15.6.2 Development environments

Embedded systems demand specialized development environments to streamline the intricacies of programming for constrained devices. Integrated development environments (IDEs) tailored for embedded development, such as Keil, IAR Embedded Workbench, and Eclipse, offer tools and features specific to the challenges of developing on resource-limited platforms.

These environments provide functionalities like cross-compilers, debugging tools, and real-time operating system (RTOS) integration, crucial for embedded systems. RTOS plays a vital role in managing concurrent tasks, ensuring the timely execution of critical functions in EML applications.

15.6.3 Integration of machine learning libraries

The integration of machine learning libraries is a critical step in incorporating intelligence into embedded systems. TensorFlow Lite and MicroML are examples of libraries designed explicitly for resource-constrained devices.

TensorFlow Lite allows developers to deploy machine learning models on microcontrollers and other embedded platforms, striking a balance between efficiency and accuracy.

In contrast, MicroML is tailored for extremely constrained devices, emphasizing minimal memory footprint and low latency. Its lightweight nature makes it suitable for applications where resources are at a premium, such as in Internet of Things (IoT) devices and wearable technology (Chhaya et al. 2017).

The seamless integration of these libraries into the development environment facilitates the deployment of machine learning models onto embedded systems. This integration is crucial for harnessing the power of artificial intelligence in applications like predictive maintenance, gesture recognition, and edge computing.

In conclusion, the software and programming models in the realm of embedded machine learning play a pivotal role in shaping the capabilities of intelligent embedded systems. The choice of programming language, the utilization of specialized development environments, and the integration of machine learning libraries collectively define the landscape of embedded systems development. Navigating this intricate terrain requires a careful balance between performance optimization and resource efficiency, ensuring that embedded machine-learning solutions can flourish in diverse applications across various domains.

15.7 EDGE COMPUTING AND IOT

Edge computing represents a paradigm shift in the way data is processed, moving computation closer to the data source rather than relying solely on centralized cloud servers. In the context of EML, edge computing becomes a cornerstone for real-time, context-aware decision-making. The importance of edge computing lies in its ability to alleviate latency issues, enhance privacy by processing sensitive data locally, and reduce the burden on network bandwidth. The edge encompasses devices such as IoT sensors, gateways, and edge servers, forming a distributed computing network that enables efficient data processing at the source. This decentralized approach is particularly advantageous in scenarios where immediate insights and actions are paramount, as seen in applications ranging from autonomous vehicles to smart cities.

15.7.1 The connection between embedded machine learning and IoT

Embedded machine learning and IoT share a symbiotic relationship, with each enhancing the capabilities of the other. EML brings intelligence to edge devices, enabling them to make autonomous decisions based on data patterns. This is especially valuable in IoT, where an abundance of data is generated by interconnected devices. The integration of machine learning algorithms directly into IoT devices allows for local data analysis, reducing the need to send every piece of information to centralized servers. This not only enhances response times but also contributes to more efficient bandwidth usage. Additionally, EML enables IoT devices to adapt and learn from their surroundings, optimizing performance over time.

15.7.2 IoT devices with embedded machine learning

Numerous case studies showcase the synergy between IoT and embedded machine learning, illustrating their transformative impact across various industries.

- **Healthcare monitoring devices:** IoT-enabled wearables equipped with embedded machine-learning algorithms can continuously monitor vital signs and detect anomalies in real time. This ensures timely intervention in case of emergencies, showcasing the potential for life-saving applications. EML algorithms integrated into industrial IoT devices can predict equipment failures before they occur, enabling proactive maintenance. This not only minimizes downtime but also extends the lifespan of machinery, optimizing operational efficiency.
- **Smart agriculture:** Figure 15.1 presents smart agriculture applications IoT sensors embedded in agricultural equipment, coupled with machine learning, can analyze soil conditions, weather patterns, and crop health. This data-driven approach facilitates precision farming, maximizing crop yield while minimizing resource usage. These case studies exemplify how the integration of embedded machine learning into IoT devices transcends theoretical benefits, offering tangible solutions to real-world challenges. In conclusion, the amalgamation of edge computing, IoT, and embedded machine learning marks a transformative phase in the evolution of intelligent systems. Edge computing's immediacy, coupled with the analytical prowess of embedded machine learning, opens new frontiers of possibilities across diverse industries. As the synergy between these technologies continues to mature, the landscape of IoT will be shaped by intelligent, autonomous devices that operate seamlessly at the edge, heralding a new era of efficiency and innovation.

EML encounters a myriad of challenges that stem from the inherent constraints of embedded systems. One significant challenge lies in the delicate balance between memory and processing capabilities. Constrained by limited resources, developers often grapple with optimizing machine learning

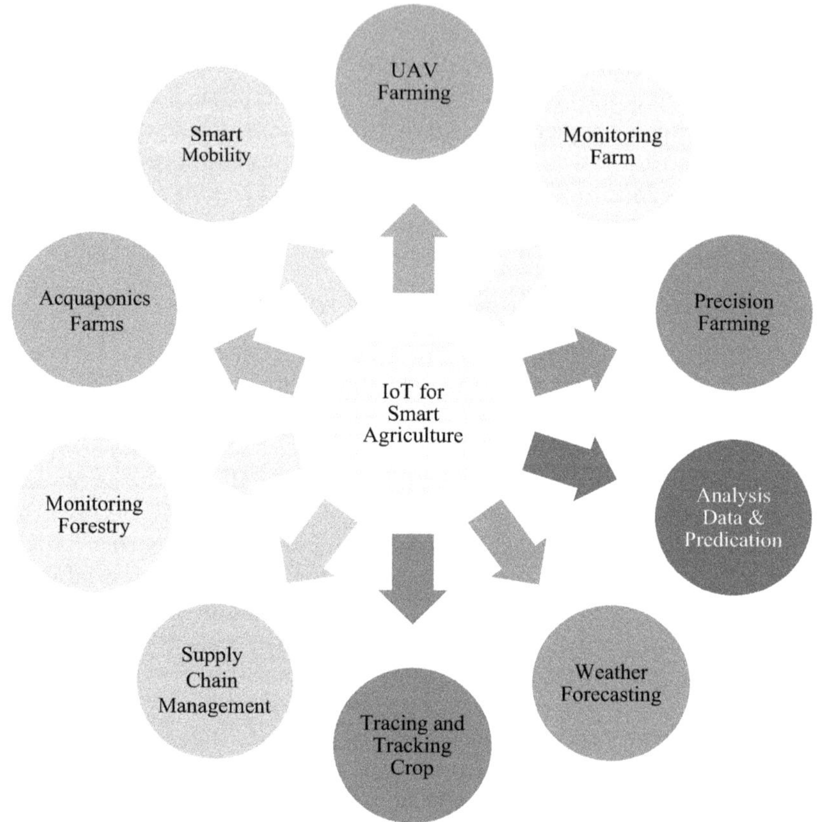

Figure 15.1 Smart agriculture applications.

models to fit within the tight confines of embedded devices. Strategies such as model quantization, which reduces precision without compromising functionality, and the use of specialized hardware accelerators aid in mitigating these constraints.

Power consumption emerges as a critical concern in the development of embedded machine-learning applications. Striking a harmonious equilibrium between computational intensity and energy efficiency is imperative. Techniques like dynamic voltage and frequency scaling (DVFS) allow devices to adapt their power consumption based on workload, effectively managing energy resources without sacrificing performance. The realm of embedded systems introduces unique security and privacy challenges. With devices operating on the edge, sensitive data may be at risk. Implementing robust encryption, secure boot processes, and incorporating hardware-based security features are essential safeguards against potential threats, ensuring the integrity of both the device and the data it processes.

In addressing these challenges, the future of embedded machine learning hinges on innovative solutions that push the boundaries of what is achievable within the constraints of embedded systems. As technology advances, the industry continues to explore novel approaches to enhance memory efficiency, minimize power consumption, and fortify security protocols, fostering a landscape where intelligent embedded systems can thrive with resilience and reliability.

15.8 FUTURE TRENDS IN EMBEDDED MACHINE LEARNING

As EML continues to evolve, envisioning the future landscape involves exploring key trends in both hardware and software advancements, identifying emerging applications, and addressing critical ethical considerations. Figure 15.2 presents the future trends in EML.

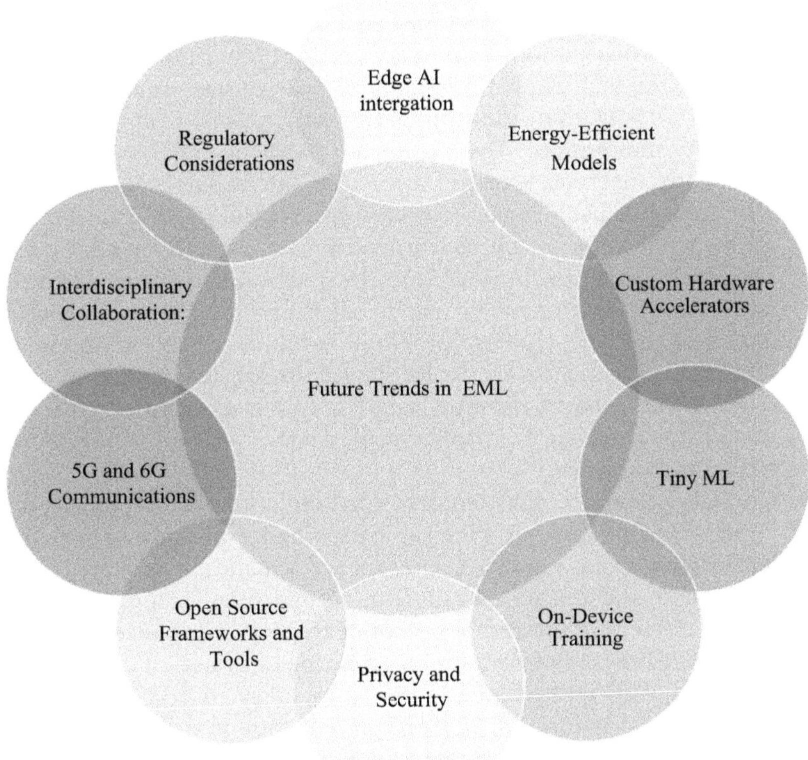

Figure 15.2 Future trends in EML.

- **Advancements in hardware and software hardware acceleration:** One of the prominent trends shaping the future of embedded machine learning is the continuous advancement in hardware technology. Specialized hardware accelerators, such as neuromorphic processors and field-programmable gate arrays (FPGAs), are gaining momentum. These accelerators are designed to handle specific machine learning workloads more efficiently than general-purpose processors, offering improved performance and energy efficiency.
- **Edge computing integration:** The integration of edge computing with embedded machine learning is another transformative trend. Edge devices, equipped with enhanced processing capabilities, enable real-time data analysis and decision-making, reducing dependence on centralized cloud servers. This shift toward edge computing aligns with the growing need for low-latency, high-throughput applications in fields like autonomous vehicles, healthcare monitoring, and industrial automation.
- **Software innovations:** On the software front, advancements in model optimization techniques and algorithmic innovations are driving the efficiency of embedded machine learning. Quantization, pruning, and knowledge distillation techniques are becoming more sophisticated, enabling the deployment of complex models on resource-constrained devices without compromising performance.
- **Edge AI integration:** The different prospects and future trends in embedded machine learning are based on the incorporation of AI. The edge has been rapidly increasing in popularity. This entails the implementation of machine learning models directly on edge devices, such as IoT devices, sensors, and other embedded systems. The continuation of this trend is probable, enabling instantaneous processing and decreased dependence on cloud resources. systems due to their restricted power supplies. Scientists are investigating methods such as quantization, model pruning, and hardware accelerators to enhance the energy efficiency of embedded devices without compromising model performance.
- **Custom hardware accelerators:** Custom hardware accelerators are being developed to optimize the performance of embedded machine learning systems. These accelerators are specifically designed to cater to the unique requirements of different machine learning workloads. These accelerators have the potential to greatly enhance the velocity and effectiveness of inference on embedded devices.
- **TinyML:** TinyML refers to the deployment of compact machine-learning models on devices with limited resources. With the growing demand for compact and energy-efficient devices, the utilization of TinyML models is expected to become more widespread. This will allow for effective machine learning inference on devices that have restricted computational capabilities.

- **On-device training:** On-device training refers to the practice of doing machine learning training directly on the device itself, rather than relying on powerful computers. This approach is gaining popularity for specific applications. Training models directly on the device enables the development of highly customized and adaptable models, eliminating the requirement for continuous connection to centralized servers.
- **Energy-efficient models:** Energy-efficient models are being prioritized in the development of EML.
- **Privacy and security:** Privacy and security will become more important as embedded machine learning is increasingly used in applications such as healthcare, surveillance, and smart homes. The protection of data handled by these devices will be a top priority. The prominence of federated learning and other privacy-preserving techniques may increase.
- **Open-source frameworks and tools:** The accessibility of open-source frameworks and tools for the development of embedded machine learning is expected to grow. By democratizing access to embedded machine learning, this initiative promotes innovation and collaboration among developers.
- **Interdisciplinary collaboration:** The implementation of embedded machine learning necessitates the cooperation of specialists in hardware design, software development, and machine learning. Interdisciplinary collaboration will be increasingly crucial in addressing the intricate difficulties related to embedded systems as the area progresses.
- **5G and 6G communications:** The implementation of 5G networks and the extensive use of edge computing can significantly augment the functionalities of embedded machine learning. Decreased delay and enhanced data transfer capacity might enable smoother connection between peripheral devices and centralized systems.

As embedded machine learning applications become more widespread, regulatory frameworks may develop to tackle ethical and legal challenges. This encompasses recommendations about data privacy, transparency, and accountability when implementing machine learning models on embedded devices.

15.9 EMERGING APPLICATIONS

- **Healthcare:** The future of embedded machine learning holds immense promise for healthcare applications. Wearable devices equipped with EML capabilities can continuously monitor vital signs, detect anomalies, and provide personalized health insights. Remote patient monitoring and early disease detection are areas where embedded machine learning is poised to make significant contributions.

- **Smart cities and infrastructure:** In the realm of smart cities, embedded machine learning is anticipated to play a crucial role in optimizing urban infrastructure. Intelligent traffic management, energy-efficient systems, and predictive maintenance of critical infrastructure are among the applications that can benefit from EML, contributing to the development of sustainable and resilient cities.

As embedded machine learning becomes more pervasive, ethical considerations become increasingly paramount.

- **Privacy concerns:** The collection and processing of data at the edge raise concerns about individual privacy. Striking a balance between leveraging machine learning for valuable insights and protecting user data is a challenge that future EML applications must address.
- **Bias and fairness:** Embedding machine learning models in decision-making processes necessitates careful consideration of biases. Ensuring fairness and equity in outcomes requires ongoing efforts to identify and rectify biases present in training data and algorithms.
- **Transparency and accountability:** The opaque nature of some machine learning models poses challenges in understanding their decision-making processes. Future trends in embedded machine learning must prioritize transparency and accountability, allowing users to comprehend the rationale behind decisions made by intelligent embedded systems.

The integration of IoT devices and embedded machine learning capabilities can augment the usefulness, efficiency, and intelligence of diverse applications. It is applied in several ways. Smart thermostats, such as the Nest Learning Thermostat, employ machine learning algorithms to comprehend human preferences and automatically modify temperature settings accordingly. Intelligent cameras equipped with machine learning technology can accurately detect and distinguish between individuals, animals, and various objects, hence minimizing the occurrence of false alarms. Wearable fitness trackers, such as Fitbit, employ machine learning to examine activity patterns, offer customized observations, and anticipate potential health concerns. ML-enabled devices can detect abnormalities in vital signs and offer timely alerts for illnesses such as arrhythmias or sleep disturbances. Predictive maintenance sensors enable machines to anticipate maintenance requirements by utilizing sensors and machine learning algorithms. This capability minimizes operational interruptions and decreases maintenance expenses. Quality control devices, such as cameras and sensors, can utilize machine learning algorithms to promptly detect and classify faults and anomalies during the manufacturing process. Precision farming devices utilize IoT sensors to gather data about soil conditions, weather patterns, and crop health in agriculture. Subsequently, machine learning algorithms can examine this data to enhance irrigation timetables and forecast agricultural output. Automobiles equipped with IoT connectivity and ML technologies can analyze up-to-the-minute traffic data, anticipate maintenance requirements, and improve safety by identifying trends in the surrounding environment. IoT devices, such as intelligent traffic lights and cameras, can utilize ML to enhance traffic flow, minimize congestion, and enhance overall transportation efficiency. Waste management can be enhanced by utilizing sensors in waste bins, which employ machine learning algorithms to optimize rubbish pickup routes based on different levels. Smart Shelves utilize machine learning technology to enable retailers to effectively track inventory levels, analyze client behavior, and strategically position products to maximize sales. Optimizing the use and allocation of energy resources in smart grids and IoT devices are used for the integrated energy grids to employ machine learning algorithms to forecast energy demand, enhance distribution efficiency, and detect regions susceptible to problems. Air quality sensors are devices that employ machine learning to analyze patterns and detect the origins of pollution, thereby enhancing environmental management. Prosthetic limbs equipped with integrated sensors and machine learning algorithms can adjust to the movements and preferences of users, resulting in a user experience that is more authentic and instinctive. The combination of IoT with machine learning facilitates the acquisition and examination of data by these devices, allowing them to make instantaneous judgments and enhance their performance progressively. It improves the flexibility, independence, and cognitive abilities of diverse systems in numerous sectors.

In conclusion, the future of embedded machine learning promises a convergence of hardware and software innovations, opening new frontiers in diverse applications while demanding a steadfast commitment to ethical principles. Striking the right balance between technological advancements and ethical considerations will be pivotal in realizing the full potential of embedded machine learning in shaping the future of intelligent, connected systems.

15.10 SUMMARY

In the culmination of this exploration into EML, it is imperative to revisit key insights that underscore the transformative potential of integrating intelligence into embedded systems. Throughout this journey, we delved into the intricacies of embedded machine learning, examining programming languages, development environments, and the integration of machine learning libraries. The symbiosis of traditional embedded systems with the cognitive capabilities of machine learning has paved the way for innovations in diverse domains. We dissected the challenges and opportunities, showcasing real-world applications and case studies that exemplify the impact of EML. As we stand at the precipice of technological evolution, the importance of embedded machine learning becomes increasingly apparent. The convergence of intelligent algorithms with the ubiquity of embedded systems promises to redefine how we interact with our surroundings. From enhancing healthcare diagnostics to optimizing energy consumption in smart cities, the applications are boundless. EML catalyzes unlocking unprecedented efficiency, responsiveness, and autonomy in devices that permeate our daily lives.

The study provides a superficial understanding of the extensive domain of embedded machine learning. The ongoing development of machine learning algorithms and embedded systems offers a promising opportunity for further investigation. Researchers, developers, and innovators are urged to explore the limits, uncover fresh opportunities, and tackle obstacles. The future of embedded machine learning offers the potential for increased integration, efficiency, and adaptability, motivating us to enthusiastically pursue knowledge and innovation.Essentially, the exploration of embedded machine learning showcases not just a technological advancement but also a fundamental change in how we perceive and utilize the capabilities of intelligent embedded systems. We eagerly invite you to further explore the limitless area of embedded machine learning, as it presents several prospects for growth and advancement.

BIBLIOGRAPHY

Chen, X., Lin, X., & Zhao, Z. (2017). Machine learning at the network edge: A survey. *IEEE Access*, 5, 4582–4599.

Chen, Y., Yang, Z., & Wang, X. (2018). Embedded machine learning for real-time image analysis. *Journal of Embedded Systems*, 20(4), 345–362.

Chhaya, L., Sharma, P., Bhagwatikar, G., & Kumar, A. (2017). Wireless sensor network based smart grid communications: Cyber-attacks, intrusion detection system, and topology control. *Electronics*, 6(1), 5.

Dhyani, S., Kumar, A., & Choudhury, S. (2023). Arrhythmia disease classification utilizing ResRNN. *Biomedical Signal Processing and Control*, 79, 104160.

Garcia, M., & Martinez, J. (2021). Embedded machine learning in healthcare: A systematic review. *Journal of Biomedical Informatics*, 105, 103429.

Goel, A., Goel, A. K., & Kumar, A. (2023a). Performance analysis of multiple input single layer neural network hardware chip. *Multimedia Tools and Applications*, 1–22.

Goel, A., Goel, A. K., & Kumar, A. (2023b). The role of artificial neural network and machine learning in utilizing spatial information. *Spatial Information Research*, 31(3), 275–285.

Gupta, R., & Chhokra, A. (2018). Energy-efficient machine learning algorithms for embedded systems. *Journal of Green Engineering*, 8(3), 303–318.

Gupta, S., HariPrasad, S. A., & Kumar, A. (2018, May). Bionics-based emergency contact for treatment of epilepsy. In *Proceedings of the 2018 10th International Conference on Bioinformatics and Biomedical Technology* (pp. 46–50). Association for Computing Machinery, New York.

Gupta, S., & Kumar, A. (2018). Bionic functionality of prosthetic hand. In Rajesh Singh, Sushabhan Choudhury, & Anita Gehlot (Eds.), *Intelligent Communication, Control and Devices: Proceedings of ICICCD 2017* (pp. 1177–1190). Springer, Singapore.

Gupta, S., & Kumar, A. (2023). Study on early accurate diagnosis and treatment of COVID-19 with smartphone tracking using bionics. *Security and Privacy*, 6(5), e303.

Johnson, M., & Smith, A. (2020). Programming languages for embedded machine learning: A comprehensive review. *ACM Transactions on Embedded Computing Systems*, 19(3), 1–24.

Kim, J., Park, H., & Lee, S. (2018). Real-time gesture recognition using embedded machine learning on microcontrollers. *Sensors*, 18(9), 2892.

Kulkarni, M., & Kagal, V. (2017). Integration of machine learning in embedded systems for smart home applications. *Procedia Computer Science*, 115, 141–148.

Kumar, A. (2023). Study and analysis of different segmentation methods for brain tumor MRI application. *Multimedia Tools and Applications*, 82(5), 7117–7139.

Kumar, A., Chauda, P., & Devrari, A. (2021). Machine learning approach for brain tumor detection and segmentation. *International Journal of Organizational and Collective Intelligence (IJOCI)*, 11(3), 68–84.

Li, H., Zhang, Q., & Lee, B. (2019). Integration of TensorFlow Lite in embedded systems for edge computing applications. *IEEE Transactions on Embedded Systems*, 14(2), 567–580.

Lu, Y., Li, W., & Wu, D. (2020). Design and implementation of embedded machine learning systems for autonomous vehicles. *IEEE Transactions on Intelligent Transportation Systems*, 21(3), 1077–1088.

Rawat, A. S., Rana, A., Kumar, A., & Bagwari, A. (2018). Application of multi layer artificial neural network in the diagnosis system: A systematic review. *IAES International Journal of Artificial Intelligence*, 7(3), 138.

Reddy, S., & Kumar, R. (2020). Edge computing and embedded machine learning for smart cities: A comprehensive review. *Sustainable Cities and Society*, 55, 102004.

Sankar, A., & Varshney, P. (2019). Security challenges in embedded machine learning: A comprehensive review. *Computers and Security*, 82, 207–226.

Wang, J., Liu, Y., & Zhang, Y. (2016). FPGA-based accelerators for embedded machine learning: A survey. *Journal of Signal Processing Systems*, 84(2), 189–202.

Wang, L., Li, M., & Zhang, S. (2017). Challenges and opportunities in hardware acceleration for embedded machine learning. *Journal of VLSI Signal Processing Systems*, 88(1), 45–62.

Zhang, L., Wang, Y., & Wang, Y. (2018). Real-time facial recognition using embedded machine learning on Raspberry Pi. *Journal of Real-Time Image Processing*, 15(4), 843–854.
Zhou, Y., Zhang, W., & Liu, Q. (2019). A survey on machine learning in the Internet of Things: Processing, communication, and security. *IEEE Access*, 7, 66989–67006.

Chapter 16

Applications of machine learning in side-channel and fault analysis

Naga Durga Saile K and Rohit Tanwar

16.1 INTRODUCTION

An attack is an offensive target made on the computers, networks, and peripherals of the system [1]. The intruder attacks the restricted areas of the system and tries to gain access to the data without authorization by passing malicious content to the system. Based on the attacks that occurred worldwide in different domains, the definition of an attack has been defined in different ways.

The Internet Engineering Task Force defines an attack [2] as a system security breach caused by a purposeful threat. This is a deliberate attempt to circumvent security safeguards and violate a system's security regulations.

According to the Committee on National Security Systems of the United States of America, an attack [3] is defined as a type of malicious behavior intended to acquire, disrupt, impede, diminish, or obliterate the resources of an information system or the information stored within it.

One of the main reasons for the increase in cyberattacks is that we rely on computers and the internet for most day-to-day activities and a huge amount of data is being transmitted over the internet daily.

As per the statistics of 2023, the data that is being transferred is expected to reach 103 billion of which 8.3 billion are for Google, and the WhatsApp data exchange is 65 billion per day. It is also expected that there will be 180 zettabytes of data by 2025 [4]. With the impact of huge data transfers happening all over the world, intruders are finding different mechanisms to access confidential data in the best ways possible, which is leading to disastrous data leakage in all confidential arenas. In the next sections, we will discuss the various kinds of attacks.

16.2 TYPES OF ATTACKS

Attacks are primarily categorized into two kinds, namely active attacks and passive attacks. Active attacks are cybersecurity attacks in which an adversary tries to change, demolish, or interrupt the normal operation of a system or network. These attacks are more damaging than passive attacks

DOI: 10.1201/9781003510420-16

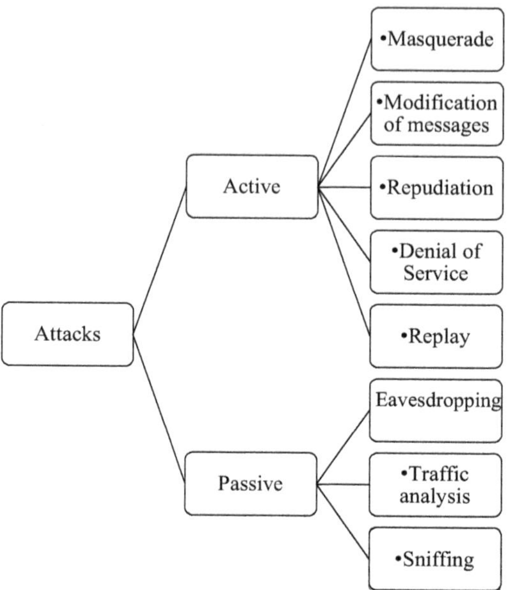

Figure 16.1 Types of attacks.

because the attacker actively interacts with the target system or network. Passive attacks, on the other hand, are aimed at acquiring or utilizing information from the system without affecting its resources. These attacks simulate eavesdropping or monitoring transmissions, with the goal to collect communicated information. In passive attacks, the attacker observes or collects data without altering or destroying it. Figure 16.1 shows examples of active and passive attacks.

16.3 CLASSIFICATION OF ATTACKS

An attack is the act of stealing confidential information through different modes and finally gaining access to the data. Attacks can be classified into three types, namely, software attacks, network attacks, and hardware attacks, as shown in Figure 16.2, through which the intruder peeks through the confidentiality of the data.

16.3.1 Software attacks

The attacks that compromise the core application of the data and gain access are called software attacks. The intruder induces malicious pieces of code into the system and hence hacks the device data. Using mechanisms to affect the system at the application level can be defined as software attacks. Out of many software attacks possible, some are discussed next.

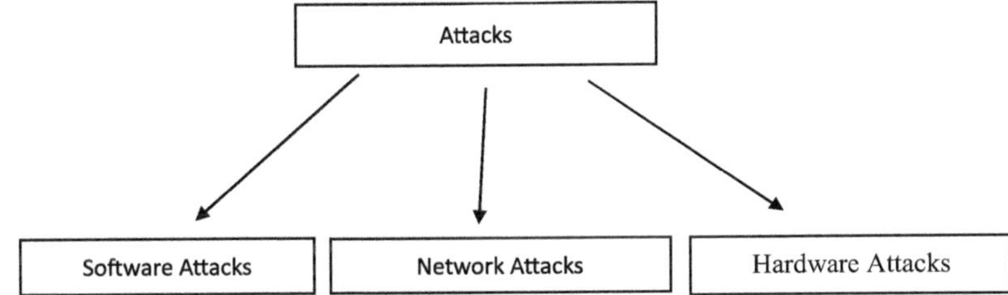

Figure 16.2 Classification of attacks.

16.3.1.1 Malware attack

Malware is a malicious software attack where the intruder intervenes in the system by passing a malicious piece of code through fake URLs, updates of the system, security updates, etc. They seem to appear trustworthy to the users, but in reality, this malware is one of the methods through which the intruder attacks the system and gains access to the system.

16.3.1.2 Brute force attack

This is a type of attack where the user guesses the passwords of the user by trial-and-error methods. Particularly this happens when you are using networks on public domains. Multiple login attempts and using a password without proper security measures are some of the methods of a brute force attack. This is one of the easiest types of attacks because the hacker is aware of the pattern of the passwords that are used or any names used.

16.3.1.3 Memory buffer overflow attack

In this attack, the intruder takes control of the memory and overwrites the memory addresses [5] with a piece of code such that the intruder can gain access to the data by manipulating the memory.

16.3.2 Network attacks

Network attacks are prone to occur when the network infrastructure of any organization is compromised by the attacker. The attacker monitors the internet traffic that is being transmitted and exploits the confidentiality of network systems. A few examples of these attacks are as follows.

16.3.2.1 Man-in-the-middle attack

In this attack, the intruder intervenes between the sender and receiver end and tries to gain access to the communication. All the messages sent between the sender and receiver will be seen by the hacker and the hacker will pretend to be the receiver and send a reply pretending to be the receiver.

16.3.2.2 Domain Name Service (DNS) poisoning

DNS plays an important role in the services provided over the internet. DNS poisoning is the type of attack in which the name of the service is being attacked [6], i.e., the hacker attacks the DNS and migrates to a different location of his own finally leading to loss of confidentiality of the data.

16.3.2.3 Denial of service (DoS)

In this attack, the system is crashed or taken down by the attackers making the system unusable for the users. DoS attacks are targeted mainly at organizations with high-profile turnovers like finance, banking, media, and government websites, where there is a greater chance of data exploitation.

Session hijacking and signal jamming are a few other network attacks possible.

16.3.3 Hardware attacks

Hardware attacks are those that try to access the data from computers by targeting the hardware components like integrated circuits (ICs), semiconductors, CMOS batteries, and other hardware devices that are connected to the system [7]. One of the major reasons for these hardware attacks is because of the backdoor activities of manufacturing and security validations. With the increase in economic crises across the globe, the scope for backdoor manufacturing activities [8] has increased. These kinds of activities have provided scope for the attackers to induce malicious hardware devices into the systems, which include radio-frequency identification (RFID) sensors. The impact of hardware attacks is very dangerous, as software attacks can be identified easily with the malfunction of the system, but tracing a hardware attack is always difficult.

The following is a list of hardware attacks that are possible [9]:

1. Side-channel attacks
2. Timing attacks
3. Row hammer attacks
4. Eavesdropping attacks
5. Attacking by fault triggering.
6. Modification attacks

In the next sections, we will discuss side-channel attacks and their consequences.

16.4 Side-channel attacks

16.4.1 What are side-channel attacks?

A side-channel attack (SCA) is a way of hacking a cryptographic algorithm that is based on an examination of the auxiliary systems employed in the encryption process. These can be accomplished by utilizing a variety of signals emitted by devices, such as electromagnetic waves, power usage, mobile sensors, and sound from keyboards and printers to target devices. Once captured, these signals are utilized to interpret signals that can then be used to breach a device's security.

16.4.2 History

It was during the year 1945 when a side-channel attack first took place in the United States when the Soviet Union presented a great seal to US Ambassador W. Averell Harriman. This seal contained a hearing device inside it and was placed in the ambassador's residence. This device transmitted audio when set on a certain radio frequency, which was handled by the Soviet Union. This hearing device was a passive device designed by Theremin and could not be identified for seven years until 1952. This device was called "The Thing" and was known to be a great political spy gadget of the times [10]. Using the radio frequency, the signals were analyzed and the Soviet Union took over the secrets of US discussions.

During World War II, the Bell telephone company also encountered this kind of glitch where it found that there was a leakage of electromagnetic signals from one-time pad printers, which was later secured by keeping a secure distance of 100 ft radius away from the printer. It was in the year 1964 that the US found a large set of earphones under the ceiling of its embassy that later leveraged the inception of Tempest, which is an intelligence system that worked on electromagnetic signals and came up with countermeasures for side-channel attacks that are observed [11].

16.4.3 Method

A side-channel attack is a form of hardware attack that undermines system security by attacking the hardware components. Rather than targeting the code of the system, the hardware of the system like the electric emissions from the system monitor, hard drives, varying the power of the system through a different channel, and trying to gain access over the system are targeted. These side-channel attacks can create a security breach to the level that even the cryptographic keys of the system can also be trapped. Side-channel attacks are also called sidebar attacks or implementation attacks [12].

A side-channel attack [13] is defined as an assault that allows information to leak from a physical cryptosystem. Timing, power usage, and

Figure 16.3 General communication between sender and receiver.

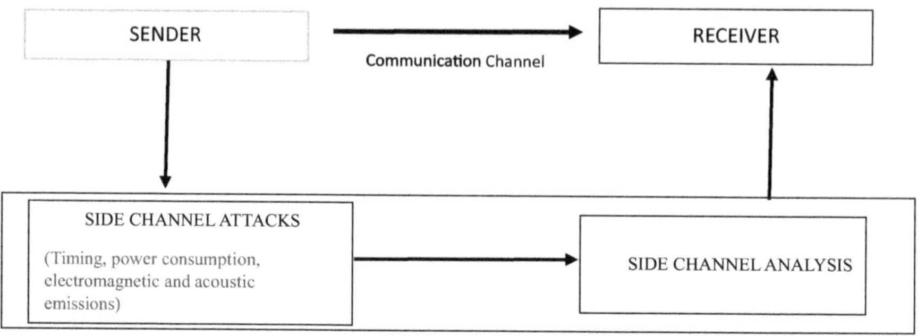

Figure 16.4 Side-channel attacks.

electromagnetic and acoustic emissions are all potential vulnerabilities in a side-channel assault. The general flow of information, from source to destination is through an encrypted channel, as shown in Figure 16.3.

The source sends the data and the receiver at the destination will decipher the text and get the code. But in a side-channel attack, as the name itself says there are other methods in which the intruder can get track of the system. Figure 16.3 shows the regular data transmission of data, whereas Figure 16.4 depicts side-channel attacks. Generally, during the design of an algorithm, only the input and output of the algorithm are discussed and taken care of. But apart from the software, there is a lot to talk about the physics associated with the peripherals of the system. Out of many physical features [14], time, sound, and power are the major ones that are affected during side-channel attacks.

16.4.4 Types of side-channel attacks

These attacks gain access to the system data by analyzing the different hardware components and other resources such as power, electromagnetic sources, and sound. The following are types of side-channel attacks that are observed.

1. **Timing attacks:** These are attacks that are likely to occur due to a computation of the time required to complete a specific job. These attacks evaluate how long it takes a machine to run a cryptographic method, hence the name timing attacks.

2. **Attack with electromagnetic (EM) radiation:** Analyzes and quantifies the electromagnetic radiation emitted by a gadget and investigates how long a system takes to run cryptographic methods.
3. **Simple power analysis (SPA):** During operation, a cryptographic system's power and electromagnetic EM variations can be immediately observed using SPA.
4. **Differential power analysis (DPA):** DPA is the process of gathering and analyzing specific statistical data from various processes.
5. **Template attack (TA):** This attack uses a "template" device to recover cryptographic keys by comparing side-channel data techniques.

16.5 FAULT ANALYSIS

Fault analysis refers to the study and exploitation of information leaked through unintended channels during the execution of cryptographic algorithms or other secure processes. Fault analysis in side-channel attacks is a cat-and-mouse game between attackers and defenders. A few fault analysis techniques are listed next.

1. **Fault injection techniques:** Researchers studied a range of fault injection methods, including voltage spikes, clock manipulations, and laser-induced faults, in order to break into cryptographic systems.
2. **Error-based attacks:** Techniques involve inducing errors in cryptographic computations to reveal sensitive information.
3. **Combining fault and side-channel attacks:** Certain research has focused on combining fault attacks with side-channel analysis to improve efficiency.

Each attack performs in a different mechanism finally compromising the data and gaining access to the confidential information.

Technology advancements and improvements in machine learning have transformed a wide variety of sectors and applications. The importance of machine learning has been increasing significantly because of its ability to forecast the future, automate processes, and find patterns in data. Some important areas of machine learning are automation, natural language processing, medical diagnosis, and scientific research, with many more areas to explore. Machine learning techniques are widely used in the field of side-channel analysis and fault analysis.

The next section delves into many categories of side-channel assaults and their accompanying responses. The applications include the following.

1. Side-channel attack detection
2. Side-channel attack profiling
3. Detection of fault attack

4. Selecting countermeasures
5. Attribution of attacks
6. Key recovery
7. Hardware Trojan detection
8. Data preprocessing and feature selection
9. Detection of anomalies

Now let us see how the machine learning algorithms work on the preceding applications for side-channel detection and fault analysis.

16.6 UNDERSTANDING MACHINE LEARNING

Machine learning (ML) algorithms are widely used as they generate increased performance output in prediction concerning any task. In any classification problems, the machine learning algorithms are fed with labeled data for training the models, which is known as supervised learning techniques. When the machine learning algorithm is not labeled, then it is said to be unsupervised learning, and in semisupervised machine learning, only some part of the data is labeled. Different machine learning algorithms constitute the supervised and unsupervised learning techniques that we will discuss in the later part of the chapter. Now we will discuss the standard architecture of machine learning, the algorithms, and the terminology used in machine learning.

The following are the basic steps in machine learning (also see Figure 16.5):

1. Data collection
2. Preprocessing
3. Feature engineering
4. Modeling
5. Parameter tuning
6. Results

The first step of the process is data collection. The data that is required for the problem to be solved is gathered from various sources and kept ready for preprocessing. Next, data preprocessing is a phase where the unnecessary data, outliers, and missing values are preprocessed to make the data ready for giving to the algorithm.

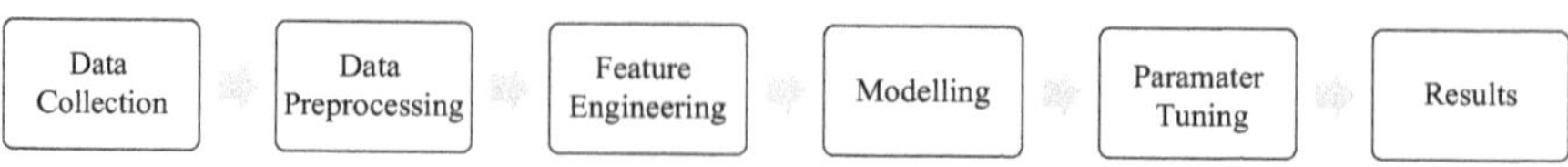

Figure 16.5 Machine learning process.

The third step involves feature engineering, which plays a major role in the ML process. The attributes that are required for the problem statement are identified. Dimensionality reduction techniques like principle component analysis, linear discriminant analysis, and singular value decomposition can reduce the data.

Now we have an appropriate dataset ready for modeling. The modeling phase is also known as selecting the algorithm and training. In this phase, the data is fragmented into training and testing and given to the appropriate model for performing the prediction. If the results are inappropriate due to overfitting or underfitting, which means there is too much training done or no training done then the parameter tuning is done. In parameter tuning, the hyperparameters are fine-tuned so that the model generates an accurate result. This summarizes a brief discussion on the machine learning process.

16.7 RELATED WORK

A recent line of research employs approaches from the machine learning domain to challenge cryptography implementations. Because no comprehensive assessment of this burgeoning topic has been conducted thus far, the purpose of this section is to assess the cutting-edge technology and explore the different machine learning techniques involved in securing side-channel attacks.

Paul Kocher [15] spoke about the side-channel attacks on public key cryptosystems which drew the attention of researchers toward learning and analyzing different timing attacks, electromagnetic radiations [16], power attacks [17], and sound attacks [18]. An example of sound attacks is typing on the keyboard. An example of a power attack is understanding the consumption of power of the target device by differential power analysis.

Rivest [19] was the first to speak about machine learning in cryptography where he observed that there are a lot of similarities between the identification of the secret key in cryptography to the target variable in machine learning. He has identified a considerable range of accurate outcomes that helped machine learning be adapted to side-channel attacks. The timeline in Figure 16.6 shows the major research on how machine learning has evolved.

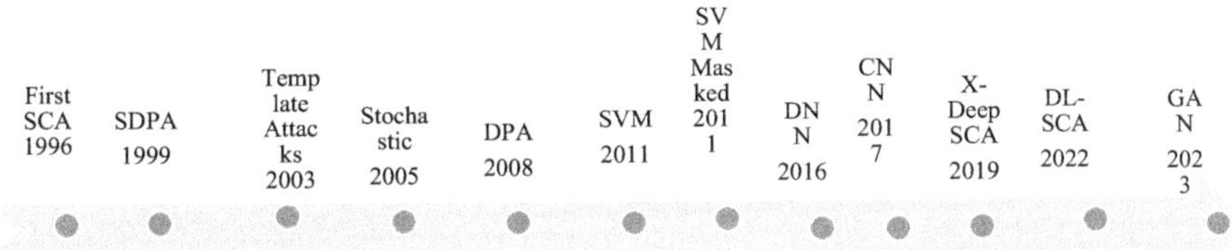

Figure 16.6 Timeline showing machine learning on SCA.

The first article on side-channel attacks was published in the year 1996, followed by differential power analysis in 1999. The research work on template attacks came in 2003 and stochastic attacks in 2003. Later with the incorporation of machine learning, the first research work was done in the year 2011 using support vector machines (SVMs) followed by masked SVM techniques. The advent of machine learning led to research in the area of deep learning using convolutional neural networks and deep learning techniques like long short-term memory (LSTM). Research in 2023 using generative adversarial networks (GANs) has taken over the implementation of side-channel attacks.

Successful side-channel attacks have been implemented on the Advanced Encryption Standard (AES) algorithm using SVMs [20]. SVM is a widely used supervised learning technique. It's also known as the max-margin classifier since it creates an ideal binary hyperplane between data points from two linearly separable categories. The attack strategy for masked implementation algorithms is given in Figure 16.7 [21]. The same SVM algorithm has been used to crack the Data Encryption Standard (DES) algorithm, which is much more sophisticated [21]. The random forest algorithm has been used to break the AES encryption and intrude into the system [22].

Triple DES was also attacked using the random forest technique [23]. The research was further extended to attack AES using random forest and support vector machines to get over the target machine [24]. The algorithms that use autoencoders and neural networks have the following architecture of hidden layer input layer and output layers [27] as shown in Figure 16.8.

The use of naïve Bayes using a 2:1 ratio of split in the traces has been studied [25]. It was further extended using correlation feature engineering using SVM and unprotected AES [26].

Maghrebi et al. [28] discuss the use of machine learning techniques to break cryptographic implementations. The authors suggest a new profiling strategy built on deep learning that is more complex than the current

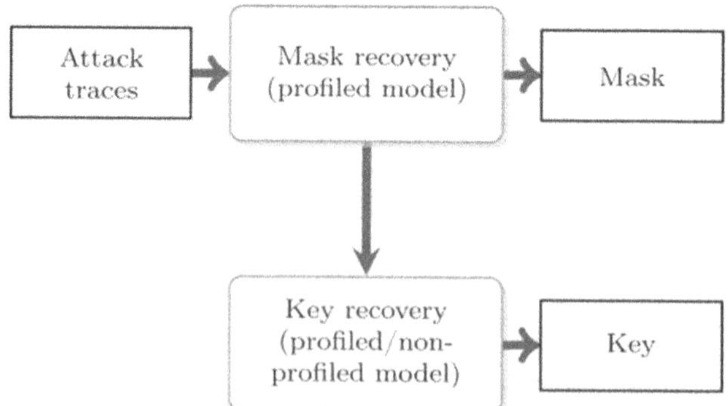

Figure 16.7 Masked implementation strategy for masked algorithms.

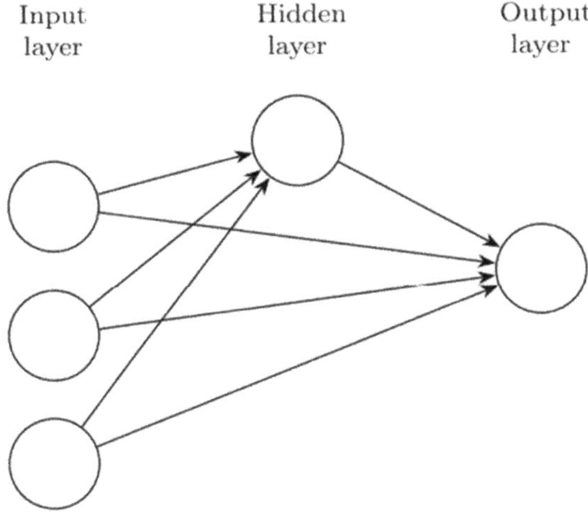

Figure 16.8 Neural networks for identifying SCA.

template attack method. The machine learning algorithms used include autoencoders. The experimental results show that the proposed approach is more effective in breaking both unprotected and protected cryptographic implementations.

He et al. [29] explore the application of machine learning techniques to side-channel cryptanalysis. The authors suggest using an SVM to recover DES keys from an 8-bit smart card. The proposed approach uses measurements during the DES key schedule to exploit the side-channel leakage. After training with at least 1,000 samples from their set of 11,000, they attain 100% accuracy using characteristics collected from measurements taken during the DES key schedule.

Zdene and Zeman [30] offer a new power analysis method that uses a neural network to examine the AES algorithm's power consumption. The suggested method combines the benefits of basic and differential power analysis and may calculate the secret key value of a cryptographic module based on measured power consumption. The proposed method is unique and effective in breaking trusted cryptographic devices such as RFID and touch smart cards.

In their study, Prouff et al. [31] investigate the use of deep learning algorithms for side-channel research and present an open database known as ASCAD. The authors provide a detailed assessment of deep learning algorithms within the context of side-channel analysis, drawing parallels with traditional template assaults. Furthermore, the study discusses hyperparameter selection for multilayer perceptron networks and convolutional neural networks (CNNs). The authors evaluate a masked version of the AES algorithm using different benchmarks and reasons. Their findings imply that deep learning algorithms demonstrate excellent efficiency in evaluating the security of embedded systems, giving various advantages over alternative methods.

Timon [32] presents a novel way for applying deep learning techniques in a non-profiled scenario, in which an attacker can only gather a finite amount of side-channel traces from a closed device for a fixed unknown key. The study demonstrates that by combining key guesses with deep learning measures, it is possible to extract information about the secret key. The research also emphasizes that masked implementations can be targeted without leaks via combination pre-processing and with fewer assumptions than traditional high-order attacks. The report includes a series of experiments done on simulated data and real traces obtained from the ChipWhisperer board and the ASCAD database to demonstrate these properties.

Specht et al. [33] present a unique algorithmic strategy for improving non-profiled attacks based on clustering against exponentiation-based implementations. This technique uses principal component analysis (PCA) to identify a few mid-ranked components with exploitable, low-variance leakage. Practical tests, including single-channel high-resolution magnetic field measurements, demonstrated a significant increase in the number of successful attacks, as stated in the study.

Zhang et al. [34] propose a new method for analyzing electromagnetic side-channel signals using wavelet transform and PCA. The proposed method is used to extract features from the electromagnetic signals and reduce the dimensionality of the data. The paper reports that the proposed method is effective in identifying the leakage information of cryptographic devices and can be used to improve the success rate of side-channel attacks.

Perin et al. [35] present a new method to attack randomized exponentiations using unsupervised learning. The proposed method is based on clustering algorithms and PCA to extract the leakage information from electromagnetic side-channel signals.

Chakraborty [367] investigates a template attack on a countermeasure provided by Joye et al. to implement the RSA-Chinese remainder theorem in embedded devices like smart cards and RFIDs. Initially thought to be robust to both simple power analysis and fault analysis attacks, the authors demonstrate its vulnerability to a template attack using a small number of power traces. The paper describes the experimental results of the suggested attack against Joye's approach, which was performed on a Xilinx Microblaze soft-core processor with the SASEBO-W standard side-channel analysis board. The authors assess the obtained power traces using binary classifiers based on the least squares support vector machine (LS-SVM). The study also covers the possible threat posed by cache timing attacks on Joye's ladder, particularly in the presence of a concurrently running spy process, and suggests a viable countermeasure to such assaults.

Duan et al. [37] propose a novel side-channel attack based on the particle swarm optimization (PSO) algorithm and directed acyclic graph support vector machine (DAG-SVM) for electromagnetic analysis. The authors demonstrate that the proposed method can effectively extract the features of electromagnetic signals and improve the accuracy of side-channel attacks. The authors also compare the proposed method with other existing methods and show that it outperforms them in terms of accuracy and efficiency. The authors also provide experimental results to validate the effectiveness of the proposed method.

Heyszl et al. [38] present a unique clustering-based approach for recovering hidden exponents that does not need previous profiles, manual tweaking, or leakage models. They show that this strategy may successfully use any available single-execution leakage and offers a simple way to combine contemporaneous observations to increase the available leakage. Furthermore, practical findings are shown, demonstrating the effectiveness of attacking an FPGA-based elliptic curve scalar multiplication with the k-means clustering method. By effectively utilizing location-based leakage generated from high-resolution electromagnetic field measurements, the suggested approach considerably reduces the secret exponent's brute-force difficulty. Furthermore, the authors compare their unique method to current alternatives, demonstrating that it is more accurate and efficient.

Perin et al. [35] propose a new unsupervised learning-based assault on randomized exponentiations. The authors demonstrate that the proposed technique may exploit the remaining leaks caused by conditional control checks and memory addressing in an RNS-based RSA implementation. The authors further demonstrate that the suggested approach recovers the whole exponent using a single trace.

16.8 FEATURES OF MACHINE LEARNING APPROACHES

When compared with the traditional approach, it is observed that machine learning provides better accuracy in cracking the security mechanisms in SCA. A few key considerations of machine learning in side-channel attacks and fault analysis are as follows.

16.8.1 Pattern recognition and adaptability

Machine learning techniques have shown a remarkable ability to recognize subtle patterns in side-channel leakage and fault-induced variations. Their adaptability across diverse scenarios and devices enhances their applicability.

16.8.2 Nonlinear analysis

The nonlinear nature of side-channel leakage and fault behavior is effectively captured by machine learning models, surpassing the limitations of traditional linear analyses. This capability enables a more comprehensive understanding of system vulnerabilities.

16.8.3 Fault detection and classification

In fault analysis, machine learning excels in both fault detection and classification. By learning the normal behavior of a system, these models can accurately identify deviations indicative of faults and classify the types of faults present.

Table 16.1 highlights key points for both side-channel attacks and fault analysis.

16.9 SUMMARY

In this chapter, we discussed attacks and their types. We have seen that there are active and passive attacks and each of them intrudes in the system and exploits the data. Side-channel attacks are the ones that intrude on the system through different hardware network sources and exploit the data. Machine learning is an emerging field in computer sciences that deals with the prediction of identification of a certain task with accurate results. Hence machine learning is used in side-channel attacks to counteract the problem. We also discussed the different algorithms that are used in SCA, including support vector machines, masked SVMs, naïve Bayes, K-means, CNNs, autoencoders, LSTM, and GANs. It is observed that a wide range of research is being conducted in power differential analysis and acoustic-based attacks, where based on the power consumption, the usage is being tracked and attacked. In acoustics, based on the key press and sound of the keyboard, the machine learning algorithms can target the encryptions.

There are other scenarios like electromagnetic radiation that are also leading to SCA. There is a lot of future scope that can be done in this domain. Along with the earlier two cases, there is a huge need for research to be carried out to provide much more inference in the SCA using machine learning. The application of machine learning in side-channel attacks and fault analysis has demonstrated promising results in enhancing the efficiency and accuracy of security analyses. Ongoing research efforts are essential to address emerging challenges and to develop advanced, robust, and privacy-aware solutions that can effectively secure systems against evolving threats.

Table 16.1 Summary

S. No.	Title	Author	Year	Technique used
1	A Practical Implementation of the Timing Attack	Paul C. Kocher	1996	This paper introduces the concept of timing attacks, demonstrating how variations in the time taken to execute cryptographic operations can leak information about secret keys.
2	Differential Power Analysis	Paul C. Kocher, Joshua Jaffe, and Benjamin Jun	1999	This seminal paper presents the differential power analysis (DPA) attack, which leverages power consumption variations to extract cryptographic keys.
3	Template Attacks. Cryptographic Hardware and Embedded Systems	S. Chari, J. R. Rao, and P. Rohatgi	2002	This paper introduces the concept of template attacks, a powerful new class of side-channel attacks. It describes the various types of template attacks, their impact on cryptographic security, and the countermeasures that can be taken to mitigate these attacks.
4	Survey of Fault Attacks	Adi Shamir	2003	This survey provides an overview of fault attacks, discussing various techniques for inducing faults in cryptographic implementations.
5	Cache Attacks and Countermeasures: The Case of AES	Dag Arne Osvik, Adi Shamir, and Eran Tromer	2005	The authors explore cache-based side-channel attacks, demonstrating vulnerabilities in the implementation of the Advanced Encryption Standard (AES).
6	Fault Attacks on RSA: Breaking CRT-RSA Through Invasive Attacks	Stefan Mangard, Elisabeth Oswald, and Thomas Popp	2005	The authors explore fault attacks on RSA, demonstrating vulnerabilities in the widely used cryptographic algorithm.

(Continued)

Table 16.1 (Continued) Summary

S. No.	Title	Author	Year	Technique used
7	Power Analysis Attacks: Revealing the Secrets of Smart Cards	S. Mangard, E. Oswald, and T. Popp	2007	This book provides a comprehensive overview of power analysis attacks on smart cards. It describes the various types of attacks, their impact on smart card security, and the countermeasures that can be taken to mitigate these attack.
8	A Survey of Elliptic Curve Cryptography Implementations	M. Joye and B. Libert	2011	This paper provides a survey of elliptic curve cryptography implementations. It describes the various types of implementations, their impact on cryptographic security, and the countermeasures that can be taken to mitigate side-channel attacks and fault injection attacks.
9	Efficient Algorithms for Generic Side-Channel Attacks	M. Joye, P. Paillier, and D.V. Stam	2011	This paper presents efficient algorithms for generic side-channel attacks. It describes the various types of attacks, their impact on cryptographic security, and the countermeasures that can be taken to mitigate these attacks.
10	Attacking RSA Private Keys Generated by the Dual_EC_DRBG Algorithm	G. Bertoni, L. Breveglieri, I. Koren, P. Maistri, and V. Piuri	2014	The paper explores fault attacks on RSA private keys generated by the Dual_EC_DRBG random number generator.
11	A Combined Power Analysis and Fault Injection Attack on an FPGA Implementation of RSA	Kasper Bonne Rasmussen, Jakob Illeborg Pagter, and Jakob Salomonsen	2014	This paper presents a combined power analysis and fault injection attack on an FPGA implementation of the RSA algorithm.

(Continued)

Table 16.1 (Continued) Summary

S. No.	Title	Author	Year	Technique used
12	A Survey of Side-Channel Analysis and Fault Injection Attacks on Field-Programmable Gate Arrays	M. Tunstall, E. Oswald, and T. Popp.	2015	This paper provides a comprehensive survey of side-channel analysis and fault injection attacks on field-programmable gate arrays (FPGAs). It describes the various types of attacks, their impact on FPGA security, and the countermeasures that can be taken to mitigate these attacks.
13	The Spy in the Sandbox: Practical Cache Attacks in JavaScript	Yossef Oren, Vasileios P. Kemerlis, Simha Sethumadhavan, and Angelos D. Keromytis	2015	This paper investigates cache-based side-channel attacks in web browsers, highlighting the security implications of JavaScript code execution.
14	Deep Learning for Side-Channel Analysis	Lichao Wu, Dirmanto Jap, Lejla Batina, and Ingrid Verbauwhede	2016	The authors investigate the use of deep learning techniques, specifically convolutional neural networks (CNNs), for side-channel analysis.
15	A Deep Learning Approach to Unsupervised Ensemble Learning in Side-Channel Analysis	Michael Pehl, Oliver Wittmann, Amir Moradi, and Christof Paar	2017	The paper explores the use of deep learning for unsupervised ensemble learning in side-channel analysis.
16	Machine Learning for Side-Channel Attack Detection: Can Small Data Convince Skeptics?	Christopher Hachenberger, Amr M. Youssef, and Thomas Eisenbarth	2017	This paper explores the application of machine learning for side-channel attack detection and addresses the challenge of dealing with limited training data.
17	Combined Side-Channel and Fault Analysis Attacks on a Hardware AES Implementation	Lei Zhang, Shujin Chai, Yu Hu, and Leibo Liu	2017	The authors investigate combined side-channel and fault analysis attacks on a hardware implementation of the Advanced Encryption Standard (AES).
18	Machine Learning in Cache-Based Side-Channel Attack: A Comparative Study	Shaza Zeitouni, Adel Anis, Amr M. Youssef, and Michel Kadoch	2018	This paper provides a comparative study of machine learning techniques in the context of cache-based side-channel attacks.

(Continued)

Table 16.1 (Continued) Summary

S. No.	Title	Author	Year	Technique used
19	Statistical Properties of Side-Channel and Fault Injection Attacks on Cryptographic Devices	J.-P. Seifert, M. Stöttinger, and F. Regazzoni	2018	This article analyzes algorithmic protections combining both side-channel prevention and fault injection detection. It surveys security models for a given set of security parameters. In general, several such models can be defined, each addressing a particular kind of attacker.
20	Deep Learning for Fault Analysis Attack on White-Box Cryptography	Bokhan Zhang, Lei Zhang, Shize Guo, Shujin Chai, and Leibo Liu	2019	The authors apply deep learning techniques to fault analysis attacks on white-box cryptography implementations.
21	Machine Learning Attacks on the Smart-Card Power Analysis Countermeasure	Chiheb Chebbi, Laurent Sauvage, Olivier Potin, Sylvain Guilley, and Jean-Luc Danger	2019	This paper investigates machine learning attacks on a specific countermeasure designed to protect against power analysis attacks on smart cards.
22	A Survey on Side-Channel Attacks of Strong PUF	Y. Zhang, X. Li, and Y. Zhang	2020	Based on the research of strong PUF attacks, this paper classifies the existing side-channel analysis methods. According to the unified symbol rules, the principles of PUF error injection, reliability attack, and power analysis are analyzed. Finally, the future development prospects of PUF side-channel attacks are discussed.
23	A Survey of Side-Channel Analysis and Fault Injection Attacks on Modern Automotive Security	M. Kasper and C. Paar	2021	This paper provides a comprehensive survey of side-channel analysis and fault injection attacks on modern automotive security. It describes the various types of attacks, their impact on automotive security, and the countermeasures that can be taken to mitigate these attacks.

(Continued)

Table 16.1 (Continued) Summary

S. No.	Title	Author	Year	Technique used
24	Side-Channel and Fault-Injection Attacks over Lattice-Based Post-Quantum Cryptography	A. Barenghi, M. Bertoni, L. Breveglieri, and G. Pelosi	2021	This work presents a systematic study of side-channel attacks (SCA) and fault injection attacks (FIA) on structured lattice-based schemes, with the main focus on the Kyber Key Encapsulation Mechanism (KEM) and Dilithium signature scheme, which are leading candidates in the NIST standardization process for Post-Quantum Cryptography (PQC).
25	A Comprehensive Survey on the Non-Invasive Passive Side-Channel Analysis	Petr Socha, Vojtěch Miškovský, and Martin Novotný	2022	This paper provides a comprehensive survey of non-invasive passive side-channel analysis. It describes both non-profiled and profiled attacks, related security metrics, countermeasures against such attacks, and leakage-assessment methodologies, as available in the literature of more than 20 years of research.

REFERENCES

1. Cyberattack. Wikipedia. (Accessed on 2023, May 30). https://en.wikipedia.org/wiki/Cyberattack
2. Internet Security Glossary, Shirey, R. (2000, May) (Accessed on 2023, May 30) https://doi.org/10.17487/rfc2828
3. Committee on National Security Systems. (2010, April 26). CNSS instruction no. 4009. (Accessed 2023, June 9)https://www.niap-ccevs.org/Ref/CNSSI_4009.pdf
4. 25+ Impressive Big Data Statistics for 2023. Techjury (Accessed on 2023, August 31).https://techjury.net/blog/big-data-statistics/
5. Alhusayni, S., Alsuwat, Dr.E. (2020). The buffer overflow attack and how to solve buffer overflow in recent research. *Academic Journal of Research and Scientific Publishing* 2, 5–11.
6. Alharbi, F., Zhou, Y., Qian, F., Qian, Z., Abu-Ghazaleh, N. (2022, July–August 1). DNS poisoning of operating system caches: Attacks and mitigations. *IEEE Trans. Dependable Secure Comput.* 19(4), 2851–2863. https://doi 10.1109/TDSC.2022.3142331.

7. Moein, S. M., Aaron, G. T., Fayez, G., Alkandari, A. (2017). Hardware attack mitigation techniques analysis. *Int J Cryptography Secur.* 7. https://10.5121/ijcis.2017.7102.

8. Paganini, P. (2013, October 11). Hardware attacks, backdoors and electronic componentqualification (Accessed on September 15). https://resources.info-sec institute.com/topic/hardware-attacks-backdoors-and-electronic-component-qualification/

9. Global, V.. (2022, August 26). Different Types of Hardware attacks. https://www.linkedin.com/pulse/different-types-hardware-attacks-vgics-global/

10. Fabio, A. (2015, December 8). Theremin's Bug: How The Soviet Union Spied On The US Embassy For *7 Years. Hackaday.* (Accessed on 2023, May 30). https://hackaday.com/2015/12/08/theremins-bug

11. Gray-Fow, E. (2021, December 10) A brief peek into the fascinating world of side channel attacks. *Medium.* (Accessed on 2023 June 3). https://medium.com/swlh/a-brief-peek-into-the-fascinating-world-of-side-channel-attacks-809f96eabea1

12. Wright, G., & Gillis, A. S. (2021, April 6). Side-channel attack. *Security.* (Accessed on 2023 June 15). https://www.techtarget.com/searchsecurity/definition/side-channel-attack

13. Franklin, J. M., Howell, G., Boeckl, K., Lefkovitz, N., Nadeau, E., Shariati, B., Ajmo, J. G., Brown, C. J., Dog, S. E., Javar, F., Peck, M., Sandlin, K. F. (2020, September 15). Mobile device security: Corporate-Owned Personally-Enabled (COPE). https://doi.org/10.6028/nist.sp.1800-21

14. What is a side-channel attack? Pedro Tavres (2020, November 30). (Accessed on 2023 June 15) https://resources.infosecinstitute.com/topic/what-is-a-side-channel-attack/

15. Kocher, P.C. (1996). Timing attacks on implementations of Diffie-Hellman, RSA, DSS, and other systems. In: Neal Koblitz *Advances in Cryptology–CRYPTO '96*, 16th Annual International Cryptology Conference Santa Barbara. California,USA, August 18–22, 1996 Proceedings, pp. 104–113. Springer, Berlin.

16. Quisquater, J.J., Samyde, D. (2001). Electromagnetic analysis (EMA): measures and counter-measures for smart cards. In: Isabelle Attali and Thomas Jensen. *Smart Card Programming and Security*, International Conference on Research in Smart Cards, E-smart 2001 Cannes, France, September 19–21, 2001. Proceedings, pp. 200–210. Springer, Berlin.

17. Kocher, P., Jaffe, J., Jun, B. (1999). Differential power analysis. In: Michael Wiener, *Advances in Cryptology–CRYPTO' 99*, 19th Annual International Cryptology Conference Santa Barbara, California, USA, August 15–19, 1999. Proceedings, pp. 388–397. Springer, Berlin.

18. Genkin, D., Shamir, A., Tromer, E. (2017). Acoustic cryptanalysis. *J. Cryptol.* 30(2), 392–443.

19. Rivest, R.L. (1993). Cryptography and machine learning. In: Imai, H., Rivest, R.L., Matsumoto, T. (eds.) *Advances in Cryptology — ASIACRYPT '91. ASIACRYPT 1991. Lecture Notes in Computer Science*, vol. 739. Springer, Berlin, Heidelberg. https://doi.org/10.1007/3-540-57332-1_36.Michael

20. Backes, M., Dürmuth, M., Gerling, S., Pinkal, M., Sporleder, C. (2010). Acoustic side-channel attacks on printers. In *Proceedings of the 19th USENIX conference on Security (USENIX Security'10).* USENIX Association, 20.

21. Lerman, L., Bontempi, G., Markowitch, O. (2015). A machine learning approach against a masked AES. *J. Cryptogr. Eng.* 5(2), 123–139.

22. Zeng, Z., Gu, D., Liu, J., Guo, Z. (2014). An improved side-channel attack based on support vector machine. 10th International Conference on Computational Intelligence and Security, Kunming, China, pp. 676–680. https://doi: 10.1109/CIS.2014.80.

23. He, H., Jaffe, J., Zou, L. (2012). *Side Channel Cryptanalysis Using Machine Learning Using an SVM to Recover DES Keys from a Smart Card.*

24. Patel, H., Baldwin, R. O. (2014, June). Random Forest profiling attack on advanced encryption standard. *Int. J. Appl. Cryptol.* 3(2), 181–194. https://doi.org/10.1504/IJACT.2014.062740Markowitch.

25. Markowitch, O., Lerman, L., Bontempi, G. (2011). *Side Channel Attack: An Approach Based on Machine Learning.* Second International Workshop on Constructive Side-Channel Analysis and Secure Design, COSADE.

26. Picek, S., Heuser, A., Guilley, S. (2017). Template attack versus Bayes classifier. *J. Cryptogr. Eng.* 7(4), 343–351.

27. Picek, S., Heuser, A., Jovic, A., Legay, A. (2017). Climbing down the hierarchy: Hierarchical classification for machine learning side-channel attacks. In: Joye, M., Nitaj, A. (eds.) *Progress in Cryptology—AFRICACRYPT 2017*, 9th International Conference on Cryptology in Africa, Dakar, Senegal, May 24–26, 2017. Proceedings, pp. 61–78. Springer, Cham.

28. Maghrebi, H., Portigliatti, T., Prouff, E. (2016). Breaking cryptographic implementations using deep learning techniques. In: Carlet, C., Hasan, M.A., Saraswat, V. (eds.) *Proceedings*, Security, Privacy, and Applied Cryptography Engineering: 6th International Conference, SPACE 2016, Hyderabad, India, December 14–18, 2016, pp. 3–26. Springer, Cham.

29. He, H., Jaffe, J., Zou, L. (2012). *Side Channel Cryptanalysis Using Machine Learning.* Standford University, CS229 Fall Project.

30. Zdenek, M., Zeman, V. (2013). Innovative method of the power analysis. *Radioengineering* 22(2), 586–594.

31. Prouff, E., Strullu, R., Benadjila, R., Cagli, E., Dumas, C. (2018). Study of deep learning techniques for side-channel analysis and introduction to ASCAD database. Cryptology ePrint Archive, Report 2018/053. https://eprint.iacr.org/2018/053.

32. Timon, B. (2018). Non-profiled deep learning-based side-channel attacks. Cryptology ePrint Archive, Report 2018/196. https:// eprint.iacr.org/2018/196

33. Specht, R., Heyszl, J., Kleinsteuber, M., Sigl, G. (2015). Improving non-profiled attacks on exponentiations based on clustering and extracting leakage from multi-channel high-resolution EM measurements. In: Mangard, S., Poschmann, A.Y. (eds.) *Constructive Side-Channel Analysis and Secure Design*, 6th International Workshop, COSADE 2015, Berlin, Germany, April 13–14, 2015. Revised Selected Papers, pp. 3–19. Springer, Cham.

34. Zhang, H., Han, G., Li, J. (2015). Wavelet transform-principal component analysis in electromagnetic attack. 7th Asia-Pacific Conference on Environmental Electromagnetics (CEEM), pp. 420–423.

35. Perin, G., Imbert, L., Torres, L., Maurine, P. (2014). Attacking randomized exponentiations using unsupervised learning. In: Prouff, E. (ed.) *Constructive Side-Channel Analysis and Secure Design*, 5th International Workshop, COSADE 2014, Paris, France, April 13–15, 2014. Revised Selected Papers, pp. 144–160. Springer, Cham.
36. Chakraborty, A. (2016). Template attack on SPA and FA resistant implementation of montgomery ladder. *IET Inf. Secur.* 10(6), 245–251.
37. Duan, L., Hongxin, Z., Qiang, L., Xinjie, Z., Pengfei, H. (2015). Electromagnetic side-channel attack based on PSO directed acyclic graph SVM. *J. China Univ. Posts Telecommun.* 22(5), 10–15.
38. Heyszl, J., Ibing, A., Mangard, S., De Santis, F., Sigl, G. (2013). Clustering algorithms for non-profiled single-execution attacks on exponentiations. In: Francillon, A., Rohatgi, P. (eds.) *Smart Card Research and Advanced Applications*, 12th International Conference, CARDIS, Berlin, Germany, November 27–29, 2013. Revised Selected Papers, pp. 79–93. Springer, Cham.

Chapter 17

Embedding learning models with pedagogy recommendation in adaptive intelligent tutoring system

Amit Kumar, Devesh Pratap Singh, Ninni Singh, and Neeraj Kumar Pandey

17.1 INTRODUCTION

The term "adaptive intelligent tutoring system" (AITS) describes a computer-based learning environment that uses artificial intelligence techniques to customize learning materials and how they are presented to students based on their requirements and preferences. The delivery of training via e-learning has advanced rapidly in tandem with advancements in web innovation (Chrysafiadi and Virvou 2013). E-learning isn't only to convey the learning material on the web to the nascent student, but it is also for educators and students who are building their knowledge of a particular subject (Sheeba and Krishnan 2019).

The traditional e-learning framework offers guidance to diverse learners, without looking at their learning inclinations, needs, and level of competencies (Amit et al. 2017; Bernard et al. 2017). Every individual is different and the same learning instructions can't be useful to everyone. Flexibility is essential to the learning process since it allows instructors to provide and monitor customized courses for each student, watching and interpreting students' activities according to their preferences and needs. The adaptable component of the e-learning system matches the learner's identity with the course content and activities (Lee et al. 2008).

Learning style is one of the most essential characteristics utilized distinction while building an adaptive framework for individuals (Graf et al. 2009). A learner's attributes, perspectives, and methods of perceiving, understanding, and interpreting information can be categorized as their learning style (Kinshuk 2004). It implies that each person has a unique set of approaches, systems, or procedures for learning (Rasheed and Wahid 2021). Learning style is defined as a combination of intellectual aptitude, affective characteristics, and mental habits that serve as largely consistent indicators of how students perceive, interact with, and react to the learning environment.

In order to support the integration of adaptivity in ITS, this study aims to investigate the learning style models applied in cognitive psychology and education. We begin with a brief review of existing learning style models,

DOI: 10.1201/9781003510420-17

definitions, their learning dimensions, and an instrument used for identifying learning styles. We also discuss the strengths and weaknesses of learning style models along with their applications. Then we present the Felder model of learning styles and the Index of Learning Styles. Last, we present our recommendations from the study and our conclusions.

17.2 METHODS

17.2.1 Criteria

In light of a comprehensive survey of 71 intellectual and learning styles conducted by Coffield et al. (2004), 10 of the best intellectual and learning styles are examined in this review. The top 13 learning style models that are evaluated in this review met the following measures: the theoretical importance of the study in the fields of education and business, the influence on learners' outcomes; their usage; cited in other studies; the hypothesis that underpinned the learning style model was clear and successful; and the instruments, surveys, or inventories are extensively used by professionals, instructors, or guides.

17.2.2 Search approach

There are uncommon developments in the field of versatile tutoring frameworks beginning around 1990. A broad archival survey was conducted using online information databases including ScienceDirect, Google Scholar, Papers, Interscience, proposal, and SpringerLink. The following keywords were used: adaptable learning systems, adaptable tutoring system, adaptation, learning style, adaptability, adaptive hypermedia, learning style model, thinking model, adaptive instruction, adaptive behavior, learning system, and modeling/student modeling. Only these keywords are included in this research.

17.3 CLASSIFICATION OF LEARNING STYLE MODELS

Coffield et al. (2004) surveyed the 13 most compelling learning style models and organized them according to criteria like adaptability, impact on the specific situation and condition, assurance by natural and psychological limitations to limit the people experience due to their mental and biological processing, and relationship to other ideas and hypotheses. In this study, the learning style models are mapped into five different groups according to the learners' characteristics and cognitive skills used in the pedagogy (Figure 17.1).

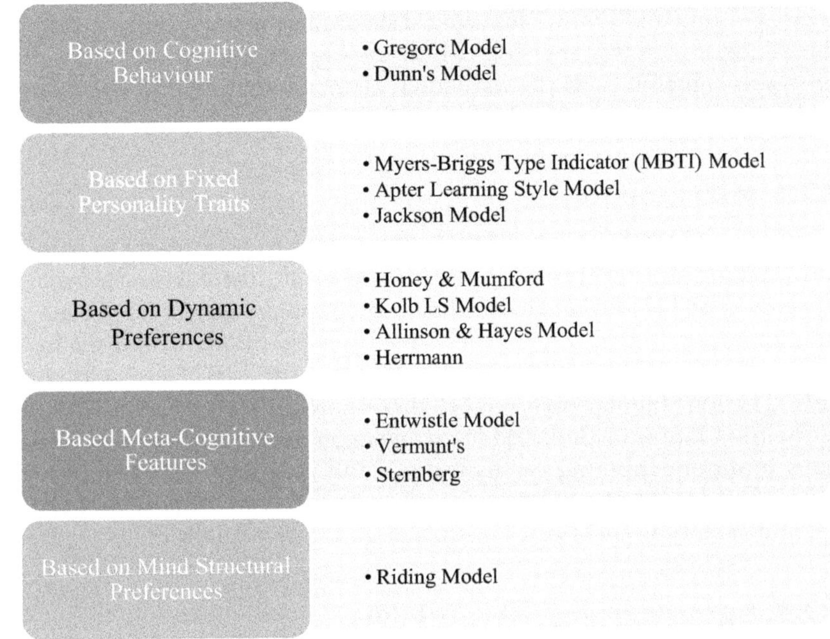

Figure 17.1 Classification of learning style model.

The current research focuses on the literature-based current learning style models. The various delineations and representations of learning measurements offered by each learning style model have been examined. Coffield et al. (2004b) identified 13 of 71 intellectual and learning style models as noteworthy or compelling. What's more, a great deal of examination has been finished over the most recent 30 years concerning specific pieces of these learning style models.

17.3.1 Gregorc learning style model

Gregorc's mental learning style model (Gregorc 1985) is one of the most valuable models surveyed by Coffield et al. The concrete–abstract and sequential–random dimensions characterize Gregorc's model. Gregorc doesn't think of people as only using one style and ignoring others. Gregorc claims that his self-test "style delineator" inventory and retests on the same people are highly consistent and independent.

17.3.2 Dunn learning style model

Dunn et al. (1990) discussed that many students fail not due to curriculum but because of the wrong instructional approaches. The Dunn and

Dunn model has been researched in over 60 educational instructions in America. Dunn and Dunn's learning style models worked on four different dimensions: emotional, sociological, psychological, and physiological. Each dimension contributes to the individual elements that map to the different learning characteristics and skills.

17.3.3 MBTI model

The Myers model (Myers and McCaulley 1985) is more about personality than learning styles. According to Coffield et al., the Myers–Briggs Type Indicator (MBTI) inventory is extremely popular and widely used and found that more than 2000 articles have been distributed about the instrument, and a detailed 2,000,000 duplicates of the MBTI are sold consistently. The four bipolar dichotomies that represent stable personality types are the foundation of the MBTI: extraversion versus introversion, sensing versus intuition, thinking versus feeling, and judging versus perceiving. A lattice of all potential combinations produces 16 types of potential characters, and each type constitutes positive and negative qualities.

17.3.4 Apter learning style model

The Apter (2001) Motivational Style Profile (MSP) is within the median range between fixed mental elements and methodologies. It suggests that the different elements of learning play in different states and motivate them to control the learner. Apter's reversal theory of personality is the foundation of Apter's MSP model. Learning can be effective if people are aware of their qualities. This knowledge can be improved and build confidence in them and give them more control over their learning.

17.3.5 Jackson learning style model

The Jackson (2002) learning style model is focused on biological and neurological theories and concepts. Jackson suggests four learning styles: the initiator, the thinker, the analyst, and the one who puts things into action. The Jackson Learning Style Profiler (LSP) is comprised of 80 inquiries and organized in four arrangements of 20 elements mapped for each learning style.

17.3.6 Honey and Mumford learning style model

Honey and Mumford (2000) suggested that a learner's preferred learning style is a collection of attitudes and behaviors. Honey and Mumford identified four different ways or approaches to learning: activist, reflector, theorist, and pragmatist. Figure 17.2 presents the four measurements of learning

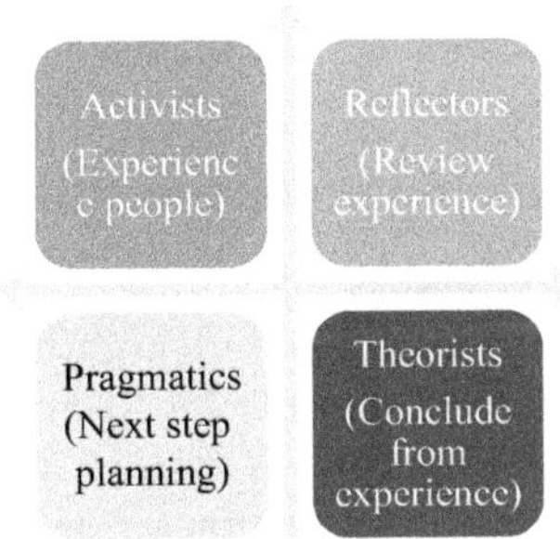

Figure 17.2 Honey and Mumford learning style model.

styles portrayed as those of activists, reflectors, theorists, and realists. In their view, learners generally stick to one learning style approach or can move in different styles based on situations. Each learning style approach is mapped to the different educational activities appreciated by learners. Their Learning Styles Questionnaire has 80 items and is divided into four blocks of 20 questions, each of which corresponds to one of the four learning styles mentioned earlier. Honey and Mumford hence view their psychometric device in a rigorously down-to-earth manner, pointed toward giving far-reaching criticism to mentors, coaches, and students about their assets and shortcomings given specific tendencies to some learning styles.

17.3.7 Kolb learning style model

Kolb (1999) has given a theory of learning style and created an inventory of questions based on psychometric theory. The Kolb learning style theory is comprised of a four-stage cycle. First is concrete experience, which may be comprised of a new or existing experience of real scenarios. Second, reflective observation, is how learners reflect based on the existing experience. Third, in abstract conceptualization, learners generate new ideas based on their learning experience. Fourth is active experimentation in which the learner creates new ideas, experiments with them, and learns from them. The Kolb Learning Style Inventory has been used in the education and business contexts.

17.3.8 Allinson and Hayes model

Allinson and Hayes (1996) developed a Cognitive Style Index (CSI) for learning style. The CSI is based on the intuition of the cognitive characteristics of a learner. As a result, Coffield et al. treat the learning styles model developed by Allinson and Hayes as a stable but adaptable collection of learning preferences that differ from person to person based on cognition and the aforementioned factors.

17.3.9 Herrmann learning style model

Coffield et al. (2004a) classified Herrmann's Brain Dominance Instrument (HBDI) as a fixed and modality-based cognitive learning style model (Herrmann 1989). The HBDI consists of 120 questions and is categorized into four groups. The theorist type is known as cerebral–left, the organizer type is known as limbic–left, the innovator type is known as cerebral–right, and the last one is the humanitarian type known as limbic–right. It is hypothesized that people who exhibit characteristics of a category will have trouble connecting with opposite types (for instance, a theorist or cerebral and left-dominant type would have trouble connecting with limbic and right-dominant types or humanitarians).

17.3.10 Entwistle learning style model

Entwistle (1998) developed the ASSIST learning style model based on the mind's characteristics, behaviors, preferences, and learning experience. Entwistle's model distinguishes three distinct teaching styles, each of which can be dominant (Mockford et al. 1998). The first is deep learning, which pursues new ideas and innovations. The second, surface apathy, is based on an assessment and maximizes output by applying minimal effort. The last one is nonapathetic, in which learners aim to learn more rather than attain good grades.

17.3.11 Vermunt learning style model

According to Vermunt, a student's learning style is a holistic approach to learning activities (Vermunt 1998). Vermunt gives the following definitions of learning styles or methods: significance-coordinated, application-coordinated, propagation-coordinated, and undirected. Vermunt developed the Inventory of Learning Styles, which consists of 120 questions, five rating scales, and subscales for each factor.

17.3.12 Sternberg learning style model

Sternberg developed the Thinking Styles Inventory (TSI). This theory connects the thinking style with the pedagogy in a very explicit way (Sternberg et al. 1999). Sternberg makes a lot of strong claims about how his thinking styles model can help students do better in school. As per Sternberg, three different levels control the mental process: functions of mental governments, forms of mental self-control, and stylistic preferences.

17.3.13 Riding learning style model

Riding (1997) developed the Cognitive Style Analysis (CSA) inventories. This model is based on the cognitive style and considers the assumptions that the cognitive style of learning is not changed. The Riding model focuses on the cognitive style of the learner, mainly the dimensions of wholistic–analytic and verbalizer–imager. According to the CSA approach, a learner's cognitive style refers to how they prefer to represent and organize information, and a learner's learning strategy is how they respond to their unique learning context-specific weights. The cognitive style of a learner is defined by two assessments, according to the CSA: the first, called verbalizer–imager, measures an individual's mental structure, and the second, called wholistic–analytic, measures an individual's cognitive organization. However, the two dimensions are unrelated to one another.

17.4 COMPARATIVE STUDY

Table 17.1 presents a summary of the aforementioned learning style models, along with their strengths, weaknesses, and applications.

17.5 FELDER MODEL OF LEARNING STYLE

The Felder learning style model was developed by Richard M. Felder and Linda K. Silverman. This model helps us to categorize individuals based on their needs and preferences. The Index of Learning Styles (ILS) is an instrument used to evaluate inclinations on four measurements of a learning style (Felder and Solomon 2006). The 13 learning style models discussed in Section 17.2 classify learners into groups, whereas Felder describes the learner's learning style in a more comprehensive way by separating them into four different dimensions. The important thing in this model is that it indicates the learning preferences from low to high or high to low.

The Felder model is the most appropriate learning style model adapted by the educational tutoring system. The model is often used in adaptive tutoring systems comparison with other learning style models. The Felder model is also an appropriate learning style model for e-learning systems.

Felder and Silverman's model defines the four dimensions to differentiate the learning styles of learners. Each learner is characterized by particular dimensions with a specific preference. Figure 17.3 presents the four different dimensions of the Felder and Silverman learning style model.

The first dimension is the active/reflective way of processing the information. Active learners are those who actively participate in learning activities, materials, and group discussions; are interested in communication; and discuss the learned material with their groups or individuals. In contrast,

Table 17.1 Summary of 13 learning style models

S.N.	Learning style models	Strengths	Limitations	Model application	Year
1.	Riding model	Excellent implication for teaching. Valid proof of the relationship between educational choices and cognitive style.	Model validity issues and bias toward the model's two poles.	Academics Business	1991
2.	Apter model	For steady personality types, it works well.	Learning style is not measured by it. Not a single educational research.	Business	1998
3.	Dunn and Dunn model	Easy and basic model, with significant effects on instruction.	Fewer evaluations. Neuroscience and physiological parameters have an erratic relationship.	Academics Business	1979 1975 2003
4.	Sternberg model	Lots of learning styles or thinking styles proposed.	Not evaluated independently.	Academics	1998
5.	Gregorc model	Has the advantage of the oblivious subjective procedures associated with the digestion and mix of data: discernment and requesting.	Learning style appears to be a fixed characteristic. Theory is ambiguous.	Academics	1977
6.	Vermunt model	It can apply of the learner thinking to enhance learner's analytical and critical thinking. Best learning strategies.	Weak predictor of learner. Can't distinguish between learning style and personality traits.	Academics	1996
7.	ASSIST	Aims to envelop ways to deal with learning, think about systems, scholarly advancement.	Very complex and instrument is not for the novice.	Academics	1979 1995 2000
8.	Jackson model	Ideal for computerized and theoretical formats. Computer-recommended for the development of personality. It is intended for academic and commercial use.	Types of personalities have steady learning styles. The reliability of the reasoner dimensions is low.	Academics Business Distance learning	2002

(*Continued*)

Table 17.1 (Continued) Summary of 13 learning style models

S.N.	Learning style models	Strengths	Limitations	Model application	Year
9.	MBTI model	Face legitimacy is uncontroversial; limited evidence demonstrating the benefits of learning style coordination for pupils' academic performance.	Hazy ramifications for instructional method, not an execution indicator.	Academics Business	1962
10.	Allinson and Hayes	Best tool for improving the effectiveness of learning. Evidence for reliability, despite the fact that the academic ramifications of the model have not been completely investigated.	The proposed single measurement is exceptionally wide and comprised of assorted, inexactly related attributes.	Business	1996
11.	HBTI model	Psychometric instruments are a fantastic way to assess learning style because they are theoretically sound and best for business.	Lack of application on pedagogical research	Academics Business	1995
12.	Honey and Mumford	It is not a psychometric tool and can be used for organizing and personal growth.	Not focused to cognitive ability. Not a very good learner predictor.	Business Distance learning	1982
13.	Kolb model	It is a fair model that is built on specific assumptions and is subject to ongoing changes.	Learning cycle is controversial.	Academics Business	1976 1985 1999

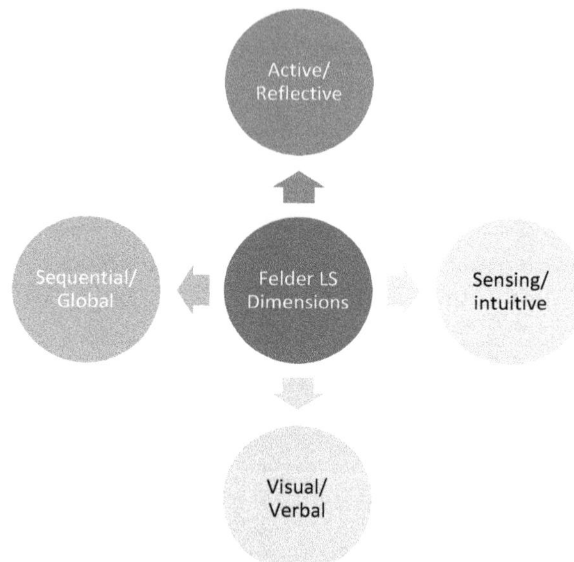

Figure 17.3 Felder learning style dimensions.

reflective learners think and reflect on the learning materials. These learners don't prefer to work in groups and are not interested in communication.

The second dimension is sensing/intuitive. Sensing learners learn from the facts and use actual learning materials. They are more patient and try to solve the problem with the standard methodology and look more genuine. In contrast, intuitive learners focus on abstract concepts and prefer to learn about theories and their meanings. They apply some pioneering and inventive ways to solve problems.

The third dimension is visual/verbal. It differentiates how learners remember things through visuals like images, flow diagrams, charts, etc., or more through spoken words.

The fourth dimension is sequential/global. Sequential learners follow the incremental approach to learning and finish tasks step by step. These learners have patience and follow the stepwise approach to find the desired solution. In contrast, global learners use the holistic approach to learning and finish the task in one go. They absorb the learning material randomly and sometimes they don't care about the connection or relation between the content. These types of learners are known to solve complex problems and try to find the relationship between the theories, areas, and materials.

The ILS model contains 44 questions and is divided into four groups. Each group constitutes 11 questions with a scale comparing to one of the four measurements (Felder and Silverman 1988). Each of the four measurements contains an arrangement of two inverse classifications. The thought

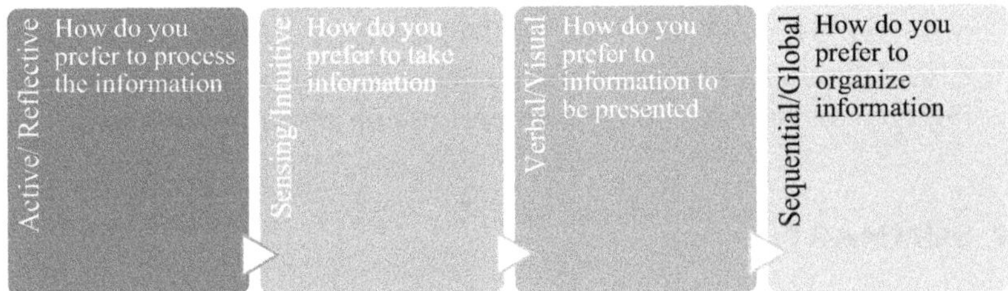

Figure 17.4 Felder learning style model.

behind these inverse classifications is that everybody utilizes every one of them in various circumstances, yet with shifting degrees of inclination. Felder's four learning dimensions are shown in Figure 17.4.

17.6 RECOMMENDATIONS

Finding the right learning styles for adaptive learning and clever mentoring techniques that would improve learning capacities were the goals of this study.

This chapter's first section introduced the reader to computer-based and online learning, with a focus on adaptable intelligent tutoring frameworks. The methods used to review the learning style model were specified in Section 17.2. A thorough analysis of 13 learning styles from Coffield et al. was presented in Section 17.3. These approaches have a history in education and offer possible advantages, disadvantages, and applications (Section 17.4). An item-by-item audit of the Index of Learning Styles (Felder and Silverman 1988; Felder and Solomon 2006) was provided in Section 17.5.

Based on the three portions of the study that came before it, the following are several specific recommendations for the usage of the ILS inside adaptive intelligent learning and intelligent tutoring systems.

1. It is prescribed that the ILS (Felder and Solomon 2006) can be utilized as an instrument to distinguish the learning styles of students to prepare for customized, computer-based remote learning.
2. A benchmark of learning styles can be evaluated utilizing the Felder–Solomon ILS online survey (Felder and Solomon 2006) to catch the underlying esteems speaking to the student's style.
3. The ILS ought to be thought about against the style of the ITS course educating/introduction styles to check whether the individuals show mismatch among learning and showing styles, and additionally electronic introduction of data.
4. It is suggested that a comparable technique be embraced for ITS. One proposal is to create space where administrators can practice and give

things a shot, or watch examples of meetings done effectively and inaccurately.

5. Also, learning styles could be incorporated into the intelligent tutoring systems or classrooms with permission given by the tutor.

17.7 SUMMARY

The purpose of this study was to review the models of learning styles used in instructional and psychological brain research and assess how they might impact education. This survey was extended to identify novel learning style models and recommendations that had not been examined by Coffield et al. (2004a), due to the lack of independent support for the consistent quality, legitimacy, and applications of the models investigated by Coffield et al. (2004). The Felder–Solomon Index of Learning Styles was the result of the investigation. To improve students' learning adequacy, it is advised that the ILS be used to identify their learning preferences.

REFERENCES

Allinson, C. W., & Hayes, J. (1996). The cognitive style index: A measure of intuition-analysis for organizational research. *Journal of Management Studies*, 33(1), 119–135.

Amit, K., Ninni, S., & Jyothi, A. N. (2017). Learning styles based adaptive intelligent tutoring systems: Document analysis of articles published between 2001 and 2016. *International Journal of Cognitive Research in Science, Engineering and Education*, 5(2), 83–98.

Apter, M. J. (2001). *Motivational Styles in Everyday Life: A Guide to Reversal Theory*. Washington, DC: American Psychological Association.

Bernard, J., Chang, T. W., Popescu, E., & Graf, S. (2017). Learning style identifier: Improving the precision of learning style identification through computational intelligence algorithms. *Expert Systems with Applications*, 75, 94–108.

Chrysafiadi, K., & Virvou, M. (2013). Student modeling approaches: A literature review for the last decade. *Expert Systems with Applications*, 40(11), 4715–4729.

Riding, R. (1997). On the nature of cognitive style. *Educational*, 17(1–2), 29–49.

Coffield, F., Moseley, D., Hall, E., & Ecclestone, K. (2004a). *Learning Styles and Pedagogy in post-16 Learning*. London: Learning and Skills Research Centre.

Coffield, F., Moseley, D., Hall, E., & Ecclestone, K. (2004b). *Should We Be Using Learning Styles? What Research Has to Say to Practice*. London: Learning and Skills Research Centre.

Dunn, R., & Griggs., S. (2003). *Synthesis of the Dunn and Dunn Learning Style Model Research: Who, What, When, Where and So What – The Dunn and Dunn Learning Style Model and Its Theoretical Cornerstone*. New York: St John's University.

Entwistle, N. J. (1998). Improving teaching through research on student learning. In: J.J.F. Forrest (ed.) *University Teaching: International Perspectives*. New York: Garland.

Felder, R. M. (1995). A longitudinal study of engineering student performance and retention. IV. Instructional methods. *Journal of Engineering Education*, 84(4), 361–367.

Felder, R. M., & Silverman, L. K. (1988). Learning and teaching styles in engineering education. *Engineering Education*, 78, 674–681.

Felder, R. M., & Solomon, B. A. (2006). Learning styles and strategies. Retrieved on February 28, 2010, from http://www.ncsu.edu/felder-public/ILSdir/styles.htm.

Graf, S., Liu, T. C., Kinshuk, Chen, N. S., & Yang, S. J. H. (2009). Learning styles and cognitive traits – Their relationship and its benefits in web-based educational systems. *Computers in Human Behavior*, 25(6), 1280–1289.

Gregorc, A. F. (1985). *Style Delineator: A Self-Assessment Instrument for Adults*. Columbia, CT: Gregorc Associates Inc.

Herrmann, N. (1989). *The Creative Brain*. Lake Lure, NC: Brain Books, The Ned Hermann Group.

Honey, P. (2002). Why I am besotted with the learning cycle. In: P. Honey (ed.) *Peter Honey's Articles on Learning and This and That*, 115–116. Maidenhead: Peter Honey Publications Ltd.

Honey, P., & Mumford, A. (2000). *The Learning Styles Helper's Guide*. Maidenhead: Peter Honey Publications Ltd.

Jackson, C. (2002). *Manual of the Learning Styles Profiler*. Retrieved from www.psi-press.co.uk.

Kinshuk, H. H. (2004). Adaptation to student learning styles in web-based educational systems. Proceeding of EDMEDIA Conference, Chesapeake.

Kolb, D. A. (1999). *The Kolb Learning Style Inventory, Version 3*. Boston: Hay Group.

Lee, J., Park, D. H., & Han, I. (2008). The effect of negative online consumer reviews on product attitude: An information processing view. *Electronic Commerce Research and Applications*, 7(3), 341–352.

Mockford, C. D., & Denton, H. G. (1998). Assessment modes, learning styles and design and technology project work in higher education. *Journal of Technology Studies*, 14(1), 12–17. Retrieved April 21, 2008, from http://scholar.lib.vt.edu/ejournals/JOTS/Winter-Spring- 1998/mockford.html.

Myers, I. B., & McCaulley, M. H. (1985). *Manual: A Guide to the Development and Use of the Myers-Briggs Type Indicator*. Palo Alto, CA: Consulting Psychologists Press.

Rasheed, F., & Wahid, A. (2021). Learning style detection in e-learning systems using machine learning techniques. *Expert Systems with Applications*, 174, 114774.

Riding, R., & Rayner, S. (1998). *Cognitive Styles and Learning Strategies: Understanding Style Differences in Learning Behaviour*. London: David Fulton Publishers Ltd.

Sheeba, T., & Krishnan, R. (2019). Automatic detection of students learning style in learning management system. In: Al-Masri, A., Curran, K. (eds), *Smart Technologies and Innovation for a Sustainable Future. Advances in Science, Technology & Innovation*. Springer, Cham. https://doi.org/10.1007/978-3-030-01659-3_7

Sternberg, R. J. (1999). *Thinking Styles*. Cambridge: Cambridge University Press.

Vermunt, J. D. (1998). The regulation of constructive learning processes. *British Journal of Educational Psychology*, 68(2), 149–171.

Chapter 18

Security issues and challenges for IIoT in Industry 4.0 paradigm

*Neeraj Kumar Pandey, Amit Kumar Mishra,
Neha Tripathi, Piyush Bagla, Umang Garg,
and Amit Kumar*

18.1 INTRODUCTION

Industry 4.0 is a phenomenon that has had both a steady rise and a remarkable expansion because of the numerous components that make up industrial systems that are connected to information and communication technologies (ICT), creating the smart industries and companies of the future. Industry 4.0 and related techniques are propelled through a new innovation that introduces uncountable new openings and increases the current market production value. Cloud design and Internet of Things (IoT) development are some examples of these technologies.

Various studies have made confident assertions about the substantial potential in terms of linked devices, global networks, and economic effects. It is projected that by 2029, approximately 15 billion IoT devices will be integrated into enterprise setups. In terms of agility, there will be an estimated 27 million new IoT connections by 2026, representing a remarkable growth rate of 120% compared to the year 2020 (Prinsloo et al. 2019). Despite these staggering numbers, cyber security and the protection of personal information are already affecting internet technologies. These issues must be considered when executing new technologies. Industry 4.0 will also have to face special security and privacy challenges. If these issues are not adequately resolved, Industry 4.0 may never reach its full potential.

The year 2021 brought about many new technological advancements. One notable development was the introduction of free-to-use spectrum 5G over CBRS (Citizens Broadband Radio Service). By using CBRS, we can see the significant growth in open-source software-defined networking (SDN) platforms and edge cloud infrastructure technology. These advancements have significant benefits for manufacturers and contribute to the ongoing industrial revolution (4.0 and 5.0) and Industrial Internet of Things (IIoT) projects (Fernandez-Carames and Fraga-Lamas 2019).

Numerous technologies and related paradigms will be included in Industry 4.0. The IIoT and the new paradigms in product development

DOI: 10.1201/9781003510420-18

typical of the 21st century include cloud infrastructure design, cloud-based development, and large-scale innovation integration.

The chapter covers five divisions, which are arranged in the following manner: An introduction is described in Section 18.1 in detail, followed by a summary of Industry 4.0 in Section 18.2. The security issues of smart systems are elaborated in Section 18.3. Section 18.4 comprises a proposed framework for the IIoT system. In Section 18.5, security issues in Industry 4.0 are analyzed. The last section, Section 18.6, concludes the chapter.

18.2 INDUSTRY 4.0

Industry 4.0, which includes the Industrial Internet of Things, has significantly shifted the sector. It first appeared when the government of Germany stimulated the use of the computing paradigm in the manufacturing sector. The concept behind Industry 4.0 was to utilize the internet to connect all devices in order to communicate and disseminate processes and information. This idea is based on a cyber-physical system, which is a network of computer components that synchronize in a planned and organized manner. Industry 4.0 offers superior business gains and significantly increases productivity. It has affected practically all spheres of life, and Industry 4.0 proponents see it as the third innovation wave (Vaidya et al. 2018).

In the 18th century, the First Industrial Revolution came into existence with steam power and a production system. The use of steam power achieved a significant breakthrough in increasing production and human productivity. In this revolution, steam engines performed powerful work instead of human muscles (Culot et al. 2019). After the massive success of this first revolution, the Second Industrial Revolution, or Industry 2.0, came with assembly line production and electricity in the 19th century. Conveyor belts used in slaughterhouses ignited the idea of Henry Ford of the assembly line and mass production. He carried forward the idea to the automobile industry, making production significantly low-cost and faster.

In the 20th century, with memory-programmable units and automation, Industry 3.0 came into existence. It began with the computer era and the internet for faster production, partial automation, renewable energy units, and connectivity. The production process can be done without human intervention in this revolution with improved efficiency. Industry 4.0 is the enhanced version of Industry 3.0 with better connectivity, full automation, self-organized logistics, and a cyber-physical production system. Industry 4.0 allows the networking of all modules and leads to a cyber-physical production system for building smart production systems, factories, components, and automated production processes (Usländer and Batz 2018).

18.3 SECURITY ISSUES IN SMART SYSTEMS

- The Internet of Things has been shaped by the perfect combination of affordable computing and expanding networking. It encompasses various devices that merge different fields and enable them to connect to the internet for data transmission (as shown in Figure 18.1). IoT plays a crucial role in transforming businesses digitally, offering the potential to enhance labor productivity, efficiency, profitability, and overall employee experience (Hearn and Rix 2018). However, as interconnected networks and devices continue to grow, the number of entry points into these systems also increases. This expansion brings forth a wide range of vulnerabilities that intruders can exploit. Figure 18.2 illustrates the functional system of Industry 4.0 and provides a concise analysis of potential security concerns. Within this industry's structure, the following three distinct levels of security can be identified:
- IoT security – The protection of electronic devices that come into contact with the outside world, as well as the systems that make up operational technology (OT), is crucial.
- Transport medium security systems – The safety of network devices and the various aspects related to computer networks involves the use of protocols, systems, tools, and methods to ensure security and prevent malicious attacks.
- Security in the cloud – The issue of security and privacy becomes more significant when data is sent and stored across multiple jurisdictions

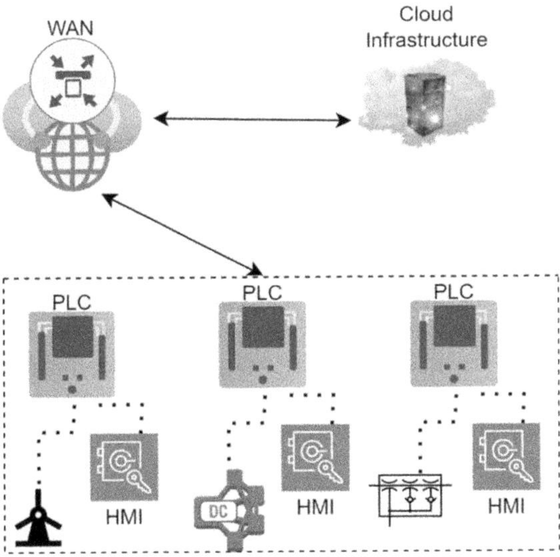

Figure 18.1 Industry 4.0 functional system.

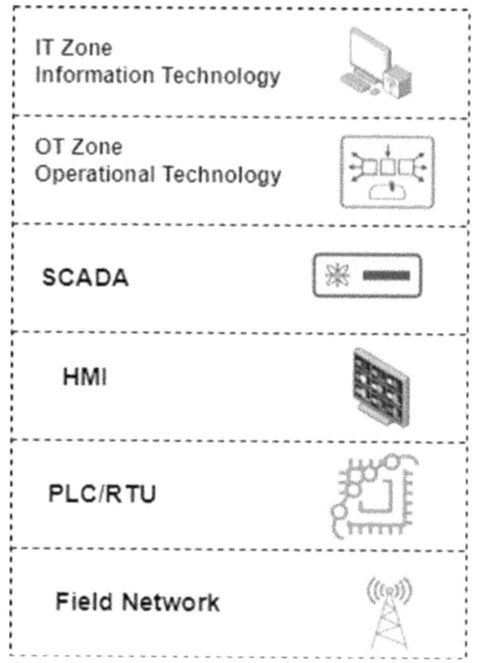

Figure 18.2 OT and IT zone convergence in SCADA.

with varying levels of security. Although there are benefits to interconnecting IoT devices, it also presents a new challenge in terms of ensuring security for unmonitored network end devices, particularly those used by enterprises. While IT departments prioritize the protection of standard network devices, the security measures for IoT devices are often overlooked or not well understood. This neglect may stem from a combination of factors (Sarker et al. 2020).

First and foremost, traditional cybersecurity systems seem to struggle with identifying specific IoT network traffic, unique risk structures, and distinct behaviors.

Moreover, IoT devices are utilized across various business sectors without being considered as fully integrated IT devices. As a result, they are often overlooked when it comes to managing and updating standard security measures. This is primarily because most IoT devices rely on different hardware types, operating systems, and antimalware software compared to traditional cyber devices. Additionally, their estimated lifespan tends to be longer than that of typical cyber devices.

Unfortunately, due to the negligence of IT departments in prioritizing IoT device security, these devices become prime targets for malicious attackers. This stands in contrast to the attention given to safeguarding personal computers (PCs), servers, laptops, or smartphones (Thames and Schaefer 2017). Without a reliable system for tracking and identifying IoT devices, they are often considered unmanaged endpoints within a network. This leaves them vulnerable to potential security risks such as password attacks or the infiltration of malware.

Traditionally, the demand for cyber security in OT was not high because their systems were not connected to the internet. In most cases, IT networks and OTs were kept separate from each other. However, this approach brings about effort when it is a matter of security and transparency issues. IT networks can only track and manage their known network devices, leaving a significant portion of the overall network unmonitored (Corallo et al. 2020). Furthermore, this separation makes it challenging to identify the source of an attack since isolated teams are unaware of all the devices connected to their network (Figure 18.2). The technology systems used in operational environments are similar to those used in IT environments but serve different purposes. They are primarily designed for machine-to-machine interaction, such as industrial control systems (ICS), ensuring that assets operate correctly according to availability or uptime requirements rather than being primarily focused on human interaction.

When it comes to this matter, the cloud emerges as an obvious solution for integrating data due to its ability to store large amounts of information for analysis. However, OT teams are not familiar with this aspect and lack knowledge about control, autonomy, and latency. Another option that businesses are leaning toward is edge computing. It seems to provide reassurance for both sides involved, allowing the IT department to have a comprehensive understanding of the system's structure and to manage and monitor it from a central level while still being controlled by the OT teams.

IIoT represents the newest iteration of SCADA (Supervisory Control and Data Acquisition) systems. These systems play a vital role in supporting critical infrastructures worldwide, such as industry, energy, transportation, and nuclear power plants. Cloud technologies have been integrated into the overall network architecture of these systems to establish their current infrastructure. A SCADA system is a broadly spread high-tech system that covers expansive physical areas and is responsible for monitoring, automating, and controlling physical tasks (Asghar et al. 2019).

IIoT, which is a crucial part of Industry 4.0, represents the latest advancement in the development of SCADA systems. By implementing IoT technology, it leverages cloud technology and its profitable benefits to enhance efficiency and reduce prices. Ensuring the safety of an organization includes data collection in real time and a thorough analysis becomes necessary. This analysis starts with understanding network protection, system management, and physical infrastructure concepts. The complexity increases when transitioning to cloud-based solutions. In the realm of cyberattacks targeting SCADA systems, there are three main types: hardware attacks, software breaches, and communication system intrusions. The SCADA control center relies on information received from remote terminal units (RTUs) to execute its operations effectively. Attacks that compromise the control process primarily focus on manipulating control data or disrupting data flow. Denial of service (DoS) attacks in various forms like distributed denial of service (DDoS) or man-in-the-middle (MITM) attacks pose significant threats to SCADA systems (Pang et al. 2021).

18.4 FRAMEWORK FOR IIOT SYSTEM

The proposed framework for the IIoT system has four basic layers. Compiling data is required to analyze the IIoT system, which can perform information flow from the sensing layer to the application layer. This section covers potential assaults and defenses aimed at the various IIoT system layers (Tian et al. 2019).

a) Sensing layer: Direct physical attacks are made against the IIoT sensing and actuator layer. Our research indicates that DoS, sensor threats, and tampering are among potential assaults that could target the sensing layer of the IIoT. Malicious code injection, node jamming, and physical damage are ways tampering attacks might be launched. Sensor threats can result from poor authentication and irresponsible deployment, while an attack like DoS may be started by changing the physical link, distorting the signal, or jamming. Additionally, this study revealed defenses against each attack. Tamper-resistant packaging and tamper-proofing and concealing are two potential solutions to the tampering issues. An intrusion detection system (IDS), public key encryption, safeguarding sensed data, and strengthening the service management system are some of the defenses against sensor assaults. On the other hand, by using a traffic monitoring system, DoS can be avoided in a significant way.

b) Network layer: Communication between nodes is essential for IIoT automation, and networking assaults can be particularly damaging. Our findings indicate that MITM and DoS are the two primary threats aimed at the network layer of the IIoT. However, as our architecture illustrates, these attacks lead to other sub-attacks. The framework asserts that various methods, including eavesdropping, routing assaults, and replay attacks, can launch MITM. However, DoS can also be initiated through flooding, spoofing, node replication, collision, sinkhole assault, exhaustion, wormhole, unfair behavior, Sybil attack, and spoofing.

c) Data/service layer: Data is collected and aggregated in the IIoT system at the layer of data processing, frequently in the cloud environment. In any online environment, the security of this data is essential. The four main threats aimed at this layer in the currently available IIoT literature are malware, session hijacking, malicious insiders, and cloud service provider difficulties. These assaults can also be grouped into three major types of malware attacks: viruses, worms, and botnets. Attacks on sessions occur at the service layer and can be either active or passive. DoS, information extraction, and privilege execution are malicious insider attacks.

Users are highly aware of session hijacking attempts, and IDS, SSL, and MAC address and CAPTCHA protection may all be checked. A monthly risk assessment, staff training, allocating less power, a rigorous security policy, and detecting disruptive behavior are some potential solutions for hostile insiders. Cloud education, input monitoring, and encrypted communication can all be utilized to mitigate the risk associated with cloud services (Pereira et al. 2017).

d) Application layer: Several security exploits target the IIoT application layer. Sniffing, phishing, malicious code injection, and DoS attacks are the four basic attack methods that predominantly target the IIoT application layer (Figure 18.3). These attacks have further sub-attacks, such as the two types of sniffing known as active sniffing and passive smelling. Floods and fatigue may result in DoS assaults at the application layer. Malware-based phishing and social engineering phishing are two types of phishing assaults. Attacks involving malicious code injection also include node and packet injection. Our research signifies that encryption and MAC filtering are the primary countermeasures utilized to defend against sniffing attempts. IP security, filtering, firewalls, proxies, and other methods can be used to prevent DoS attacks at the application layer. Using authentication and IDS can prevent malicious code injection, and client- and server-side security

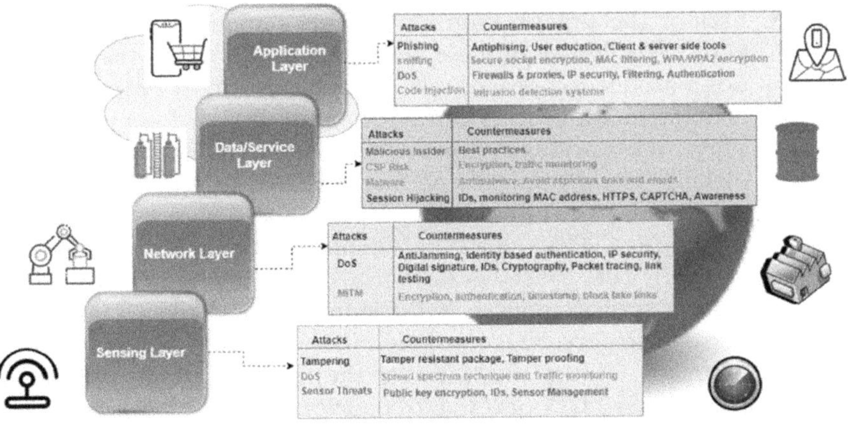

Figure 18.3 Proposed framework of IIoT system.

tools, network-level security, user education, and robust authentication mechanisms can halt phishing attempts (Borky et al. 2019).

18.5 SECURITY ISSUES IN INDUSTRY 4.0

Among various IoT-based frameworks (Ghosh et al. 2021), Sharma and Sharma (2022) discuss various existing attacks on IoT devices. Social engineering and DoS are the most prevalent assaults in both the physical and cyber categories. The study also proposed methods to mitigate them. Several security issues in Industry 4.0 have been generated due to vulnerabilities in IoT devices and sensors.

a) Inadequacy of IT/OT security knowledge and awareness: This conflict is the result of persons who are either involved in changing manufacturing processes or in the security of the digital components. In both cases, there is a serious deficit of information security knowledge.
   - People involved in the manufacturing process are usually uninformed of the security precautions that must be taken, and those responsible for the security component of the process do not completely understand the manufacturing process in order to safeguard it in a way that precludes outside intrusions.
   - To understand the full process, a person must be skilled in a variety of domains, such as IT and OT security, embedded systems, and network security. Finding suitable candidates for all of the aforementioned occupations is getting increasingly difficult.
   - Employees who work with IT/OT equipment are frequently undertrained in manufacturing organizations. When the organizations do provide training, it is insufficiently extensive or relevant to the needs of the organization's business.
   - Even if the organization can locate a solid security specialist knowledgeable in all these areas and familiar with the organization's production procedures, all employees must be aware of the manufacturing process and security pitfalls.
   - There are various reasons behind this, including employees' inability to comprehend and learn updated security measures and the expense of training. These days, there is less cybersecurity training available and it is becoming increasingly costly (Wolf and Serpanos 2017).

b) Inadequate security policies and funding: This challenge falls under the person category. This revolution in the industry requires security requirements. The main reason for the security pitfalls and privacy requirements is a lack of standard policies, technical flaws, and a lack of awareness.

Therefore, some organizations that have prioritized security before launching a new product into the market over other challenges have reduced their budgets and are hesitant to invest or spend more in research and development related to current security issues. As a result, cybersecurity technology will no longer spread and improve as swiftly as it should. For example, if a corporation chooses to move its information systems to the cloud rather than storing them locally, it fails to spend sufficiently on security. Such a transfer now occurs when a corporation realizes how much money it is wasting on local data storage. One of the critical motivations for making the switch is usually the financial savings that can be realized by using cloud services.

c) Products and liability: This security challenge that Industry 4.0 operators face falls under the second group of procedures. In the context of Industry 4.0 security, the issue we're considering is liability. Numerous stakeholders complicate creating an intelligent device that can connect to the internet through IoT. Stakeholders include everyone in the supply chain involved in manufacturing a particular smart device, including security teams, software teams, and many others. Determining responsibility in the event of a security problem involving that smart device could be a substantial barrier to industry adoption due to the vast number of players involved.

d) Lack of uniform standardization: Science has advanced enormously. To make working with these technologies more accessible, hundreds of thousands of papers, guides, and standards have been released; however, this does not include Industry 4.0 security. Industry 4.0 security does not currently have any market standards. However, IoT and other new technologies have extensive standardization, making it simple for users to adhere to these standards while implementing their projects. The few Industry 4.0 standards currently in use are either fragmented or lack consensus among the researchers developing the technology (Mentsiev et al. 2020).

e) The devices' technical constraints: Because Industry 4.0 focuses on digitizing current industrial processes rather than developing a completely new system, previously manual procedures will now be handled digitally. This means that digital platforms will be built on top of existing platforms, which is the only feasible option given the impossibility of developing a new sector from the start.

So, while this is a straightforward approach to implement Industry 4.0, it does have certain limitations. The majority of these gadgets, which are still in use or are built using antiquated procedures, have technological limitations that pose substantial security issues. Most of these gadgets have minimal operational capability, to the point where they cannot perform even the most basic of tasks. This is done for various reasons, one of the most important of which is to keep pricing stable.

The second most severe technical issue is that none of these already running devices have the infrastructure to support any safety system. All these devices were created with a single principal use in mind, and the defensive side was given little thought. Furthermore, many new gadgets are still produced without this critical defense architecture. Finally, even though some machines have some processing power and can complete straightforward tasks, like a refrigerator measuring the temperature and transmitting it to its control unit, they are unable to complete more complex security tasks, like encryption or authentication. Even though these actions seem straightforward, they demand extra processing capabilities, which increase the delay time (Alqahtani et al. 2019).

18.5.1 Security recommendations for Industry 4.0

Some of the recommendations (Lim et al. 2021) for IIoT security are discussed in this section which deals with security flaws in the technology itself or one of its foundational technologies:

- Industry 4.0 can employ SDN to reduce security risks and make security rules more manageable.
- Small-scale manufacturing: To reduce money and increase security, Industry 4.0 solutions can benefit from centralized security solutions. Adopting security frameworks and policies as standards can be highly beneficial. Different standards are used for strengthening critical infrastructure cybersecurity.
- Ensure that all systems are patched and up to date. This can defend systems against a variety of zero-day threats.
- Always have appropriate policies in place. The system's use is governed by security and other regulations.
- The deployment of modern IoT gateways can significantly decrease security threats. When employing smart cards, be sure they've been tested for security and that you're using the right cryptographic methods.
- Industrial systems that have existed for a long time have undoubtedly been the target of numerous attacks over the years. This should push businesses to use newer technologies or, at the very least, existing technologies that have been thoroughly evaluated.
- To strengthen security in the smart industrial 4.0 environment, it was proposed that trusted platform modules be deployed. Although this affects reliability, this method looks to be promising.
- In smart industrial contexts, safeguarding endpoints is as vital as protecting central systems.

- Investing in cybersecurity is critical to an organization's survival.
- An organization must show that it can safeguard its technology assets. Your clients will have more faith in you because of this.
- Don't try to solve new security problems with old solutions. It's critical to stay on top of new threats and not fall back on outdated security safeguards.

18.5.2 IP camera case study

This section will analyze an IP camera's current security features to identify vulnerabilities (Pandey et al. 2023) To implement a testbed, we select an IP camera with an internet connection and vulnerabilities present in it. We have performed some of the steps for the exploration of attacks and vulnerabilities.

a) Exploration of camera: Only mobile systems like Android and iOS can access an IP camera (Figure 18.4). The end-user's account in the IP camera must be associated with the devices and connected through the internet to establish the connection. An SSID and camera password are required, which can provide a secure connection. This can be achieved using the Android application for the IP camera.

   A dedicated server provides all types of communication after establishing the connection. To view the video broadcast from anywhere on the internet, a remote server must be at the center of all network activity. When using the camera, it's fascinating to notice what happens when the mobile client and camera are connected to the same LAN. Once it detects a similar external IP address for both devices, it informs the application. An ARP (Address Resolution Protocol) message is sent to the LAN during the transmission of the packet. This kind of situation will improve video quality and video streaming.

b) Security perspective of camera: A password is required whenever a new device is added to the IP camera. Therefore, since the server

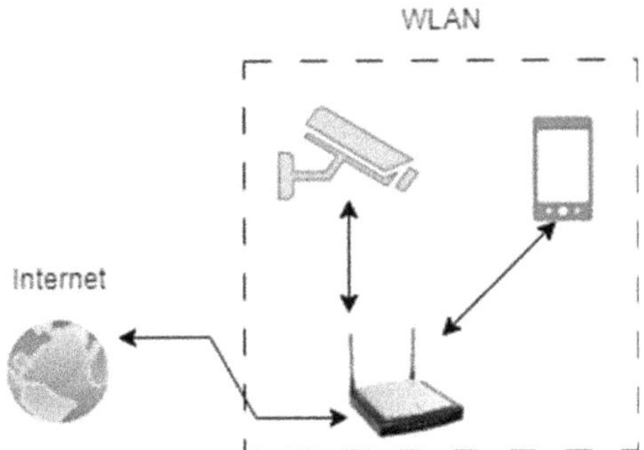

Figure 18.4 Exploration of IP camera.

requires authentication, adding any device to the mobile is not possible to access the video stream. So, it requires analyzing the application and network simultaneously. It would grant access to programs that are now executing on the device and occasionally even allow locating configuration files that contain usernames and passwords.

c) Inspection of network services: In order to proceed with additional testing, the NMAP (Network Mapper) network scanning software can be used to locate the IP address of the camera. This can be achieved by searching for all active hosts on the network. However, since the camera doesn't have a specific name, it might be difficult to differentiate it from other devices. By conducting LAN scanning with the NMAP tool and using the "A" flag command, you will be able to determine which port is being used and identify any open ports on the network.

1) Port: 554/TCPRTSP

2) Scan port: TCP 5000

RTSP is used to monitor the real-time scanning of the network, which is directly connected to the LAN.

d) Communication inspection among devices: There are some tools like Wireshark and tcpdump used to extract the packet details and record all data for further analysis. Wireshark was employed in our situation to complete this operation. There are numerous operating modes for Wireshark, including monitor, promiscuous, and non-promiscuous modes. The utility usually operates in a non-promiscuous mode. This mode records all data coming into and leaving the device. Any data entering the network interface is recorded in promiscuous mode, even if the Ethernet layer filters it out. If the promiscuous mode operates correctly on the host system, the process becomes straightforward; if not, a specific network configuration is needed to capture all the data.

In order to understand how these two sets of data are functioning and detect any confidential information being sent to the server or other devices, it is important to closely monitor their flow. Capturing data flow from mobile devices is relatively simple by setting up a proxy in the settings of an Android or iOS phone or by installing an emulator. We were able to complete this task using Android-x86 because it provides full control over the operating system. To achieve this, a network tap is utilized, as shown in Figure 18.5 of the setup.

In Figure 18.6, the network taps serve to capture and store all the communication that occurs between two endpoints in a network. These endpoints, referred to as A and B in the diagram, can be either devices or separate networks themselves. In this specific scenario, A represents a device while

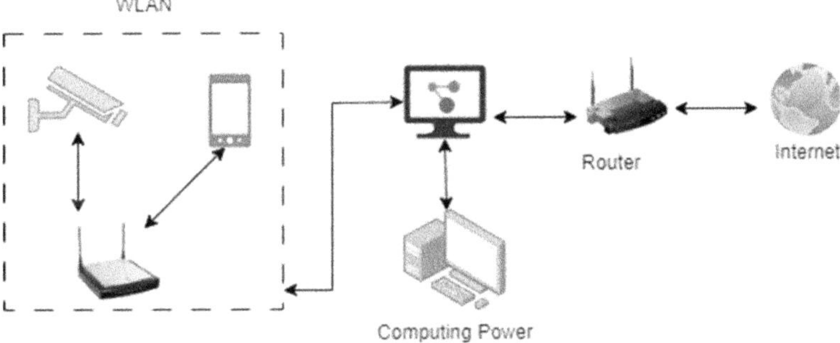

Figure 18.5 Setup for packet capture.

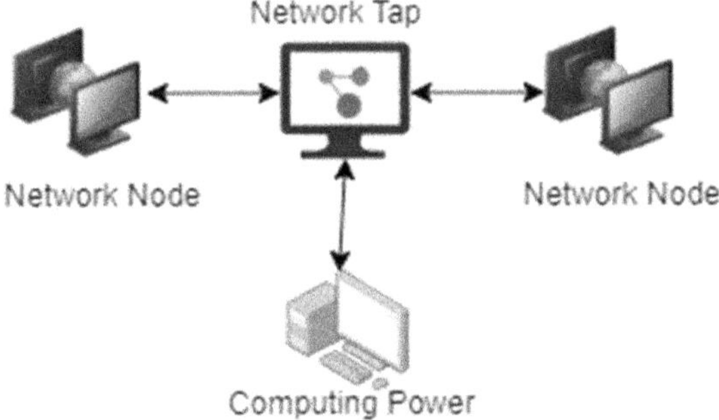

Figure 18.6 Network tap operations.

B represents an independent network. Additionally, it's worth noting that every network flow between A and B is also duplicated to C.

In the network setup depicted in Figure 18.6, we utilize the network tap feature of the Wireshark tool on device C. This allows us to capture and analyze all data transmitted within the network. However, for our specific needs, we are only interested in the information exchanged between devices A and C. To isolate this data, we can apply filters within Wireshark that focus solely on capturing data from the IP camera.

$$"src" = [IP] \text{ or } dst = [dst\ IP]"$$

To gather information from IP cameras and mobile applications, you can combine the filter and operator. Once you have configured this correctly, Wireshark will start capturing a large amount of data. During our investigation, we observed that the cameras shifted toward LAN-based video streaming instead of using IP camera servers. Additionally, we discovered that all IP camera media streaming servers utilize the domain name p*p* .videoipcamera.cn (as shown in Figure 18.7). It's worth noting that any

```
Standard query response 0x6f26 A p2p2.videoipcamera.com A 218.30.35.92
Standard query response 0x6f26 A p2p2.videoipcamera.com A 218.30.35.92
Standard query response 0x6f26 A p2p2.videoipcamera.com A 218.30.35.92
Standard query response 0x6f26 A p2p2.videoipcamera.com A 218.30.35.92
Standard query response 0xdcae A p2p3.videoipcamera.cn A 220.231.142.137
Standard query response 0xdcae A p2p3.videoipcamera.cn A 220.231.142.137
Standard query response 0x6f26 A p2p2.videoipcamera.com A 218.30.35.92
Standard query response 0x6f26 A p2p2.videoipcamera.com A 218.30.35.92
Standard query response 0xdcae A p2p3.videoipcamera.cn A 220.231.142.137
Standard query response 0xdcae A p2p3.videoipcamera.cn A 220.231.142.137
Standard query response 0x6f26 A p2p2.videoipcamera.com A 218.30.35.92
Standard query response 0x6f26 A p2p2.videoipcamera.com A 218.30.35.92
Standard query response 0xdcae A p2p3.videoipcamera.cn A 220.231.142.137
```

Figure 18.7 Wireshark results.

number between 1 and 6 can replace the "*" in the URL, so a valid IP camera server could be p2p3.ipcamera.com. If these IP camera servers are vulnerable, it is likely that all cameras across different organizations would be using the same servers, which creates a potential single point of failure for all cameras. Considering this, it is prudent to not overlook server security when dealing with IP cameras.

18.6 SUMMARY

When it comes to analyzing the cybersecurity of Industry 4.0 systems, we can break it down into three major security categories; IoT security, security of transport medium, and cloud security. Among these, the security of IoT devices is of utmost importance.

In terms of cyberattacks on SCADA systems, they can be categorized into hardware attacks, software attacks, and communication system attacks. The control center of SCADA plays a vital role in processing all the information collected from RTUs. Attacks targeting this command-and-control process usually aim to manipulate control data or disrupt data transfer. Common examples of such attacks include DoS, DDoS, and MITM attacks.

Cloud-based systems face similar security risks, as they are also part of the same network infrastructure. When discussing SCADA systems integrated into public clouds, several vulnerabilities need to be considered. Firstly, cloud-based SCADA systems are more susceptible to common attacks mentioned earlier due to sharing network infrastructure with unknown parties.

Firstly, it's important to note that the connections between SCADA systems and the cloud can be quite risky, potentially impacting the overall industrial control process. Additionally, the application protocols utilized by SCADA systems, such as Modbus or DNP3, are known to have significant security vulnerabilities as they lack support for authentication and encryption algorithms.

Moreover, relying on commercial solutions that fail to address a wide range of security issues instead of proprietary ones can have severe consequences for the cyber security of the entire system. There are multiple methods through which sensitive information can be obtained. One such method involves passive traffic analysis where malicious code is introduced into operational programs with the intention of modifying control measures, stealing the virtual identities of end users, and gaining unauthorized access to information. Another concerning approach is disrupting the normal functioning of routing protocols by causing interference in communication channels. Exploiting software vulnerabilities, depleting resources, taking control of devices, or infecting them with malware are some ways in which this interference could occur. Manipulating routing information and affecting network traffic opens doors for attacks like DoS or black holes.

Lastly, hackers may employ side-channel attacks to scan and identify vulnerable devices within a network infrastructure. By analyzing factors like battery or memory conditions, device interconnections, or routing information, hackers gain insights into device vulnerabilities. Hackers also manipulate network nodes to inject malicious data, infiltrate various network devices, and carry out eavesdropping attempts. Neglecting proper access control measures can result in unauthorized individuals gaining access to valuable resources. Another potential risk for organizations is the reliance on cloud computing. While many enterprises opt to store their data using cloud infrastructure, sharing this infrastructure with numerous other customers can present a range of significant challenges, including:

- Performing DDoS attacks involves exploiting vulnerabilities within the scheduler component of certain hypervisors to disrupt services and render them unavailable.
- Deploy malicious software within the cloud environment to create a virtual machine replica, thus gaining unauthorized access to the data that is usually transmitted through the genuine one.
- By engaging in side-channel attacks, the perpetrator can observe the electromagnetic field encompassing the devices and obtain unauthorized access to their resources.
- By examining and researching various clients who have access to other applications, the attacker gains the ability to gather information about account names and passwords, as well as introduce harmful services into the cloud. It's important to understand that most of these attacks can only be executed by experienced hackers with significant resources at their disposal.

REFERENCES

Y. Alqahtani, S. M. Gupta, and K. Nakashima, "Warranty and maintenance analysis of sensor embedded products using internet of things in industry 4.0," *International Journal of Production Economics*, vol. 208, pp. 483–499, 2019.

M. R. Asghar, Q. Hu, and S. Zeadally, "Cybersecurity in industrial control systems: Issues, technologies, and challenges," *Computer Networks*, vol. 165, p. 106946, 2019.

J. M. Borky, T. H. Bradley, J. M. Borky, and T. H. Bradley, "Protecting information with cybersecurity," *Effective Model-Based Systems Engineering*, pp. 345–404, 2019.

A. Corallo, M. Lazoi, and M. Lezzi, "Cybersecurity in the Context of Industry 4.0: A structured classification of critical assets and business impacts," *Computers in Industry*, vol. 114, p. 103165, 2020.

G. Culot, F. Fattori, M. Podrecca, and M. Sartor, "Addressing industry 4.0 cybersecurity challenges," *IEEE Engineering Management Review*, vol. 47, no. 3, pp. 79–86, 2019.

T. M. Fernandez-Carames and P. Fraga-Lamas, "A review on the application of blockchain to the next generation of cybersecure industry 4.0 smart factories," *IEEE Access*, vol. 7, pp. 45201–45218, 2019.

K. Ghosh, S. Sharma, P. Bagla, and K. Kumar, "Applications of IoT based frameworks in Industry 4.0: Applications of IoT based frameworks." In: >Vikram Bali, Vishal Bhatnagar, Deepti Aggarwal, Shivani Bali, Mario José Diván(Eds.). *Cyber-Physical, IoT, and Autonomous Systems in Industry 4.0.* CRC Press, 2021, pp. 163–178.

M. Hearn and S. Rix, "Cybersecurity considerations for digital twin implementations," *IIC Journal of Innovation*, vol. 10 pp. 107–113, 2019.

H. Lim et al., "A review of industry 4.0 revolution potential in a sustainable and renewable palm oil industry: HAZOP approach," *Renewable and Sustainable Energy Reviews*, vol. 135, p. 110223, 2021.

U. Mentsiev, E. R. Guzueva, and T. R. Magomaev, "Security challenges of the Industry 4.0." *Journal of Physics: Conference Series*, vol. 1515, no. 3, p. 032074, 2020.

N. K. Pandey, K. Kumar, G. Saini, and A. K. Mishra, "Security issues and challenges in cloud of things-based applications for industrial automation," *Annals of Operations Research*, pp. 1–20, 2023.

T. Y. Pang, J. D. Pelaez Restrepo, C.-T. Cheng, A. Yasin, H. Lim, and M. Miletic, "Developing a digital twin and digital thread framework for an 'Industry 4.0'Shipyard," *Applied Sciences*, vol. 11, no. 3, p. 1097, 2021.

T. B. Pereira, L. Barreto, and A. Amaral, "Network and information security challenges within industry," *Procedia Manufacturing*, vol. 4, pp. 1253–1260, 2017.

J. Prinsloo, S. Sinha, and B. von Solms, "A review of industry 4.0 manufacturing process security risks," *Applied Sciences*, vol. 9, no. 23, p. 5105, 2019.

H. Sarker, A. S. M. Kayes, S. Badsha, H. Alqahtani, P. Watters, and A. Ng, "Cybersecurity data science: An overview from machine learning perspective," *Journal of Big Data*, vol. 7, pp. 1–29, 2020.

R. Sharma and N. Sharma, "Attacks on resource-constrained IoT devices and security solutions," *International Journal of Software Science and Computational Intelligence (IJSSCI)*, vol. 14, no. 1, pp. 1–21, 2022.

L. Thames and D. Schaefer, *Cybersecurity for Industry 4.0.* Springer, 2017.

S. Tian, K. Tang, P. Yang, A. Jia, and H. Melvin, "Secure cloud computing model for communication network management," *Journal of Intelligent and Fuzzy Systems*, vol. 37, no. 1, pp. 27–34, 2019.

T. Usländer and T. Batz, "Agile service engineering in the industrial Internet of Things," *Future Internet*, vol. 10, no. 10, p. 100, 2018.

S. Vaidya, P. Ambad, and S. Bhosle, "Industry 4.0–a glimpse," *Procedia Manufacturing*, vol. 20, pp. 233–238, 2018.

M. Wolf and D. Serpanos, "Safety and security in cyber-physical systems and internet-of-things systems," *Proceedings of the IEEE*, vol. 106, no. 1, pp. 9–20, 2017.

Index

For Product Safety Concerns and Information please contact our EU
representative GPSR@taylorandfrancis.com
Taylor & Francis Verlag GmbH, Kaufingerstraße 24, 80331 München, Germany